Sustainable Bioprocessing for a Clean and Green Environment

Sustainable Bioprocessing for a Clean and Green Environment

Concepts and Applications

Edited by M. Jerold, A. Santhiagu, Rajulapati Sathish Babu and Narasimhulu Korapatti

CRC Press
Taylor & Francis Group
Boca Raton London New York

CRC Press is an imprint of the
Taylor & Francis Group, an **informa** business

First edition published 2022
by CRC Press
6000 Broken Sound Parkway NW, Suite 300, Boca Raton, FL 33487-2742

and by CRC Press
2 Park Square, Milton Park, Abingdon, Oxon, OX14 4RN

Library of Congress Cataloging-in-Publication Data
Names: Jerold, M., editor.
Title: Sustainable bioprocessing for a clean and green environment : concepts and applications / edited by M. Jerold, A. Santhiagu, Rajulapati Sathish Babu, and Narasimhulu Korapatti.
Description: First edition. | Boca Raton, FL : CRC Press, 2021. | Includes bibliographical references and index. | Summary: "This book highlights the importance of waste to health in which waste is safely converted to value-added products via bioprocess technologies. Providing fundamental concepts and applications, this book also offers readers the methodology behind the operation of a variety of biological processes used in developing valuable products from waste. This interdisciplinary book is essential for researchers in chemical, environmental, and bioprocess engineering"-- Provided by publisher.
Identifiers: LCCN 2020055297 (print) | LCCN 2020055298 (ebook) | ISBN 9780367459086 (hardback) | ISBN 9781003035398 (ebook)
Subjects: LCSH: Green chemistry. | Recycling (Waste, etc.) | Biochemical engineering. | Fuel. | Chemicals.
Classification: LCC TP155.2.E58 S855 2021 (print) | LCC TP155.2.E58 (ebook) | DDC 660.6/3--dc23
LC record available at https://lccn.loc.gov/2020055297
LC ebook record available at https://lccn.loc.gov/2020055298

ISBN: 978-0-367-45908-6 (hbk)
ISBN: 978-0-367-76236-0 (pbk)
ISBN: 978-1-003-03539-8 (ebk)

DOI: 10.1201/9781003035398

Typeset in Times
by SPi Technologies India Pvt Ltd (Straive)

Contents

Preface

Global sustainability is gearing up for the dynamic harmony of humans and nature, societies and the ecosystem, as well as the earth and the world. The goal of sustainable development is an avenue for human development in the context of world climate change and earth resilience to support present and future generations.

One of the key areas of sustainability is waste management. Waste management is a global challenge in maintaining a pollution-free environment. Due to the huge hurdles in waste management, it is essential to practice the concepts of "reuse and resource recovery" from the waste generated by humans, plants, and animals for prospective heath and wealth as well as for socio-economics. Bioprocessing is a novel and advanced technique for the management of biowaste from various sources. In particular, bioprocessing facilitates the efficient conversion of organic waste into value added products using microorganisms as biocatalyst. Sustainable bioprocessing is a key platform in addressing the valorization of biowaste and establishing circular bioeconomy. Biorefinery is an innovative and advanced concept in bioprocessing explored in the field of biotechnology for bioremediation. Biowaste is used as the renewable feedstock for the recovery of bioproducts, biochemicals, and bioenergy, using sustainable bioprocessing techniques.

This book, *Sustainable Bioprocessing for a Clean and Green Environment: Concepts and Applications,* is proposed with a goal of delivering all the up-to-date concepts on bioprocessing related to bioremediation and bioconversion. The whole book is a net outcome of eminent academicians and researchers working in multidisciplinary areas on sustainable clean and green environments.

This book has 15 chapters on emerging concepts related to waste management and sustainable development. Indeed, several researches are undertaken today for the proper disposal of solid and liquid waste. Interestingly, resource is recovered from the waste which is a successful milestone in waste management. This book gives deep understanding about bioplastic from biomass, bioelectricity generation, bioethanol from biomass waste, and so on. Further, importance is given to wastewater treatment using biological methods. A couple of chapters discuss liquid waste management using nanomaterials. Today, algae are widely explored in various ways for sustainable development. Therefore, we have included a few chapters on algal technology for the production of bioenergy and nutraceuticals. This book also gives information about the development of biosurfactants and corrosion inhibitors from biomass.

Essentially, the idea behind the writing of this book is to deliver information for multidisciplinary researchers. So, we believe this book gives diversified information related to pollution abatement. In a nutshell, this book is an interdisciplinary book highly focused on a research-based solution for the concept of a green and clean environment and will provide sparkling information for those readers working in the cutting-edge research areas of sustainable bioremediation and pollution management. Moreover, the chapters are written by highly esteemed researchers from various technical universities utilizing their expertise and research outcomes. Therefore, we earnestly believe this book will cater to the needs of the present generation of researchers and engineers striving to develop green technologies for sustainable development.

Enjoy reading the book!

Best wishes

Dr. M. Jerold
Prof. A. Santhiagu
Dr. R. Rajulapati Sathish Babu
Dr. K. Narasimhulu Korapatti

Editor Biographies

Dr. Jerold is an academician and researcher in the field of biochemical engineering. His areas of interest include biorefineries, waste valorization, and algal bioremediation. He has embarked on his research potential for a clean and green environment. His research is mainly focused on sustainable bioconversion; in particular, his idea is to produce wealth from waste. He is a reviewer of many national and international journals. He has industrial, teaching and research experience. His research areas encompass fermentation, phytoremediation, biological wastewater treatment, biosoprtion, microbial fuel cells, nanotechnology, biosurfactants, biocataysts, biofuels, and adsorption of heavy metals, to list a few. His research is highly focused on problems related to environmental issues, especially liquid and solid waste management. He was awarded a seed grant to execute research on biodiesel production using enzymatic catalysts by the National Institute of Technology, Warangal, India. He has identified macroalgae as feedstock for the production of bioethanol at low cost. He has extensively worked on dye removal using algae for the textile industry. He has developed a biocomposite for the removal of dye with maximum biosorption capacity. Being an interdisciplinary area researcher, he is working on hot topics related to sustainable energy production from microbial fuel cells, algal lipid, biohydrogen from wastewater, and so on. He is also involved in the extraction of various phytochemicals from algae.

In addition to his research, he is involved in teaching for undergraduate engineering students. He has handled various subjects related to biotechnology and biochemical engineering bioreactor design and analysis, downstream processing, protein engineering, bioprocess heat transfer, biofuel technology and engineering, bioprocess instrumentation, environmental biotechnology, herbal biotechnology, introduction to life science, engineering biology, transport phenomena in the bioprocess system, and biochemical thermodynamics. He has handled various laboratory courses for Bachelor of Technology in Biotechnology students, such as bioprocess and bioreaction engineering, transport phenomena in the bioprocess system, heat and mass transfer, instrumental methods of analysis, fluid mechanics, and environmental and pollution control. He is a highly committed teacher and researcher who works enthusiastically on cutting-edge research areas of biotechnology. He has been with Dr. D.G.S. Dhinakaran fellowship for his postgraduate education and the MHRD Fellowship by Government of India for his doctoral studies. In addition, he has been honored with gold medals for his higher academic credentials by the university of his higher studies. He was the state rank holder during his Pre University level.

Jerold has published 24 research and review papers as well as reviewing articles in international and national peer-reviewed journals. In addition, Jerold has authored six book chapters and has guided 15 project students for their dissertation work. He is currently guiding one doctoral student in the area of bioprocessing and bioseparation. To his credit he has edited a couple of books and published by CRC Press/ Taylor & Francis and Springer Nature. He is a life member of the Biotech Research Society of India (BRSI) and the International Society for Technical Education (ISTE), India.

Dr. A. Santhiagu completed his Bachelor of Pharmacy (B.Pharm) from Tamil Nadu Dr. M.G.R Medical University, Chennai in 1998. Later, he achieved his Master of Technology in Biochemical Engineering from Banaras Hindu University (IIT-BHU), Varanasi in 2001, then he completed his Doctor of Philosophy in Biochemical Engineering at Banaras Hindu University (IT-BHU), Varanasi in 2006. Following this, he joined as Assistant Professor at Kalasalingam Academy of Research and Education, Sivakasi for two years, before moving to Ultra College of Pharmacy, Madurai where he served as Head of Department for one year. He then joined NIT Calicut, School of Biotechnology in the year of 2009 as Assistant Professor where he is now Professor. In addition to teaching he has been deputed in various administrative positions like HOD, Associate Dean, Deputy Registrar and Chief Warden.

He has published over 18 research papers in various peer-reviewed international journals and around 25 research papers in various national and international conferences. He has evidenced his subject knowledge in guiding five Ph.D. students, four Master's scholars and many project students in getting their degrees awarded. At present, he has seven Ph.D. scholars pursuing their degrees under his guidance. He has drawn various funded projects from government-funding agencies like DST, DBT, and KSCSTE in the tune of 10 billion for various research projects. He has organized one international and one national conference sponsored and supported by DBT, DST, ICMR, and CSIR. His key research areas include bioprocessing, bioremediation and controlled drug delivery systems. He is a life member of the Biotech Research Society of India. He and his team have worked in the field of bioprocessing for the last decade and have attributed their research findings to the field of bioprocessing of various biological products. In their lab, students are working on different products like gellan production, biofuel production, fibrinolytic enzyme production, and so on. They are developing various strategies for commercializing the products. They are still at the experimentation stage and most of the products are in pipeline. The most interesting part of their research finding is on gellan gum production using recombinant strains and they have isolated a novel strain from a marine source for the production a novel block-buster enzyme.

Thus, he has proved his research on various perspectives on science and engineering. So, he is now interested in deliver his research finding and expertise via Books and Journals. Perhaps, he has various publications in peer reviewed international journals. He would like to step into authoring a book with the support of his colleagues who are working in the similar kind of discipline.

Dr. Rajulapati Sathish Babu is presently serving as Associate Professor and Head in the Department of Biotechnology. He completed his B.Tech in Chemical Engineering from NIT Warangal, M.Tech and his Ph.D. from Jawaharlal Nehru Technological University, Hyderabad, India. His area of research includes bioprocess engineering, environmental biotechnology modeling, and simulation of bioprocesses metabolic engineering bioinformatics. He has more than 14 years' teaching experience handling various subjects like bioprocess engineering, biochemical engineering, biochemical reaction engineering, downstream processing, bioinformatics, heat transfer in bioprocess, mass transfer operations in bioprocess, transport phenomena in bioprocess, process engineering principles, microbial bioreactor design, and microbial engineering. He has also handled various labs like bioprocess engineering, bioreaction engineering, bioinformatics, and downstream processing for B.Tech students.

In addition to teaching he has been deputed in various administrative positions like HOD, Warden, and Security officer. He has published over 26 research papers in various peer-reviewed international journals and many research papers in various national and international conferences. He has evidenced his subject knowledge in guiding three Ph.D students, several Master's scholars, and project students in getting their degrees awarded. He has organized various workshops and national conferences to deliver his expertise to academicians, researchers, and students. He received funding for a project from the Council of Scientific and Industrial research (CSIR).

His research is in the development of low-cost methods to remove pollutants from ground water. His research group is working on the isolation of novel enzymes and its application cancer cells studies. They have developed a low-cost ultrafiltration membrane for the removal of fluoride from ground water. For more than a decade, his research group has done significant work for the benefit of the research community.

He has published a chapter in the book *Chemical and Bioprocess Engineering: Trends and Developments* (2015), published by CRC Press, and he has one patent to his credit (continuous solution phase production of beta peptide using spiral copper channel microreactor, Application No.: 635/CHE/2015, Publication Date: 9/02/2015). He received Best Poster award at the 3rd International Conference on Desalination using Membrane Technology organized by Elsevier in Gran Canaria, Spain. He has honored by the Venus International Society in 2016 for his outstanding performance as Bioprocess Engineering Faculty. He is an academic fellow at Telangana Academy of Sciences.

Dr. Narasimhulu Korapatti is currently at the National Institute of Technology Warangal as Associate Professor in the Department of Biotechnology. He has 20 years of teaching experience. He was Visiting Researcher at Rice University, USA in the Department of Chemical and Biomolecular Engineering has completed his B.Tech in Chemical Engineering at NIT Warangal and his M.Tech at Jawaharlal Nehru Technological University, Hyderabad, India. Following this, Dr. Korapatti completed his Doctor of Philosophy at NIT Warangal, and, in addition, he has his Diploma in Sugar Technology.

His areas of research are environmental biotechnology, bioprocess engineering, modeling and simulation of bioprocesses, biofuels, and systems biology. He is an expert in teaching various subjects like biochemical thermodynamics, bioprocess calculations, bioprocess engineering, bioreaction engineering, modeling and simulation of bioprocesses, bionanaotechnology, and bioprocess plant design. He is actively guiding Ph.D. research students and has successfully guided three Ph.D students in the field of biotechnology. He has also guided several undergraduate and Master's students in their dissertation work. He has received various sponsored funded projects from DST, SERB, MHRD for conducting research works and other academic activities. He has published 28 research papers in various peer-reviewed international journals and many in various national and international conferences. In addition to teaching and research, he has been deputed in various administrative positions for the prospective of institute development. His research group studies the production of biofuel from ligocellulosic feedstock using enzymatic hydrolysis, and his team is working on the bioremediation of organic pollutants.

Dr. Korapatti has conducted various workshops, FDPs, and conferences to deliver his expertise to the young students, researchers, and engineers. He is a senior member of Asia Pacific Chemical, Biological & Environmental Engineering Society (APCBEES) and life member of various professional bodies. He has visited the USA, Thailand, Australia, and Dubai attending conferences and other academic works and received the International travel award by the Department of Biotechnology (DBT), Government of India, for attending the International Conference on Tissue Science and Engineering. He has also been the recipient of the Young Faculty and Young Scientist Award held by the Venus International Foundation, Chennai, India.

Contributors

Chanchpara Amit
Analytical and Environmental Science
    Division & Centralized Instrument
    Facility
CSIR-Central Salt & Marine Chemicals
    Research Institute
Bhavnagar, India

Madhava Anil Kumar
Analytical and Environmental Science
    Division & Centralized Instrument
    Facility
CSIR-Central Salt & Marine Chemicals
    Research Institute
Bhavnagar, India

and

Academy of Scientific and Innovative
    Research
Ghaziabad, India

N. Arunkumar
Department of Microbiology
Central University of Tamil Nadu
Thiruvarur, India

Deep Bhattacharya
Department of Biotechnology
School of Bioengineering
SRM Institute of Science and
    Technology
Chennai, India

K. M. Meera Sheriffa Begum
Department of Chemical Engineering
National Institute of Technology
Tiruchirappalli, India

Ramesh Desikan
Department of Vegetable Science
Horticultural College and Research
    Institute for Women
Tamil Nadu Agricultural University
Tiruchirappalli, India

**Radhakrishnan Edayileveetil
Krishnankutty**
School of Biosciences
Mahatma Gandhi University
Kerala, India

Tamil Elakkiya Vadivel
Department of Biotechnology
Bharathidasan Institute of Technology
Anna University
Tiruchirappalli, India

Elangovan Elakkiya
Department of Biotechnology
PSG College of Technology
Coimbatore, India

Tholan Gajendran
Department of Biotechnology
Bharathidasan Institute of Technology
Anna University
Tiruchirappalli, India

Carlin Geor Malar
Department of Biotechnology
Rajalakshmi Engineering College
Thandalam, India

Samuel Jacob
Department of Biotechnology
Faculty of Engineering and Technology
School of Bioengineering
College of Engineering and Technology
SRM Institute of Science and Technology
Chennai, India

Mansi Kikani
Analytical and Environmental Science
    Division & Centralized Instrument
    Facility
CSIR-Central Salt & Marine Chemicals
    Research Institute
Bhavnagar, India

Santhosh Kumar Kookal
International Centre for Genetic
  Engineering and Biotechnology
New Delhi, India

Erudayadhas Lavanya
Department of Biotechnology
PSG College of Technology
Coimbatore, India

Sunaina Nag
Department of Biotechnology
Faculty of Engineering and
  Technology
School of Bioengineering
College of Engineering and Technology
SRM Institute of Science and Technology
Chennai, India

Samsudeen Naina Mohamed
Department of Chemical Engineering
National Institute of Technology
Tiruchirappalli, India

Rajupalepu S. Monish
Department of Biotechnology
School of Bioengineering
SRM Institute of Science and Technology
Chennai, India

Subramaniapillai Niju
Department of Biotechnology
PSG College of Technology
Coimbatore, India

R. Nithya
Department of Industrial Biotechnology
Government College of Technology
Coimbatore, India

Amisha Pani
Department of Biotechnology
Faculty of Engineering and Technology
School of Bioengineering
College of Engineering and Technology
SRM Institute of Science and Technology
Chennai, India

Riasha Pal
Department of Biotechnology
School of Bioengineering
SRM Institute of Science and Technology
Chennai, India

Divya Palaniswamy
Department of Chemistry
Avinashilingam Institute for Home
  Science and Higher Education for
  Women
Coimbatore, India

Jishma Panichikkal
School of Biosciences
Mahatma Gandhi University
Kerala, India

Nilanjan Paul
Department of Biotechnology
School of Bioengineering
SRM Institute of Science and Technology
Chennai, India

Velusamy Priya
Department of Civil Engineering
SNS College of Engineering
Coimbatore, India

Karuppusamy Priyadharshini
Department of Biotechnology
PSG College of Technology
Coimbatore, India

Suchitra Rakesh
Department of Microbiology
Central University of Tamil Nadu
Thiruvarur, India

Doddabhimappa Ramappa Gangapur
Plant Omics Division
CSIR-Central Salt & Marine Chemicals
  Research Institute
Bhavnagar, India

and

Academy of Scientific and Innovative
    Research
Ghaziabad, India

Krishnan Ravi Shankar
Department of Biotechnology
Bharathidasan Institute of Technology
Anna University
Tiruchirappalli, India

Joseph George Ray
School of Biosciences
Mahatma Gandhi University
Kerala, India

Rajalakshmi Reguramnan
Department of Chemistry
Avinashilingam Institute for Home Science
    and Higher Education for Women
Coimbatore, India

Shreya Sadukha
Department of Biotechnology
Faculty of Engineering and Technology
School of Bioengineering
College of Engineering and Technology
SRM Institute of Science and Technology
Chennai, India

Renganathan Sahadevan
Department of Biotechnology
Anna University
Chennai, India

Prasanthkumar Santhakumaran
School of Biosciences
Mahatma Gandhi University
Kerala, India

Muthulingam Seenuvasan
Department of Chemical Engineering
Hindusthan College of Engineering &
    Technology
Coimbatore, India

Suganya Subburaj
Analytical and Environmental Science
    Division & Centralized Instrument
    Facility
CSIR - Central Salt & Marine Chemicals
    Research Institute (CSMCRI)
Bhavnagar, India

and

Academy of Scientific and Innovative
    Research
Ghaziabad, India

Karthikeyan Subburamu
Department of Renewable Energy
    Engineering
Agricultural Engineering College and
    Research Institute
Tamil Nadu Agricultural University
Coimbatore, India

A. Thirunavukkarasu
Department of Industrial Biotechnology
Government College of Technology
Coimbatore, India

Saloni Tripathy
Department of Biotechnology
Faculty of Engineering
    and Technology
School of Bioengineering
College of Engineering and Technology
SRM Institute of Science and
    Technology
Chennai, India

Theresa Veeranan
Department of Biotechnology
Anna University
Chennai, India

Sivakumar Vivek
Department of Civil Engineering
Hindusthan College of Engineering and
    Technology
Coimbatore, India

1 Alternative Plastics from Wastes through Biomass Valorization Approaches

Amisha Pani, Saloni Tripathy, Shreya Sadukha,
Sunaina Nag and Samuel Jacob
SRM Institute of Science and Technology, India

CONTENTS

1.1 INTRODUCTION

Synthetic polymers, almost without exception, are non-biodegradable. Industries such as packaging, healthcare, textiles, and so on are a few of the major consumers of the petroleum-derived synthetic polymers (polypropylene (PP), polyvinylchloride (PVC), high-density polyethylene (HDPE), polystyrene (PS), etc.) (USA Energy Information Administration, 2012). Plastics and polymers form an integrated part of our daily routine.

1

Also, food packaging is one of the most important requirements in food industries. The major concern faced by them is how to preserve and protect all types of foods, and for this petroleum-derived plastics are predominantly being used. On the other hand, these plastics pose both health and serious environmental hazards.

The depletion of the finite petrochemical resources, supply security, and the negative effects on the environment has necessitated the development of eco-friendly polymers. For the last decade, great effort has been incorporated into producing biopolymers and has attracted considerable attention because of their environmental advantages. Industrial ecology, green chemistry, eco-efficiency, and sustainability are guiding the next generations of materials, processes, and products. Polymers are a chemical compound consisting of discrete building blocks linked together in a long, repeating chain. Biopolymers are defined as the polymer formed under the natural processes and hence is also known as a natural polymer.

The majority of the biopolymers are extracted from agricultural products such as starch, cellulose, and protein. However, with comparison to thermoplastic based on synthetic polymers, biopolymers pose difficulty when processed with conventional technologies and show subservient performances in terms of functional and structural properties (Mensitieri et al., 2011, Bahram et al., 2020). The blending of different biopolymers has been considered as an alternative to this problem. The most familiar and potential biopolymers are starch, gelatin, chitosan, alginate, poly lactic acids, poly hydroxy alkanoates (PHAs), poly hydroxy butyrates (PHBs), and so on. Being extracted from nature they are biodegradable and hence can be consumed by microorganisms and converted into simple, eco-friendly compounds. Biopolymers are used to form biobased plastic which can be reusable and are biodegradable (Elisabeta-Elena et al., 2014). The raw materials, chemical composition, and the structure of the finally produced bioplastic, along with the environment under which the bioplastic is expected to degrade, determines its biodegradability. Some of the sources of polymers available in natural resources have been schematically represented in Figure 1.1.

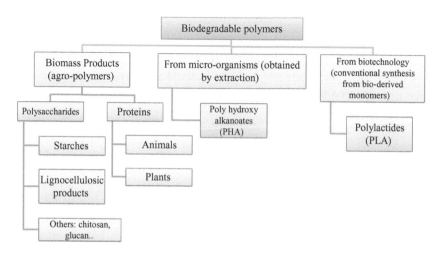

FIGURE 1.1   Sources of bio-polymers used for plastics synthesis.

1.2 CLASSIFICATION OF BIOPLASTIC

1.2.1 BASED ON BIOLOGICAL MACROMOLECULES

1.2.1.1 Starch

Starch is the most commonly used raw material which is renewable and biodegradable in nature. Starch can be obtained from potato, rice, wheat, and cassava (Whistler and BeMiller, 2007). It is thermoplastic in nature. It can be plasticized by using specific amounts of plasticizers (sorbitol, mannitol, xylitol, glycerol) and heat and then it is obtained. The main reason for using starch as a material for packaging is low cost, easy availability, and biodegradable in nature.

The use of starch in the plastics industry reduces the use of synthetic polymers in these industries. The structure of starch is made up of amylose and amylopectin. Biodegradable forms of plastics are mostly made of amylopectin, but films made up from pure starch showcase inferior physical properties (Akter et al., 2012). The main reason why amylopectin is used for bioplastic preparation rather than amylose is amylose constitutes only 20 percent of starch while amylopectin constitutes 80 percent of starch, and amylopectin is more soluble in water and branched as compared to amylose which has a linear chain structure.

The main limitation for using starch is that it has a poor mechanical property and a resistance to moisture. In order to improve these properties, the starch is blended with other varieties of biopolymers and certain additives (Yadav et al., 2018). Starch can be blended with PLA composites and can also be mixed with polyvinyl alcohol and chitosan (Wang et al., 2010).

1.2.1.1.1 Modifications of Starch

The native form of starch has unfavorable properties such as brittleness and is not soluble in cold water. Therefore to overcome these limitations it can be modified chemically as discussed in the subsequent section. The types of modification for starch are represented in Figure 1.2.

1.2.1.1.2 Cross-linking

Cross-linking of the starch is the most significant modification followed. A study was done by Kapelko et al. (2015) in which cross-linking weak hydrogen bonds are replaced with strong covalent bonds. Choi and Lee (1999) and Franssen and Boeriu

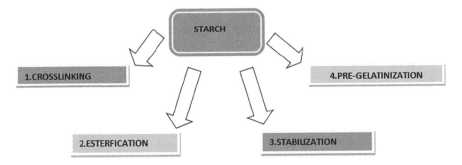

FIGURE 1.2 Ways of modification of starch.

(2014). reported that the amount of $C_6H_{12}O_6$ that exists in starch mainly consists of two 2° and one 1° hydroxyl groups that can interact with other compounds. This can lead to increase in cross-linking which causes it to be more amicable to the gelatinization process and increases the stability by reducing the solubility as stated (Ratnayake and Jackson 2008; Zhong et al., 2013). The cross-linked starch can be utilized in textile, adhesive, and food industries (Phadnis and Jadhav, 1991).

Jyothi et al. (2006) reported cross-linking of cassava starch with epichlorohydrin (an organo-chloride and epoxide) in three types of different media: water, water with N,N- dimethylformamide, and water with phase contrast catalyst. Upon characterization based on thermal, retrogradation, and physicochemical properties, the cross-linking was very high when N,N- dimethylformamide was used. Also, there was a subsequent decrease in the capacity of water-binding ability with an increase in the extent of cross-linking.

1.2.1.1.3 Esterification

In esterification, hydroxyl groups are replaced with ester groups that enhance the thermal stability (Grommers, 2009). Different types of starch esters are synthesized, namely, starch sulfates, starch phosphates, and starches of fatty acid (Wischmann et al., 2005; Vasiliadou et al., 2015).

Starch modified by esterification deciphered the disappearance of the crystalline structure of starch composites after the reaction. This led to the enhancement in physical properties and thermal stability of the native starch.

1.2.1.1.4 Stabilization

Stabilization is performed to improve the ability of the starch to tolerate the variation in temperatures. This is also done to escalate the shelf life of the products made from starch. The huge groups such as octenyl-succinate are replaced on to the starch in order to prevent the straight scattered remains to re-associate or re-crystallize the polysaccharides present in gelatinized starch (Murphy, 2007). The number of groups added to it decides the effect of stabilization. It has been seen that when potato starch was annealed with different amylose/amylopectin ratios with subsequent increases in the temperature, there was an increase in stability which might be due to thickening with amylase as observed from the X-ray crystallography results. The thickening of the crystal mainly involves amylopectin and amylose to go through co-crystallization while annealing (Gomand et al., 2012).

1.2.1.1.5 Pre-gelatinization

Pre-gelatinization is a method which was developed in order to eliminate the requirement of cooking the starch (Miyazaki et al., 2006). Two ways of acetone precipitations were employed, that is, cold and hot form, and it has been observed that the product exhibited good powder properties (Ohwoavworhua and Osinowo, 2010). However, it takes a longer time to dry a pre-gelatinized starch preparation.

1.2.1.1.6 Thermoplastic Starch

Starch in its native form is fragile and tends to be water absorbent, hence hindering its usage especially in food packaging. It is less thermally stable and has a high melting point (Wang et al., 2003). Starch can be transformed into thermoplastic form by inducing plasticizers like polyols in it. This thermoplastic starch can be mixed with a variety of polymers each having their own properties and potentiality. Thermoplastic starch can be blended with a wide range of polymers along with some plasticizers. This enhances its water-resistant and mechanical properties (Murphy, 2007). Blending of biological components such as starch and PLA with other thermoplastic has been done in the recent past and which are discussed in the subsequent sections.

1.2.1.1.6.1 Thermoplastic Starch-Polyethylene Blends Polyethylene (PE) is tolerant to chemicals and is harmless when acids, bases, or the salts are used. Apart from these advantages, other properties such as cheap, flexible, and toughness of PE makes it suitable candidate as a blending agent. Hua et al. (2014) were able to prepare thermoplastic-PE blends by using glycerol as a plasticizer. A vulcanized way was chosen in which the thermoplastic starch-PE blends were compounded in one non-stop course in a twin-screw extruder of co-rotating type which was delivered by a single-screw extruder. In order to prevent the premature evaporation of water, high pressure was maintained throughout the process. These blends demonstrated high elongation properties, as high elongation property is very important for a plastic to be used as a packaging material (van den Broek et al., 2015). A material which has elevated elongation property will eventually soak up a huge quantity of energy before it breaks.

1.2.1.1.6.2 Thermoplastic Starch-Polypropylene Blends Polypropylene (PP) is an extremely superior insulator and even exhibits good electrical properties. Reddy et al. (2003) stated that it is tolerant to varieties of chemicals such as alcohols, glycerol, some bases, and so on, when exposed at elevated temperatures which are commonly used in the wrapping of consumer goods (Imre and Pukanszky, 2013). Kaseem et al. (2012) did research in which they blended PP with thermoplastic starch by a melting and extrusion (single) process which was further shaped by using an injection molding process. The change in rheological and mechanical properties of blends was analyzed and it was observed that the firmness of thermoplastic starch reduced as the glycerol content increased. In regard to the mechanical property results, analysis indicated that the strain present at the break ends was found to be less than PP, also the blends had a higher Young's modulus than that of PP. This blend has a high crystalline nature that improves its tensile strength and stiffness which are the most significant factors required for food packaging.

1.2.1.1.6.3 Thermoplastic Starch-PLA Blends Polylactic acid (PLA) being a polymer of organic acid is biologically compatible and eco-friendly. It can be used in the field of pharmacology and textile industries. Furthermore, its monomer mixture using appropriate microorganisms and sustainable resources decreases the implications in global warming to an extent. Therefore, researchers have performed experiments to utilize natural sources to transform native PLA to improve its functional properties (Bishai et al., 2014). PLA has a lot of characteristics equivalent to those of PP and polystyrene, such as rigidity and tensile strength. PLA has also been made up into a large range of customer products, which includes degradable bags, paper covering and has also been spun into cloth. Polyethene oxide mixed with PLA is used as blended plastic material for secondary packaging material to improve the average shelf life of the products due to its enhanced barrier property (Iotti et al., 2009). It has been concluded that the Young's modulus and tensile strength was more for sorbitol plasticized blends as compared to other blends.

1.2.1.2   Chitosan

Chitosan is the most preferred polymer next to starch which is primarily ideal because of the properties such as biodegradability, biocompatibility, non-toxicity, antioxidant, anticancer, and antimicrobial properties. As chitosan is mainly extracted from the aquatic waste resources like crustaceans (crab shells, prawn exoskeleton, etc.) it renders cost-effectiveness (Abdou et al., 2008; Dayarian et al., 2014; Vilela et al., 2017; Xie et al., 2017). Alkaline deacetylation of chitin produces chitosan which is a derivative of chitin. Chitin is present in the exoskeleton of insects and crustaceans (Siripatrawan and Vitchayakitti, 2016; Darbasi et al., 2017; Kalaycıoğlu et al., 2017; De et al., 2018). Chitosan is highly soluble in contrast to chitin (Elsabee and Abdou, 2013). Due to solubility, chitosan has a large number of applications in industrial, medicine, and agriculture. The degree of deacetylation, molecular weights, and functional properties of chitosan depends on the source from which it has been obtained (Leceta et al., 2013; Akyuz et al., 2017). The capability of chitosan to form an exceptional plastic property makes it an appropriate polymer for a huge number of applications. The physical property, such as the tensile strength of the bioplastic derived from chitosan, is directly proportional to the MW (molecular weight) and extent of deacetylation of chitosan (Park et al., 1999, 2002; Nunthanid et al., 2001; Ziani et al., 2008; Fernández-Pan et al., 2010; Kerch, 2015).

It has been reported that an increase in tensile strength of films made from chitosan was observed upon the addition of natural phenolic groups on the backbone of chitosan (Liu et al., 2017). On incorporation of gallic acid into films made from chitosan, increases in the tensile strength was observed (Rui et al., 2017). Bioplastics of chitosan produced using acetic acid had tensile strength much higher as compared to those prepared by using lactic, malic, or citric acid. The physical properties were further enhanced by the addition of glycerol as a plasticizer. When silicon carbide nano-composites are incorporated, there is an increase in the physical properties (like tensile strength) of chitosan and silicon merged films (Pradhan et al., 2015; Giannakas et al., 2016). Chitosan-based films are well known to have antimicrobial properties and hence used to cover vegetables and fruits by applying antimicrobial substances through using polyvinyl alcohol/chitosan blends, which are toughened with cellulose nano-crystals and multifunctional nano-sized fillers (Kerch and Korkhov, 2011; Azizi et al., 2014).

1.2.1.2.1   Chitosan Incorporation with Other Polymers
As chitosan comprises amine functional group, these polymers can go through chemical alterations. Structural changes leading to significant changes in

biological and physicochemical properties are observed on interaction of chitosan with other polymers. Presently, research is mostly done on blending the chitosan with a wide range of other polymers. Deng et al. (2017) did a study to ensure compatibility of cellulose nano-fiber with methylcellulose, carboxy-methylcellulose, and chitosan. The results showcased that this type of polymer can play a significant role to alter the physicochemical properties of carboxy-methylcellulose films. A decrease in hydrophilicity was observed when chitosan was added to these films, which showed an enhanced compatibility with carboxy-methylcellulose in contrast to other polymers used in the research. The carboxy-methylcellulose-chitosan-based films can be utilized as an on-the-go, edible wrapping material by applying to patties of fresh beef. These bioplastics showcased their accomplishment by preventing wetness.

1.2.1.2.2 Chitosan-based Films

Chitosan-based films (Figure 1.3) are processed by physical interaction, solvent evaporation, and chemical cross-linking with a wide range of co-mixtures. Nevertheless, the very poor permissibility and mechanical properties of polymeric films is obtained by using physical methods as compared to those films obtained by adopting chemical reactions. The problem is that toxicity is induced when chemical cross-linking agents are used, which affects the materials. Hence, enzymatic methods are being employed to improve the films (Kumar et al., 2000).

1.2.1.2.3 Chitosan/Whey Protein Conglomerated Films

Recent research fabricated whey protein containing chitosan films and casted the polymeric solutions in the absence and presence of mTGase (mTGase stands for Transglutaminase enzyme, that catalyzes the formation of isopeptide bond amid the carboxamide group of glutamine and the amino group of lysine causing release of ammonia) (Di Pierro et al., 2006). A good optical characteristic and edible properties were observed for all those films obtained from chitosan and whey proteins in contrast to those fabricated in the presence of TGase. The results of the study concluded that cross-linking of protein covalently, catalyzed by TGase, and was able to enhance the mechanical resistance of the films.

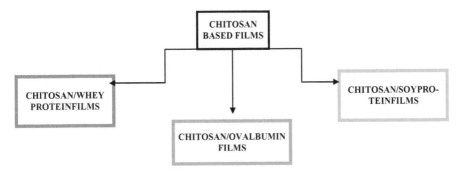

FIGURE 1.3 Types of chitosan-based films.

1.2.1.2.4   Chitosan/Ovalbumin Films

Di Pierro et al. (2006) obtained a slightly yellowish, smooth, flexible, and transparent chitosan/ovalbumin films in the absence and presence of mTGase. These films which were enzymatically cross-linked were insoluble even after incubating for 24 hours which was done at 25°C and kept at a diverse range of pH values, but there was a discharge of huge quantity of soluble proteins when they were treated for 20 hours with trypsin at the same temperature. The research concluded that protein obtained from these films has the capability to act as a protease substrate even when chitosan is present.

1.2.1.2.5   Chitosan/Soy Protein Films

A new form of fit for human consumption film was procured containing a blend of Konjac glucomannan (a water-soluble polysaccharide that is mainly well-thought-out as a dietary fiber), soy protein, and chitosan (Brandenburg et al., 1993).The weight ratio of soy protein, Konjac glucomannan, and chitosan was obtained as 1:1:1 and the film also displayed very low water vapor permeability.

Chitosan films are fairly high, but the utilization of chitosan films in industries is restricted. The foremost cause is for this is that these chitosan blend films undergo ruthless chemical processes which elevate the price of processing of the films. Research can be done to take on new procedures of casting these chitosan films which can be economical and using ingredients which cause its production to be low cost. Also, another concern is less sources of chitosan, which confines its large-scale manufacture. So, further research should be done to discover fresh or freely available sources of chitosan.

1.2.1.3   Proteins

1.2.1.3.1   Plant-based Sources

1.2.1.3.1.1   Corn Zein   The chemical properties and its applications have been thoroughly studied (Shukla and Cheryan, 2001). The two units of corn zein are α-zein and β-zein. The β-zein is formed by the group of α-zein connected by disulphide bonds and has a relatively lower tendency for coagulation and precipitation as compared to α-zein (Shukla and Cheryan, 2001). Zein has poor solubility in water thus the drying of the alcoholic aqueous dispersions to fabricate the film are relatively brilliant and grease resistant. The property for film formation of corn zein has been thoroughly reviewed (Takenaka et al., 1967; Park and Chinnan, 1990; Aydt et al., 1991; Herald et al., 1996). Hence the corn zein proteins have been used for packaging materials, preservation of fresh food, retention of enriching vitamins and for controlled delivery of medicinal drugs. Biodegradable plastic production and manufacturing has benefited from mixtures of starch and zein. Plastics prepared by injection molding portray high sensitivity to water, while plastics produced by cross-linking have shown lower absorption of water and higher mechanical property (Jane et al., 1994).

1.2.1.3.1.2   Wheat Gluten   The gluten basically consists of two fractions of protein which differ in structure and aqueous alcohol solubility as insoluble glutenins and soluble gliadins (Wieser, 2007). Glutenins form the elastic component of the wheat gluten while the gliandin forms the viscous component. Extensive study has been conducted on the wheat gluten protein's ability to form film (Park and Chinnan, 1990; Aydt et al., 1991; Gennadios et al., 1993; Gontard et al., 1996). The traditional procedure to obtain the wheat gluten film involves casting of a thin layer and then drying of aqueous alcoholic proteic solution (under acidic or basic condition) in the presence of disruptive agents such as sulphite. Wheat gluten-based films are water resistant and have similar properties and applications to those of zein films. They have been used for encapsulation of additives, improving cereal product quality and retention of antimicrobial and antioxidant additives on the surface of food. They also exhibit remarkable gas barrier properties thus having a potential application in preservation of fresh vegetables and fruits (Tanada-Palmu and Grosso, 2005; Xing et al., 2016).

1.2.1.3.1.3 Soy Proteins The classification of soy protein is conducted on the basis of their ultracentrifugation rate (Hernandez-Izquierdo and Krochta, 2008). The fractions of the protein which are obtained vary in terms of molecular weight such as 2S, 7S (conglycinin), 11S (glycinin), and 15S. They possess potential film-forming properties and traditionally have been used in Asia by collecting the lipoproteic skin of boiled soy milk to obtain edible films. These films not only constitute protein, but also polysaccharides and lipids. The soy-based films exhibit good mechanical strength and are generally hydrophobic in nature. Synthesis of these films can be performed by casting thin layers and drying of aqueous alcoholic solution (Gennadios et al., 1993; Stuchell and Krochta, 1994). These films are commonly applied in the coating and preserving of food. Apart from this, biodegradable plastic has also been prepared using soy isolate protein (Jane et al., 1994).

1.2.1.3.1.4 Peanuts and Cotton Seed Protein collected from the lipoproteic skin of peanuts is used for the formation of films and water-soluble bags (Aboagye and Stanley, 1985). The cotton seed protein solution treated with various cross-linking agents is used to prepare biodegradable bags (Marquié et al., 1997).

1.2.1.3.1.5 Milk Proteins The two major protein portions in milk are caseins and whey protein. Caseins are low in cysteine and have 300–350 kDa molecular weight whereas whey protein is significantly abundant in cysteine. Casein forms the majority portion of milk protein (Wu and Bates, 1973). Both fractions possess film-forming properties. Transparent and flexible films can be obtained. Films fabricated on processing equipment surfaces and at air-water interfaces by heating non-fat milk are used to obtain casein-based films. Whey protein is used to form the whey-based films by the heating and boiling of whey dispersions and collecting the lipoproteic skin. Transparent, odorless, colorless, and flexible film can be obtained. Network stabilization provided by disulphide bonds partly cause whey protein-based films to be insoluble in water. Casein-based films are capable of enhancing the appearance of food, generating water-soluble bags, the manufacturing of identification labels used for pre-cut cheese, and encapsulation of polyunsaturated lipids used for animal feed (Guilbert 1988; Avena-Bustillos and Krochta 1993).

1.2.1.3.2   Animal-based Sources

1.2.1.3.2.1   Collagen and Gelatin

Collagen and gelatin are both obtained from animals. The most abundant protein found in animals is collagen and it comprises of three cross-linked α-chains, whereas collagen's denatured derivative is gelatin. Collagen is high in amino acids like glycine and proline/hydroxyproline and low in methionine. Collagen has various applications in the meat industry, the pharmaceutical industry, and edible coatings (Tryhnew et al., 1973). Gelatin-derived films have been found to be flexible, transparent, water insoluble, and impermeable to oxygen. The gelatin-derived films also have several applications such as fabrication of tablets and capsules, used as raw materials in photographic films, and encapsulation of vitamins, aroma, and sweeteners (Balassa et al., 1971).

1.2.1.3.2.2   Keratin

Keratin is known to contain approximately 20 proteins with a 10kDa molecular weight. They are rich in amino acids which are hydrophobic in nature and cysteine residues (Gomez-Guillen et al., 2009) and hence not soluble in water. The presence of cysteine bonds provides stiffness to keratin. Prior to film formation, solubilization is required through chemical modification (Tanabe et al., 2004). Keratin-derived water-insoluble films were fabricated by casting and removing moisture from alkaline dispersions.

1.3   WASTES AS SOURCE OF BIOPLASTIC

1.3.1   Sugar Refinery Waste (Cane Molasses)

Bioplastics that are produced from agricultural waste are not only biodegradable and biocompatible, but have also mimicked conventional plastics. For the production of PHAs (polyhydroxyalkonoates), various bacterial genera like *Bacillus, Alcaligens, Pseudomonas*, and *Azetobactor* have been incorporated in studies (Schübert et al., 1988; Anderson et al., 1990). Several studies have found that microorganisms produced maximum PHA. Soil bacterium such as *Pseudomonas aeruginosa* is seen to produce a good concentration of PHA using substrate extracted from sugarcane molasses. PHAs can be synthesized using bacterial species under minimal growth conditions and these polymers provide an alternative source to petrochemical plastics (Choi and Lee, 1999; Dias et al., 2006). There are various physical parameters that affect the production of PHAs like temperature, pH, agitation speed, and so on. From the study it was found that PHA production using *Pseudomonas aeruginosa* under optimum growth and operating conditions gave the following values: Sugarcane molasses 40g/L, urea 0.8g/L, temperature at 36°C, pH at 7.0 and agitation speed at 175 rpm which had undergone fermentation in a 7.5L fermentor (Tripathi and Srivastava, 2011). As nitrogen source was essential, urea was a better option because it existed as a uncharged polar molecule and *Pseudomonas aeruginosa* could assimilate it easily (Kulpreecha et al., 2009). It was analyzed that the

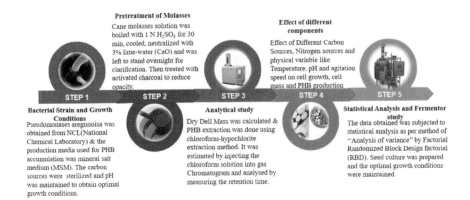

Pretreatment of Molasses
Cane molasses solution was boiled with 1 N H₂SO₄ for 30 min, cooled, neutralized with 3% lime-water (CaO) and was left to stand overnight for clarification. Then treated with activated charcoal to reduce opacity.

Effect of different components
Effect of Different Carbon Sources, Nitrogen sources and physical variable like Temperature, pH and agitation speed on cell growth, cell mass and PHB production

STEP 1   STEP 2   STEP 3   STEP 4   STEP 5

Bacterial Strain and Growth Conditions
Pseudomonases aregunoisa was obtained from NCL(National Chemical Laboratory) & the production media used for PHB accumulation was mineral salt medium (MSM). The carbon sources were sterilized and pH was maintained to obtain optimal growth conditions.

Analytical study
Dry Dell Mass was calculated & PHB extraction was done using chloroform-hypochlorite extraction method. It was estimated by injecting the chloroform solution into gas Chromatogram and analysed by measuring the retention time.

Statistical Analysis and Fermentor study
The data obtained was subjected to statistical analysis as per method of "Analysis of variance" by Factorial Randomized Block Design factorial (RBD). Seed culture was prepared and the optimal growth conditions were maintained.

FIGURE 1.4   PHB production from sugarcane waste industry.

concentration of PHA and maximum biomass obtained when the nitrogen source was urea, was 2.89 ± 0.08 g/L & 5.36 ± 0.08 g/L respectively. PHA that was obtained from *Pseudomonas aeruginosa* resembled commercial PHA as at 1,244 cm^{-1} wave number a strong band was observed that shows ester bond. Few other absorption bands with alkenes, carbonyl, and hydroxyl groups were also studied from FTIR analysis which were similar to that of commercial PHA (Nur et al., 2004). From a comparison with prior research it was seen that *Pseudomonas aeruginosa* gave a high yield in short fermentation periods (Saranya and Shenbagarathai, 2010). The PHA production in a batch fermentation process using an inexpensive carbon and nitrogen source from sugar refinery waste (Figure 1.4) has shown a reduction in manufacturing cost, fermentation tome, and provides for an overall economic process (Serafim et al., 2008).

1.3.2   PAPER MILL WASTE

Prior studies have shown that a huge group of bacterial species can synthesize compounds in low nutrient environments (Chua et al., 2003). They have also shown comparable thermic properties to traditional plastics and one of them is PHA (polyhydroxyalkanoate) (Steinbüchel and Füchtenbusch, 1998). PHA is biodegradable and biocompatible in nature. Since the 1980s, many studies have shown the production of PHA, but these mainly focused on industrial biotechnology using pure culture techniques and genetic manipulation (Gao et al., 2012). Sterilization of the equipment, as well as the batch process methods that were adopted, leads to the high cost of PHA. Prior investigations have shown that agro-industrial waste can be a suitable substrate for PHA production as compared to artificial substrates (Albuquerque et al., 2011). A dry weight of 55% of PHA content has been reported in a majority of the studies and real wastewater gave a cell dry weight of 75% which accounts for the best PHA storage capacity (Coats et al., 2007; Pijuan et al., 2009). The best production result obtained from agro-industrial waste streams that the enriched microbial

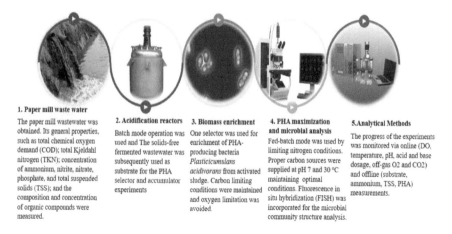

1. Paper mill waste water

The paper mill wastewater was obtained. Its general properties, such as total chemical oxygen demand (COD); total Kjeldahl nitrogen (TKN); concentration of ammonium, nitrite, nitrate, phosphate, and total suspended solids (TSS); and the composition and concentration of organic compounds were measured.

2. Acidification reactors

Batch mode operation was used and The solids-free fermented wastewater was subsequently used as substrate for the PHA selector and accumulator experiments

3. Biomass enrichment

One selector was used for enrichment of PHA-producing bacteria *Plasticicumulans acidivorans* from activated sludge. Carbon limiting conditions were maintained and oxygen limitation was avoided.

4. PHA maximization and microbial analysis

Fed-batch mode was used by limiting nitrogen conditions. Proper carbon sources were supplied at pH 7 and 30 °C maintaining optimal conditions. Fluorescence in situ hybridization (FISH) was incorporated for the microbial community structure analysis.

5.Analytical Methods

The progress of the experiments was monitored via online (DO, temperature, pH, acid and base dosage, off-gas O2 and CO2) and offline (substrate, ammonium, TSS, PHA) measurements.

FIGURE 1.5 PHA production from paper mill wastewater.

biomass could assemble was found to be 77% PHA dry cell weight in 5 hours. This showed the efficiency of the specific biomass. There were several challenges in this process. First was feedstock preparation. Batch mode was incorporated due to simplicity but continuous mode would have been a better choice. Second, oxygen mass transfer was a challenge due to the low solubility of oxygen which was a limiting step in the process design of the reactor. The third challenge was the utilization of paper mill waste water stream as a substrate. The formation of the bioplastic could be done by the usage of inorganic precipitation but it deteriorates the accuracy of pH and DO sensors. The main PHA producer in this enrichment of biomass was *Plasticicumulans acidivorans* which also had another surrounding population that yielded low PHA. The biomass proportion of *P. acidivorans* was dependent on the amount of total COD and volatile fatty acids in the wastewater streams after undergoing the acidification process. The flow process of PHA production from paper mill wastewater is represented in Figure 1.5.

1.3.3 BIOPLASTIC FROM WASTE GLYCEROL

The high production costs of PHAs is mainly due to the large amount of feedstock required for the fermentation process. Feedstock for bacterial growth and biodegradable polymer like sugars or fatty acids accounts for almost 50% of the entire production cost (Choi and Lee, 1997). So, discovering alternative, inexpensive, and renewable alternatives to feedstock can reduce the overall manufacturing cost of polymers. One such alternative is waste glycerol that is obtained as a by-product during biodiesel production. Glycerol has seen its utilization in food, cosmetics, and pharma industries but it is quite an uneconomical process for the refinement and recovery of pure glycerol for these commercial applications. Glycerol fermentation has been investigated to give many value added products like ethanol, propionic acid, 1.3- propanediol, butanol, dihydroxyacetone, citric acid, glycerine, bio-surfactants, and biopolymers like PHAs (Nakas et al., 1983). These value added products can be used as monomers to chemically synthesize plastics like polyesters, polyurethanes, and polyethers (Adkins et al., 2012). In comparison to these polymers, which are chemically catalyzed and are of biological origin, PHAs are naturally occurring biodegradable plastics. Trans-esterification is a process through which biodiesel is produced from animal or vegetable fats along with methanol. During this process the alkali or acid catalysts convert methanol and triacylglycerol into glycerol and fatty acid methyl esters (biodiesel).

Crude glycerol produced during the transesterification process is approximately around 10% of the final amount of biodiesel (Pachauri and He, 2006). As the production of biodiesel has drastically increased, the manufacture of crude glycerol has also been obtained in very large quantities (Zhu et al., 2010). The transformation of glycerol in polymers like PHA depends highly on the concentration of the substrate in the medium. Reports have shown that *Zobellellade nitrificans* MW1 gave the highest yield which is 0.31g PHB/g from a glycerol concentration of 10g/L (Ibrahim and Steinbüchel, 2010). It was also observed that glycerol concentrations that were usually higher range like 20 g/L–50 g/L gave a low yield of product that was found to be 0.03, 0.12, and 0.21 PHB/g glycerol accordingly. The product yield escalated from 0.10 to 0.25 PHB/g when this same process was optimized and incorporated in the fed-batch system. Double-staged fermentation by *Ralstonia eutropha* using pure and waste glycerol, gave a yield of 0.36 and 0.34 g PHB/g glycerol, respectively (Cavalheiro et al., 2009). The activated sludge found in municipal waste water treatment plant usually contains mixed group microbe communities which can systematically utilize crude or raw glycerol and reportedly gave a yield of 0.40g PHA/g glycerol that was somewhat similar to the conversion rate of fatty acids which are used as carbon sources (Liu et al., 2010; Motralejo-Garate et al., 2011; Malaviya et al., 2012).

Thus, we can say that glycerol has a promising future as it can significantly reduce the production costs of PHA and increase the supply of value added by-products. The de novo synthesis of microbial PHA has been represented in Figure 1.6.

1.3.4 VEGETABLE WASTE

Biodegradable plastic derived from renewable sources is a great substitution for synthetic plastic (Kiser, 2016). Various food manufacturing plants trigger environmental as well as economic issues by producing waste in each step, that is, from production and supply, to its disposal. Food waste can be used as sources of raw materials for bioplastic production which will lead to significant improvement of the economy as well as effective waste utilization (Bayer et al., 2014; Perotto et al., 2018). For the development of the biorefinery concept, vegetable waste can be used as an important raw material (Clark et al., 2009). The most important macromolecule in the vegetable is cellulose that can serve as a resource for many value added products. The vegetable waste powder was directly converted into bioplastic with the help of HCl by water-based process. Precisely, the bioplastic was formed by blending cellulose crystals which are present in vegetable waste, pectin that dissolutes using HCl (May, 1990), and sugars which act as a plasticizer. Homogenous bioplastic with matrix and

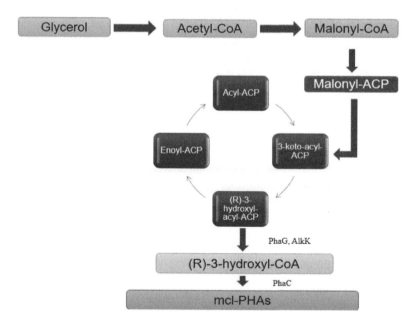

FIGURE 1.6   Fatty acid *de novo* synthesis.

crystalline cellulose was obtained by 85%–90% of the vegetable waste conversion. The vegetable waste-based bioplastic showed mechanical properties which are similar to PP and starch-based bioplastic, thus they are possible to combine with functional nanoparticles and hydrophobic macromolecules. These bioplastics are biodegradable; also, a loss of a small part of their mass when interacted with water makes them hydrophobic in nature which leads to remarkable depletion in mechanical characteristics. Biofilms have antioxidant effects indistinguishable to vegetables containing anthocyanins (known antioxidant), and acts as starch elastomeric (Tran et al., 2017). In food wrapping, appearance and longevity are the most important factors which are hindered by normal packaging in some ways. Thus, it's a good option to wrap it with vegetable-based bioplastic rich in anti-oxidants which prevents it from discoloration, thus positively improving shelf life (Kanatt et al., 2012).

1.3.5 Food Waste Valorization

An enormous amount of food, around 1.3 billion tons, is wasted across the world. It can be a threat to the environment with inappropriate disposal. A lot of factors affect food waste (FW) generation, such as production of the crop, infrastructure, distribution chains, and consumer purchase/usage habits. According to Ravindran and Jaiswal (2016), about 30% of food in the supply chain is discarded as waste. Processes like land-filling, soil amendment, fermentation, and so on, are used widely employed for safe disposal of food waste which otherwise results in groundwater contamination and greenhouse gas (GHG). Bioplastic production from food waste is an inexhaustible feasible procedure, whereupon resources are orchestrated from carbon

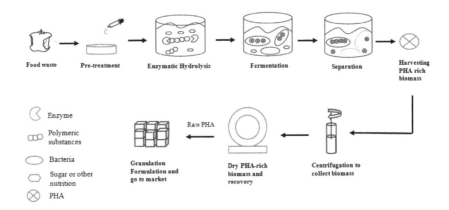

FIGURE 1.7 PHA biosynthesis process.

impartial assets. Certain biodegradable plastics are compostable by mechanical processes (Dietrich et al., 2017). Hence, usage of a biorefinery stage for food waste is a perfect alternative to source, for example, the creation of significant valuable items while diminishing the magnitude of waste. PHA has various uses, including bundling, clinical trials, vitality, and fine synthetics. The properties of PHA are perceived as better oxygen hindrance and high water vapor permeability. Microorganisms are able to store PHA in cell cytoplasm under pressured conditions. The creation strategy of PHA for the most part is experienced as substrate planning, PHA-aggregating aging, and extraction. There are 250 sorts of regular PHA makers which have been distinguished, and just a couple of microbes have been received for the business creation of PHA. These microbes, including *Alcaligeneslatus*, *Bacillus megaterium*, *Cupriavidusnecator*, and *Pseudomonas oleovorans*, are found to change over different carbon components to PHA. Specifically, *C.necator* is broadly used (Reddy et al., 2003). There are different advancements for changing over food waste to fermentable substrates, for example, mechanical and thermal transformation, chemical transformation and biological changes. Figure 1.7 shows a schematic outline of PHA production.

1.3.6 PALM TREE BIOMASS-BASED PROCESSING PLANTS

The availability of sources is the main consideration for the production of environmentally friendly, cost-effective, and efficient bioplastic. Biorefinery should be built in an area providing an adequate amount of resources from neighboring zones. Plantations as well as palm oil production generates enormous amounts of biomass (waste) which is 50–70 tons per 1ha. Hence, these can be a good raw material for biofuel production.

Expenses can be additionally decreased by enhancing the area of the creation industry and feedstock used (Shuit et al., 2009). A few activities were executed in Malaysia, such as the use of renewable energy, bioethanol, and so on, production from the waste generated to use biomass to discover savvy sustainable power resources (Ludin et al., 2004; Ong et al., 2011).

Considering the amount of biomass generated, Indonesia could be a good option for the ace generation of bioethanol and xylitol (Kresnowati et al., 2015). Based on total expenses for the professional duction of xylitol, an overall benefit of 4.3USDkg⁻¹ can be acknowledged as to whether biomass is utilized as a crude source. In a like manner, non-food sugar created from oil palm frond can be utilized for the practical creation of poly (3HB) (Zahari et al., 2015). It is evaluated that the creation cost of poly (3HB) can be diminished to (Konopka and Schnur, 1981; Kaewbai-Ngam et al., 2016) USD kg⁻¹ by utilizing inexhaustible sugars created from a palm leaf. Aside from poly (3HB) creation (Chiew and Shimada, 2013),the utilization of EFB can prompt the creation of various products including biogas, electricity, and paper (Sompong et al., 2012).

1.3.7 BANANA WASTE

Numerous amounts of banana leaf were dismissed and misused (in 2012) which was 26.46% and 6.67% separately (Quinaya and Alzate, 2014). Different products, such as PHB and biofuel can be generated from these waste products. In different papers it has been stated that 316 kg of glucose, 238 kg of ethanol, and 31.5 kg of PHB can be generated from one ton of banana waste (Naranjo et al., 2014). Additionally, banana strips are likewise significant feedstocks for the creation of differing items. Around 100 kg of banana strips can be utilized to produce 57, 2, 25, and 5 kg of glucose, acidic corrosive, and methane, separately (Quinaya and Alzate, 2014). In the main setting, PHB was an exceptional item, while banana strips were taken as the backup setting (Naranjo et al., 2014). PHB was produced from the hydrolyzed starch of banana skins.

1.4 CYANO BACTERIA AND PHB

Poly-β- hydroxybutyrate (PHB) is a storage compound found in prokaryotic organisms like cyanobacteria (Liebergesell et al., 1994; Ramaswamy et al., 2006; Mallick et al., 2007).The properties of poly-β-hydroxybutyrate are that they are thermoplastic, stable, completely biodegradable, and have absolute resistance to water, which indicates that PHB could be a better substitute to ordinary plastics and it also fits with waste management strategies.

1.4.1 PHB SYNTHESIZE

Poly-β-hydroxybutyrate is obtained from acetyl coenzyme A through three enzymatic reactions. In this, 3-Ketothiolase enzyme converts 2-acetyl-coA to 1-acetocetyl-coA molecule, then NADPH dependentacetoacetyl-coAreductase converts acetoacetyl-coA to D-3-hydroxybutyryl-coA, and the last enzyme PHB synthase catalyzes linking of the D-3-hydroxybutyryl moiety to an existing PHB molecule via an ester bond (Petrasovits et al., 2007).

Cyanobacteria can be chosen as a substitute host system for PHB production as they require a minimum amount of nutrients and utilize photoautotrophic method for the production of Poly-β- hydroxybutyrate. Cyanobacteria are known to collect the homo-polymer form of poly-β-hydroxybutyrate under photoautotrophic conditions.

Cyanobacteria are a group of oxygen developing photosynthetic bacteria; having a short generation time, they need inorganic nutrients like nitrate, phosphate, magnesium as macro nutrients and manganese, cobalt, and zinc as micronutrients for their growth and multiplication (Campell et al., 1982; Rehm and Steinbüchel, 1999, Dias et al., 2008; Fernandez-Nava et al., 2008; Jyotsana et al., 2010). The first cyanobacterial species which showed the presence of PHB is *Chlorogloeafritschii* in 1966. At present, a large number of cyanobacterial species demonstrate the presence of PHB such as *Spirulina sp.*, *Gloethece sp*, and so on (Campbell et al., 1982; Capon et al., 1983, Arino et al., 1995; Vincenzini et al., 1990; Wu et al., 2001; Melnicki et al., 2009).

They have oval structures which resemble PHB granules, which was detected by using ultra–structural analysis in *Trichodesmiumthiebautti and Microcystis aeruginosa*. Also, the PHB presence in *Oscillatorialmosasppand Gloethecespp.* were detected by Gas-liquid chromatography (Stal, 1992; Stal, 1992; Stahl et al., 1998; Miyake et al., 1996; Sabirova et al., 2006). These organisms can be easily cultured in waste waters as they use inorganic nitrogen, phosphorous, and so on (Poirier et al., 1995; Sudesh et al., 2000; Andreessen et al., 2010).

Cyanobacterium has low weight in contrast to other bacteria, due to its minute size and mass. It was also reported that the ability to synthesize PHB by cyanobacterium might be similar to most bacteria in nature. At times of large quantities of sunlight, carbon dioxide, and when growth is limited by other compounds like phosphorous and nitrogen, an excess of metabolic energy is synthesized by glycogen (a poly glucose) (Kaewbai-Ngam et al., 2016; Singh et al., 2017). During the night, an oxidative pentose phosphate pathway is used to oxidize glycogen and use it as an energy source. Previous studies reported that glycogen has more capability to store energy as compared to PHB, as cyanobacteria have an incomplete tricarboxylic acid cycle (De et al., 1992). Present studies show that a tricarboxylic cycle in cyanobacteria is still functional. Irrespective of this, the main use of PHB in the metabolism of cyanobacteria is still not clear. Most microbes can generate either glucogen or PHB but not both (Damrow et al., 2016). It is possible that PHB can be used as a carbon storage compound in the cell.

Research done by Kucho et al. (2005) on cyanobacteria *Synechocystis sp.* PCC6803, demonstrated that the circadian rhythm was found to be linked to the expression of PHB synthesis-related genes. It was observed that during the transition from light to dark cyclic expression of PHB synthesis-related genes peaked, along with other types of genes which were related to respiration (Kucho et al., 2005) (Figure 1.8). Based on this study they concluded that PHB can play an important role in providing carbon and energy during night time.

The utilization of cyanobacteria as a PHB producer in industries has the advantage that it can convert carbon dioxide which is a greenhouse gas to environmental-friendly plastics by using sunlight. Many species of cyanobacteria can accumulate PHB in large amounts. Those species of cyanobacteria which cannot accumulate PHB can

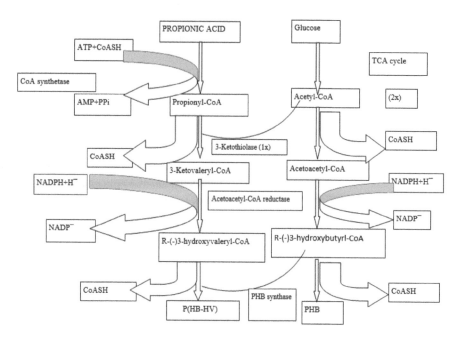

FIGURE 1.8 PHB production pathway in cyanobacteria.

now be engineered genetically by transformation with genes involved in PHB pathway (Winfred and Robards, 1973; Brandl et al., 1990; Sharma and Mallick, 2005).

1.4.2 DETECTION AND ANALYSIS OF PHB

When poly-β-hydroxybuty rate containing cyanobacteria are isolated from environment, it becomes very important to quickly screen a broad variety of cyanobacteria in a very short interval of time. Stains which show specificity to the PHB involve the detection of granules. In these techniques a viable colony gets stained. To test the cyanobacterial fat stain, Sudan black B is used (Hartman, 1940; Burdon et al., 1942; Ratledge, 2001; Haas et al., 2008). When Nile blue is used, a strong orange fluorescence is obtained in PHB granules. Nile blue stain shows more specificity for PHB than Sudan black B (Kranz et al., 1997; Salehizadeh and Van, 2004).

1.4.3 BIODEGRADABILITY AND BIOLOGICAL CONSIDERATIONS OF POLY-B-HYDROXYBUTYRATE

The main characteristics of PHB that makes it preferred to petroleum-derived bioplastics is their biodegradable nature. PHB under aerobic conditions when it undergoes biodegradation will release CO_2 and H_2O but under anaerobic conditions, the product of degradation released are CO_2 and CH4. It is compostable at variety of temperatures, such as maximum temperature 60°C. Based on present studies, 85% of PHB can get degraded within seven weeks. Research done on the biodegradability of PHB concluded that PHB does not float in water. Hence, when it is discarded, the plastic material made by using PHB gets degraded in the sediments of the surface by biogeochemical mechanisms (Mergaert et al., 1994; Porier et al., 1995; Lee, 1996; Lemos et al., 2006; Murphy, 2007).

PHB can be biodegraded in the environment. Microorganisms start to take over on the polymer surface and start to produce enzymes which degrade the P(HB-HV) into HB and HV units. These are later used by the cells as a source of carbon for the growth of biomass (Figure 1.9). Surface area, temperature, moisture, and the presence of other nutrient materials are the factors on which the rate of degradation of a polymer depends.

Cyanobacterial bioplastics manufactured using biopolymers can be obtained in two forms such as biopolymer based or obtained from polymerizable molecules (Hankermeyer and Tjeerdema, 1999; Kim and Lenz, 2001, Beccari et al., 2009). The biopolymers used can be starch, cellulose, and soya-based protein. Bioplastics obtained from cyanobacteria is the recent trend in the era of bioplastics compared to traditional methods (Madison and Huisman, 1999; Chen and Li, 2008; Castilho et al., 2009; Zhenggui et al., 2011). Cyanobacterial-based bioplastics have a lot of advantages, such as high yield in a range of environments. The use of these bioplastics neutralizes greenhouse gas emissions from power plants and factories, And hence helps in the conversion of fossil resources and reduces carbon dioxide emissions, thereafter helping sustainable development. Balaji et al. (2013) concluded that cyanobacteria has the capability to produce PHB by using CO_2 as an only source of carbon, but the technological methods for producing cyanobacterial-based bioplastics is at the research stage and will take time to be commercialized. These bioplastics are biodegradable and environmentally friendly and better than ordinary bioplastics.

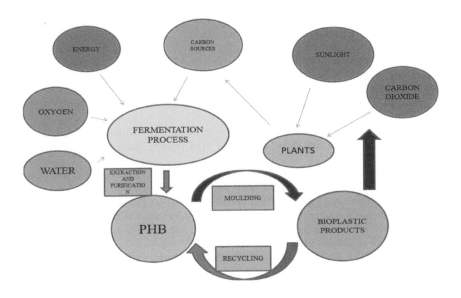

FIGURE 1.9 Biodegradabilty process of poly-β-hydroxybutyrate.

1.5 CONCLUSION

The improper discarding of pre-owned or used plastic waste leads to contamination of the environment. Long, drawn-out usage and exposure of plastics and plastic goods to high temperatures can lead to the leaching of toxic chemical constituents into food, drinks, and water. This can be a root of enormous health hazards, such as birth deformities, skin disorders, harmful to the immune system, cancer, respiratory diseases. On the level of raw materials, use of recycled materials or use of renewable resources as discussed in this chapter are two strategies to reduce CO_2 emissions and the dependency on fossil resources. The production process is another level where adjustments, for example towards a more energy-efficient process, can be made. A final level where efforts can increase sustainability is waste management.

REFERENCES

Abdou, E.S., K.S. Nagy, M.Z. Elsabee, Extraction and characterization of chitin and chitosan from local sources, *Bioresource Technology* 99(5) (2008):1359–1367.

Aboagye, Y., D.W. Stanley, Texturization of peanut proteins by surface film formation. 1. Influence of process parameters on film forming properties, *Canadian Institute of Food Science and Technology Journal* 18(1985):12–20

Adkins, J., S. Pugh, R. McKenna, D.R. Nielsen, Engineering microbial chemical factories to produce renewable "biomonomers", *Frontiers in Microbiology* 3 (2012):13.

Akyuz, L., M. Kaya, B. Koc, M. Mujtaba, S. Ilk, J. Labidi, A.M. Salaberria, Y.S. Cakmak, A. Yildiz, Diatomite as a novel composite ingredient for chitosan film with enhanced physicochemical properties, *International Journal of Biological Macromolecules* 105 (2017):1401–1411.

Albuquerque, M.G.E., V. Martino, E. Pollet, L. Averous, M.A.M. Reis, Mixed culture polyhydroxyalkanoate (PHA)production from volatile fatty acid (VFA)-rich streams: effectof substrate composition and feeding regime on phaproductivity, composition and properties, *Journal of Biotechnology* 151(1) (2011):66e76.

Andreessen, A., A.B. Lange, H. Robenek, A. Steinbuchel, Conversion of glycerol to poly (3-hydroxypropionate) in recombinant *Escherichia coli*, *Applied and Environmental Microbiology* 76 (2010):622–626.

Arino, X., J. Ortega-Calvo, M. Hernandez-Marine, C. Saiz-Jimenez, Effect of sulfur starvation on the morphology and ultrastructure of the cyanobacterium Gloeothece sp. PCC6909, *Archives of Microbiology* 163 (1995):447–453.

Avena-Bustillos, R.J., J.M. Krochta, Water vapor permeability of caseinate-based films as affected by PH, calcium cross-linking and lipid content, *Journal of Food Science* 58(1993): 904–907

Aydt, T.P., C.L. Weller, R.F. Testin, Mechanical and barrier properties of edible corn and wheat protein films, *Transactions of ASAE* 34(1991):207–211

Azizi, S., M.B. Ahmad, N.A. Ibrahim, M.Z. Hussein, F. Namvar, Cellulose nanocrystals/zno as a bifunctional reinforcing nanocomposite for poly (vinyl alcohol)/chitosan blend films: fabrication, characterization and properties, *International Journal of Molecular Sciences* 15(6) (2014):11040–11053.

Bahram, K., M. Bilal, K. Niazi, G. Samin, Z. Jahan, Thermoplastic Starch: A Possible Biodegradable Food Packaging Material—A Review, *Journal of Food Process Engineering*, 40(3) (2020):e12447.

Beccari, M., L. Bertin, D. Dionisi, F. Fava, S. Lampis, M. Majone, F. Valentino, G. Vallini, M. Villano, Exploiting olive oil mill effluents as a renewable resource for production of biodegradable polymers through a combined anaerobic–aerobic process, *Journal of Chemical Technology and Biotechnology* 84 (2009):901–908.

Brandenburg, A.H., C.L. Weller, R.F. Testin, Edible films and coatings from soy protein, *Journal of Food Science* 58 (1993):1086–1089

Brandl, H., R.A. Gross, R.W. Lenz, R.C. Fuller, Plastic bacteria? Progress and prospects for polyhydroxyalkanoate production in bacteria, *Advances in Biochemical Engineering/ Biotechnology* 41 (1990):77–93.

Burdon, K.L., J.C. Stokes, C.E. Kimbrough, Studies of the common aerobic spore-forming Bacilli staining for fat with Sudan Black B- stain, *Journal of Bacteriology* 43 (1942):717–724.

Campell, J., S. Edward Stevens, D.L. Balkwill, Accumulation of polyb-hydroxybutyrate in Spirulina platensis, *Journal of Bacteriology* 149 (1982):361–363.

Capon, R.J., R.W. Dunlop, E.L. Ghisalberti, P.R. Jefferies, Poly-3-hydroxyalkanoates from marine and freshwater cyanobacteria, *Phytochemistry* 22 (1983):1181–1184.

Castilho, L.R., D.A. Mitchell, D.M.G. Freire, Production of polyhydroxyalkanoates (phas) from waste materials and by-products by submerged and solid-state fermentation, *Bioresource Technology* 100 (2009):5996–6009.

Cavalheiro, J.M.B.T., M.D. de Almeida, C. Grandfils, M.M.R. da Fonseca, Poly (3-hydroxybutyrate) production by Cupriavidus necator using waste glycerol, *Process Biochemistry* 44 (2009): 509.

Chen, H., X. Li, Effect of static magnetic field on synthesis of polyhydroxyalkanoates from different short-chain fatty acids by activated sludge, *Bioresource Technology* 99 (2008):5538–5544.

Chiew, Y.L., S. Shimada, Current state and environmental impact assessment for utilizing oil palm empty fruit bunches for fuel, fiber and fertilizer - a case study of Malaysia, *Biomass Bioengineering* 51 (2013):109–124.

Choi, J.I., S.Y. Lee, Process analysis and economic evaluation for Poly(3-hydroxybutyrate) production by fermentation. *Bioprocess Engineering* 17 (1997):335.

Choi, J., S.Y. Lee, Factors affecting the economics of polyhydroxyalkanoate production by bacterial fermentation, *Applied Microbiology and Biotechnology* 51 (1999):13.

Chua, A.S.M., H. Takabatake, H. Satoh, T. Mino, Production of polyhydroxyalkanoates (PHA) by activated sludge treating municipal wastewater: effect of PH, sludge retention time (SRT), and acetate concentration in influent, *Water Research* 37 (15) (2003):e3602–e3611.

Clark, J.H., F.E.I. Deswarte, T.J. Farmer, Biofuels, *Bioproducts and Biorefining* 3 (2009):72–90.

Damrow, R., I. Maldener, Y. Zilliges, R.E. Blankenship, The multiple functions of common microbial carbon polymers, glycogen and PHB, during stress responses in the non-diazotrophic Cyanobacterium Synechocystissp, *Frontiers in Microbiology* 7 (June) (2016):1–10.

Darbasi, M., G. Askari, H. Kiani, F. Khodaiyan, Development of chitosan based extended-release antioxidant films by control of fabrication variables, *International Journal of Biological Macromolecules* 104 (2017):303–310.

Dayarian, S., A. Zamani, A. Moheb, M. Masoomi, Physico-mechanical properties of films of chitosan, carboxymethyl chitosan, and their blends, *Journal of Polymers and the Environment* 22(3) (2014):409–416.

De Moraes Crizel, T., A. De Oliveira Rios, V.D. Alves, N. Bandarra, M. Moldão-Martins, S.H. Flôres, Active food packaging prepared with chitosan and olive pomace, *Food Hydrocolloids* 74 (2018):139–150.

De Philippis, R., A. Ena, M. Guastiini, C. Sili, M. Vincenzini, Factors affecting polyβ-hydroxybutyrate accumulation in cyanobacteria and in purple non-sulfur bacteria, *FEMS Microbiology Letters* 103(2–4) (1992):187–194.

Deng, Z., J. Jung, Y. Zhao, Development, characterization, and validation of chitosan adsorbed cellulose nanofiber (CNF) films as water resistant and antibacterial food contact packaging, *LWT-Food Science and Technology* 83 (2017):132–140.

Dias, J.M., P.C. Lemos, L.S. Serafim, C. Oliveira, M. Eiroa, M.G. Albuquerque, Recent advances in polyhydroxyalkanoate production by mixed aerobic cultures: from the substrate to the final product, *Macromolecular Bioscience* 6 (2006):885–906.

Dias, J.M.L., A. Oehmen, L.S. Serafim, P.C. Lemos, M.A.M. Reis, R. Oliveira, Metabolic modelling of polyhydroxyalkanoate copolymers production by mixed microbial cultures, *BMC Systems Biology* 2(1) (2008):1–12.

Dietrich, K., M.J. Dumont, L.F. Del Rio, V. Orsat, Producing phas in the bio-economy towards a sustainable bioplastic, *Sustainable Production and Consumption* 9 (2017):58–70.

ElisabetaElena, T., R. Maria, P. Ovidiu, Biopolymers based on renewable resources - a review, *Bioresources Technology* 87 (2014):137–146.

Elsabee, M.Z., E.S. Abdou, Chitosan based edible films and coatings: A review, *Materials Science and Engineering: C* 33(4) (2013):1819–1841.

Fernandez-Nava, Y., E. Maranon, J. Soons, L. Castrillon, Denitrification of wastewater containing high nitrate and calcium concentrations, *Bioresource Technology* 99(17) (2008): 7976–7981.

Fernández-Pan, I., K. Ziani, R. Pedroza-Islas, J. Maté, Effect of drying conditions on the mechanical and barrier properties of films based on chitosan, *Drying Technology* 28(12) (2010):1350–1358.

Gao, X., J.-C. Chen, Q. Wu, G.-Q. Chen, Polyhydroxyalkanoates as a source of chemicals, polymers,and biofuels, *Current Opinion in Biotechnology* 22(6) (2012):e768–e774.

Gennadios, A., H.J. Park, C.L. Weller, Relative humidity and temperature effects on tensile strength of edible protein and cellulose andher films, *Transactions of ASAE* 36 (1993):1867–1872

Giannakas, A., M. Vlacha, C. Salmas, A. Leontiou, P. Katapodis, H. Stamatis, N.-M. Barkoula, A. Ladavos, Preparation, characterization, mechanical, barrier and antimicrobial properties of chitosan/PVOH/clay nanocomposites, *Carbohydrate Polymers* 140 (2016):408–415.

Guilbert, S., Use of superficial edible layer to protect intermediate moisture foods: application to the protection of tropical fruit dehydrated by osmosis. In: C.C. Seow, T.T. Teng, C.H. Quah (Eds.), *Food Preservation by Moisture Control*. London: Elsevier Applied Science, 1988, 199–219

Haas, R., B. Jin, F.T. Zepf, Production of poly(3-hydroxybutyrate) from waste potato starch, *Bioscience, Biotechnology, and Biochemistry* 72 (2008):253–256.

Hankermeyer, C.R., R.S. Tjeerdema, Polyhydroxybutyrate: plastic made and degraded by microorganisms, *Reviews of Environmental Contamination and Toxicology* 159 (1999):1–24.

Hartman, T.L., The use of Sudan Black B as a bacterial fat stain, *Staining Technology* 15 (1940):23–28.

Herald, T.J., K.A. Hachmeister, S. Huang, J.R. Bowers, Corn zein packaging materials for cooked turkey, *Journal of Food Sciences* 61 (1996):415–421

Hernandez-Izquierdo, V.M., J.M. Krochta, Thermoplastic processing of proteins for film formation—a review, *Journal of Food Science* 73(2008):R30–R39. Doi:10.1111/j.1750-3841.2007.00636.x

Ibrahim, M.H.A., A. Steinbüchel, Zobellella denitrificans strain MW1, a newly isolated bacterium suitable for poly (3-hydroxybutyrate) production, from glycerol, *Applied Environmental Microbiology* 108 (2010): 214.

Jane, J., S. Lim, I. Paetau, K. Spence, S. Wang, Biodegradable plastics made from agricultural biopolymers. In: Fishman, M.L., Friedman, R.B., Huang, S.J. (Eds.), *Polymers from Agricultural Coproducts*. Chicago, 1994, 92–100

Jyotsana, M., L. Priyangshusarma, K. Meeta, L. Ajoymandal, Evaluation of bacterial strains isolated from oil-contaminated soil for production of polyhydroxyalkanoic acids (PHA), *Pedobiologia* 54 (2010):25–30.

Kaewbai-Ngam, A., A. Incharoensakdi, T. Monshupanee, Increased accumulation of polyhydroxybutyrate in divergent cyanobacteria under nutrient-deprived photoautotrophy: an efficient conversion of solar energy and carbon dioxide to polyhydroxybutyrate by *Calothrix scytonemicola* TISTR 8095, *Bioresource Technology* 212 (2016):342–347.

Kalaycıoğlu, Z., E. Torlak, G. Akın-Evingür, İ. Özen, F.B. Erim, Antimicrobial and physical properties of chitosan films incorporated with turmeric extract, *International Journal of Biological Macromolecules* 101 (2017):882–888

Kanatt, S.R., M. Rao, S. Chawla, A. Sharma, *Food Hydrocolloids*, 29 (2012): 290–297.

Kerch, G., Chitosan films and coatings prevent losses of fresh fruit nutritional quality: a review, *Trends in Food Science & Technology* 46(2) (2015):159–166.

Kerch, G., V. Korkhov, Effect of storage time and temperature on structure, mechanical and barrier properties of chitosan-based films, *European Food Research and Technology* 232(1) (2011):17–22.

Kim, Y.B., R.W. Lenz, Polyesters from microorganisms, *Advances in Biochemical Engineering/Biotechnology* 71 (2001):51–79.

Kiser, *Nature* 531 (2016):443–446.

Konopka, A., M. Schnur, Physiological ecology of a metalimnetic *Oscillatoria rubescens* population, *Journal of Phycology* 17 (1981):118–122.

Kranz, R.G., K.K. Gabbert, M.T. Madigan, Positive selection systems for discovery of novel polyester biosynthesis genes based on fatty acid detoxification, *Applied and Environmental Microbiology* 63 (1997):3010–3013.

Kresnowati, M., E. Mardawati, T. Setiadi, Production of xylitol from oil palm empty fruits bunch: a case study on biorefinery concept, *Modern Applied Science* 9(7) (2015):206–213.

Kucho, K., K. Okamoto, Y. Tsuchiya, S. Nomura, M. Nango, M. Kanehisa, M. Ishiura, S. Pcc, Global analysis of circadian expression in the Cyanobacterium Synechocystis sp. Global analysis of circadian expression in the Cyanobacterium, *Journal of Bacteriology* 187(6) (2005):2190–2199.

Kulpreecha, S., A. Boonruangthavorn, B. Meksiriporn, Inexpensive fed-batch cultivation for high poly (3-hydroxybutyrate) production by a new isolate of *Bacillus megaterium*, *Journal of BiosciBioeng* 107(3) (2009):240–245.

Leceta, I., P. Guerrero, K. De la Caba, Functional properties of chitosan-based films, *Carbohydrate Polymers* 93(1) (2013):339–346.

Lee, S.Y., Bacterial polyhydroxyalkanoates, *Biotechnology and Bioengineering* 49 (1996): 1–14.

Lemos, P.C., L.S. Serafim, M.A.M. Reis, Synthesis of polyhydroxyalkanoates from different short-chain fatty acids by mixed cultures submitted to aerobic dynamic feeding, *Journal of Biotechnology* 122 (2006):226–238.

Liebergesell, M., K. Sonomoto, M. Madkour, F. Mayer, A. Steinbuchel, Purification and characterization of the poly(hydroxyalkanoic acid) synthase from Chromatiumvinosum and localization of the enzyme at the surface of poly(hydroxyalkanoic acid) granules, *European Journal of Biochemistry* 226 (1994):71–80.

Liu, J., C.G. Meng, S. Liu, J. Kan, C.-H. Jin, Preparation and characterization of protocatechuic acid grafted chitosan films with antioxidant activity, *Food Hydrocolloids*63 (2017):457–466.

Liu, B.F., N.Q. Ren, J. Tang, J. Ding, W.Z. Liu, J.F. Xu, G.L. Cao, W.Q. Guo, G.J. Xie, Bio-hydrogen production by mixed culture of photo and dark fermentation bacteria, *International Journal of Hydrogen Energy* 35 (2010):2858–2862

Ludin, N.A., M.A. Bakri, M. Hashim, B. Sawilla, N.R. Menon, H. Mokhtar, *Palm oil biomass for electricity generation in Malaysia.* Pusat Tenaga Malaysia, Malaysia Palm Oil Board, SIRIM Berhad, 2004.

Madison, L.L., G.W. Huisman, Metabolic engineering of poly(3-hydroxyalkanoates): from DNA to plastic, *Microbiology and Molecular Biology Reviews* 63 (1999):21–53.

Malaviya, A., Y.S. Jang, S.Y. Lee, Continuous butanol production with reduced byproducts formation from glycerol by a hyper producing mutant of *Clostridium pasteurianum*, *Applied Microbiology Biotechnology* 93 (2012):1485.

Mallick, N., L. Sharma, A.K. Singh, Polyhydroxyalkanoate (PHA) synthesis by Spirulina sub-salsa from Gujarat coast of India, *Journal of Plant Physiology* 164 (2007):312–317.

Marquié, C., A.M. Tessier, C. Aymard, S. Guilbert, HPLC determination of the reactive lysine content of cottonseed protein films to monitor the extend of cross-linking by formalde-hyde, glutaraldehyde, and N-glyoxal, *Journal of Agricultural Food Chemistry* 45 (1997):922–926

May, C.D., Industrial pectins: sources, production and applications, *Carbohydrate Polymers* 12 (1990):79–99.

Melnicki, M.R., E. Eroglu, A. Melis, Changes in hydrogen production and polymer accumula-tion upon sulfur-deprivation in purple photosynthetic bacteria, *International Journal Hydrogen Energy* 34 (2009):6157–6170.

Mergaert, J., C. Anderson, A. Woulers, T. Swings, Microbial degradation of poly (3-hydroxy-butyrate) and poly (3-hydroxybutyrate co-3-hydroxyvalarate) in compost, *Journal of Environmental Polymer Degradation* 2 (1994):177–183.

M. Miyake, M. Erata, Y. Asada, A thermophilic cyanobacterium, Synechococcus sp. MA19, capable of accumulating poly- b-hydroxybutyrate, *Journal of Fermentation and Bioengineering* 82 (1996):512–514.

Miyazaki, M.R., P.V. Hung, T. Maeda, N. Morita, Recent advances in application of modified starches for breadmaking, *Trends Food Science Technology* 17 (2006):591–599.

Motralejo-Garate, H., E. Mar'atusalihat, R. Kleerebezem, M.M. Loosdrecht, *Microbial com-munity engineering for biopolymer production from glycerol*, Applied Microbiology Biotechnology 92 (2011): 631.

Murphy, D.J., Improving containment strategies in biopharming, *Plant Biotechnology Journal* 5 (2007):555–569.

Nakas, P., M. Schaedle, C.M. Parkinson, C.E. Coonley, S.W. Tanenbaum, *Applied Environmental Microbiology* 46 (1983: 1017.

Naranjo, J.M., C.A. Cardona, J.C. Higuita, Use of residual banana for poly- hydroxybutyrate (PHB) production: case of study in an integrated biorefinery, *Waste Management* 34 (2014):2634–2640.

Nunthanid, J., S. Puttipipatkhachorn, K. Yamamoto, G.E. Peck, Physical properties and molecular behavior of chitosan films, *Drug Development and Industrial Pharmacy* 27(2) (2001):143–157

Nur, Z.Y., A. Belma, B. Yavuz, M. Nazime, Effect of carbon and nitrogen sources and incuba-tion time on poly-beta-hydroxybutyrate (PHB) synthesis by Bacillus megaterium 12, *African Journal of Biotechnology* 3 (2004):63–69.

Ohwoavworhua, F.O., A. Osinowo, Preformulation studies and compaction properties of a new starch-based pharmaceutical aid, *Research Journal of Pharmaceutical, Biological and Chemical Science* 1 (2010):255–270

Ong, H.C., T.M.I. Mahlia, H.H. Masjuki, A review on energy scenario and sus-tainable energy in Malaysia, *Renewable Sustainable Energy Review* 15(1) (2011):639–647.

Pachauri, N., B. He, *Value-added utilization of crude glycerol from biodiesel production: a survey of current research activities*. Presented at *2006 American Society of Agricultural and Biological Engineers Annual International Meeting*, Portland, Oregon, 2006.

Park, H.J., M.S. Chinnan, Properties of edible coatings for fruits and vegetables. ASAE Paper 90-6510, 1990

Park, H., S. Jung, J. Song, S. Kang, P. Vergano, R. Testin, Mechanical and barrier properties of chitosan-based biopolymer film, *Chitin and Chitosan Research* 5 (1999):19–26.

Park, S., K. Marsh, J. Rhim, Characteristics of different molecular weight chitosan films affected by the type of organic solvents, *Journal of Food Science* 67(1) (2002):194–197.

Perotto, G., L. Ceseracciu, R. Simonutti, U.C. Paul, S. Guzman-Puyol, T.N. Tran, I.S. Bayer, A. Athanassiou, Bioplastics from vegetable waste via an eco-friendly water-based process, *Green Chemistry* 20(4) (2018):894–902.

Petrasovits, L.A., M.P. Purnell, L.K. Nielsen, S.M. Brumbley, Production of polyhydroxybutyrate in sugarcane, *Plant Biotechnology Journal* 5 (2007):162–172.

Pijuan, M., C. Casas, J.A. Baeza, Polyhydroxyalkanoate synthesis using different carbon sources by two enhanced biological phosphorus removal microbial communities, *Process Biochemistry* 44 (2009):97–105.

Poirier, Y., C. Somerville, L.A. Schechtman, M.M. Satkowski, I. Noda, Synthesis of high-molecular-weight poly([R]-(−)-3-hydroxybutyrate) in transgenic Arabidopsis thaliana plant cells, *International Journal of Biological Macromolecules* 17 (1995):7–12.

Porier, Y., C. Nawarath, C. Somerville, Production of polyhydroxyalkanoates, a family of biodegradable plastics and elastomers in bacterial and plant, *Biotechnology* 13 (1995):142–150.

Pradhan, G.C., S. Dash, S.K. Swain, Barrier properties of nano silicon carbide designed chitosan nanocomposites, *Carbohydrate Polymers* 134 (2015):60–65.

Quinaya, S.H.D., C.A.C. Alzate, Plantain and banana fruit as raw material for glucose production, *Journal of Biotechnology* 185 (2014):S34.

Ramaswamy, A.V., P.M. Flatt, D.J. Edwards, T.L. Simmons, B. Han, W.H. Gerwick, The secondary metabolites and biosynthetic gene clusters of marine cyanobacteria. Applications in biotechnology. In: P. Proksch, W.E.G. Muller (Eds.), *Frontiers in Marine Biotechnology*. UK: Horizon Bioscience, 2006, 175–224.

Ratledge, B.K., *Basic Biotechnology*, 2nd edn. Cambridge: Cambridge University Press, 2001.

Reddy, C.S.K., R. Ghai, V.C.C. Kalia, Polyhydroxyalkanoates: an overview, 2003.

Rehm, B.H.A., A. Steinbüchel, Biochemical and genetic analysis of PHA synthases and other proteins required for PHA synthesis, *International Journal of Biological Macromolecules* 3 (1999):3–19.

Rui, L., M. Xie, B. Hu, L. Zhou, D. Yin, X. Zeng, A comparative study on chitosan/gelatin composite films with conjugated or incorporated gallic acid, *Carbohydrate Polymers* 173 (2017):473–481.

Sabirova, J.S., M. Ferrer, H. Lunsdorf, V. Wray, R. Kalscheuer, A. Steinbüchel, K.N. Timmis, P.N. Golyshin, Mutation in a "tesB-like" hydroxyacyl-coenzyme A-specific thioesterase gene causes hyperproduction of extracellular polyhydroxyalkanoates by Alcanivorax borkumensis SK2, *Journal of Bacteriology* 188 (2006):8452–8459.

Salehizadeh, H., M.C.M. Van Loosdrecht, Production of polyhydroxyalkanoates by mixed culture: recent trends and biotechnological importance, *Biotechnology Advances* 22 (2004):261–279.

Saranya,V., R. Shenbagarathai, Effect of nitrogen and calcium sources on growth and production of PHA of Pseudomonas sp. LDC-5 and its mutant, *Current Research Journal of Biological Science* 2(3) (2010):164–167

Schübert, P.A., A. Steinbüchel, H.G. Schlegel, Cloning of the *Alcaligenes eutrophus* genes for the synthesis of poly-β-hydroxybutyrate (PHB) and synthesis of PHB in *Escherichia coli*, *Journal of Bacteriology* 170 (1988):5837–5847.

Serafim, L.S., P.C. Lemos, M.G. Albuquerque, M.A. Reis, Strategies for PHA production by mixed cultures and renewable waste materials, *Applied Microbiology Biotechnology* 81(4) (2008):615–628.

Sharma, L., N. Mallick, Enhancement of poly-β-hydroxybutyrate accumulation in Nostocmuscorum under mixotrophy, chemoheterotrophy and limitations of gas-exchange, *Biotechnology Letters* 27 (2005):59–62.

Shuit, S.H., K.T. Tan, K.T. Lee, A.H. Kamaruddin, Oil palm biomass as a sustainable energy source: a Malaysian case study, *Energy* 34 (2009):1225–1235.

Shukla, R., Cheryan, M., Zein: the industrial protein from corn, *Industrial Crops Production* 13 (2001):171–192

Singh, A.K., L. Sharma, N. Mallick, J. Mala, Progress and challenges in producing polyhydroxyalkanoate biopolymers from cyanobacteria, *Journal of Applied Phycology* 29 (3) (2017):1213–1232.

Siripatrawan, U., W. Vitchayakitti, Improving functional properties of chitosan films as active food packaging by incorporating with propolis, *Food Hydrocolloids* 61 (2016):695–702.

Sompong, O.-T., Boe, K., Angelidaki, I., Thermophilic anaerobic co-digestion of oil palm empty fruit bunches with palm oil mill effluent for efficient biogas production, *Applied Energy* 93 (2012):648–654.

Stahl, A., D. Sayler, G. Geesey, *Techniques in Microbial Ecology*. Ronald New York, Oxford: Oxford University Press, 1998.

Stal, L.J., Poly(hydroxyalkanoate) in cyanobacteria: a review, *FEMS Microbiology Reviews* 103 (1992):169–180.

Steinbüchel, A., B. Füchtenbush, Bacterial and other biological systems for polyester production, *Trends in Biotechnology* 16(10) (1998):419e427.

Stuchell, Y.M., Krochta, J.M., Enzymatic treatments and thermal effects on edible soy protein films, *Journal of Food Science* 59 (1994):1332–1337

Sudesh, K., H. Abe, Y. Doi, Synthesis, structure and properties of polyhydroxyalkanoates: biological polyesters, *Progress in Polymer Science* 25 (2000):1503–1555.

Takenaka, H., H. Ito, H. Asano, H. Hattori, On some physical properties of film forming materials, *Gifu YakkaDaigaku* 17 (1967):142–146

Tanada-Palmu, P.S., C.R.F. Grosso, Effect of edible wheat gluten-based films and coatings on refrigerated strawberry (*Fragaria ananassa*) quality, *Postharvest Biology Technology* 36(2) (2005):199–208

Tran, T.N., A. Athanassiou, A. Basit, I.S. Bayer, *Food Chemistry* 216 (2017):324–333.

Tripathi, A.D., S.K. Srivastava, Kinetic study of biopolymer (PHB) synthesis in Alcaligenes sp. In submerged fermentation process using TEM, *Journal of Polymer Science Environment* 19 (2011):732–738.

Tryhnew, L.J., K.W. Gunaratne, J.V. Spencer, Effect of selected coating materials on the bacterial penetration of the avian egg shell, *Journal of Milk and Food Technology* 36 (1973):272–275

U.S.A. Energy Information Administration, Annual Energy Review 2011, 2012, http://www.eia.gov/totalenergy/data/annual/pdf/aer.pdf. Accessed Jan. 15, 2013.

van den Broek, L.A., R.J. Knoop, F.H. Kappen, C.G. Boeriu, Chitosan films and blends for packaging material, *Carbohydrate Polymers* 116 (2015):237–242.

Vilela, R.J., J. Pinto, M.R. Coelho, S. Domingues, P. Daina, S.A. Sadocco, C.S. Santos, Bioactive chitosan/ellagic acid films with UV-light protection for active food packaging, *Food Hydrocolloids* 73 (2017):120–128.

Vincenzini, M., C. Sili, R. Philippis, A. Ena, R. Materassi, Occurrence of poly-bhydroxybutyrate in Spirulina species, *Journal of Bacteriology* 172 (1990):2791–2792.

Wang, J., Z.B. Yue, G.P. Sheng, H.Q. Yu, Kinetic analysis on the production of polyhydroxy-alkanoates from volatile fatty acids by Cupriavidusnecator with a consideration of substrate inhibition, cell growth, maintenance, and product formation, *Biochemical Engineering Journal* 49 (2010):422–428.

Wieser, H., Chemistry of gluten proteins, *Food Microbiology* 24(2) (2007):115–119

Winfred, F.D., A.W. Robards, Ultra structural study of Poly hydroxy butyrate granules from Bacillus cereus, *Journal of Bacteriology* 114 (1973):1271–1280.

Wu, L.C., R.P. Bates, Influence of ingredients upon edible protein-lipid characteristics, *Journal of Food Science* 38(1973):783–787

Wu, G.F., Q.Y. Wu, Z.Y. Shen, Accumulation of poly-b-hydroxybutyrate in cyanobacterium Synechocystissp PCC6803, *Bioresource Technology* 76 (2001):85–90.

Xie, T., Z. Liao, H. Lei, X. Fang, J. Wang, Q. Zhong, Antibacterial activity of food-grade chitosan against *Vibrio parahaemolyticus* biofilms, *Microbial Pathogenesis* 110 (2017):291–297.

Xing, Y., Q. Xu, X. Li, C. Chen, L. Ma, S. Li, Z. Che, H. Lin, Chitosan-based coating with antimicrobial agents: preparation, property, mechanism, and application effectiveness on fruits and vegetables, *International Journal of Polymer Science* 2016 (2016): 4851730.

Zahari, M.A.K.M., H. Ariffin, M.N. Mokhtar, J. Salihon, Y. Shirai, M.A. Hassan, Case study for a palm biomass biorefinery utilizing renewable non-food sugars from oil palm frond for the production of poly(3-hydroxybutyrate) bioplastic, *Journal of Cleaning Products* 87 (2015):284–290.

Zhenggui, L., Y. Wang, N. He, J. Huang, K. Zhu, W. Shao, H. Wang, W. Yuan, Q. Li, Optimization of polyhydroxybutyrate (PHB) production by excess activated sludge and microbial community analysis, *Journal of Hazardous Materials* 185 (2011):8–16.

Zhu, C., C.T. Nomura, J.A. Perrotta, A.J. Stipanovic, J.P. Nakas, Production and characterization of poly-3-hydroxybutyrate from biodiesel-glycerol by Burkholderia cepacia ATCC 17759, *Biotechnology Progress* 26 (2010):424.

Ziani, K., J. Oses, V. Coma, J.I. Maté, Effect of the presence of glycerol and Tween 20 on the chemical and physical properties of films based on chitosan with different degree of deacetylation, *LWT-Food Science and Technology* 41(10) (2008):2159–2165.

2 Bioelectrochemical System

Waste/Wastewater to Bioenergy Conversion Technology

Samsudeen Naina Mohamed and K. M. Meera Sheriffa Begum
National Institute of Technology, Tiruchirappalli, India

CONTENTS

2.1 INTRODUCTION

Globally, fossil fuel, especially petroleum, oil and natural gas, have supported countries' economics over past century. The energy generated from solar, thermal and hydro-power are not stable sources and cannot be completely depended on for energy requirements [1]. Industries consume maximum energy for the production as well as wastewater treatment processes. Industrial operations have not increased energy demand alone and pollute the environment by releasing waste/wastewater into the aquatic environment. Wastewater contains significant amounts of chemical oxygen demand (COD) and biological oxygen demand (BOD), total dissolved solids (TDS), sulphites, nitrates, color, and so on, which contaminate soil alkalinity and pollute underground water if not properly discharged. Generally, physico-chemical and biological methods (aerobic and anaerobic) are employed to treat industrial wastewater [2]. In biological treatment, aerobic or anaerobic bacteria utilize the organic matter in the wastewater to produce alternative fuel (methane, hydrogen, volatile fatty acids). However, these are an energy consuming process and pollute the environment by releasing greenhouse gas (CO_2, N_2O etc.) during treatment. Therefore, the development of wastewater treatment technologies, that is, converting biodegradable organic matter into energy by green technologies, is one of the ideal solutions to achieve renewable energy production and environmental sustainability.

Recently, bioelectrochemical systems (BES) have been considered as novel green technology that generates bioenergy in the form of bioelectricity, biohydrogen, and other value-added products by microbial metabolism and simultaneously treating the waste/wastewater [3, 4]. The BES technology has several advantages over conventional process such as electricity and other value-added products, no aeration is needed, no intermediate processes, less production, can be operated at any temperature, used as biosensor, and so on [5]. However, the BES have several limitations such as construction and material costs are still expensive. The power generation of BES is still low as compared to Hydrogen Fuel Cell (FC). Recently, power generation has increased several-fold by various researchers by varying operational and design factors [6]. It is expected that BES technology will be a highly promising energy source in the future.

2.2   BASIC PRINCIPLES OF BIOELECTROCHEMICAL SYSTEM (BES)

BES is a promising and sustainable technology to produce bioenergy, especially using wastewater as a substrate, that generates electricity and wastewater treatment simultaneously [7, 8]. The typical BES consists of an anode and cathode compartment separated by salt bridge or membrane as shown in Figure 2.1. In anode, the microorganism degrades the inorganic and organic matter present in the waste/wastewater and releases the protons (H^+) and electrons (e^-) under an anaerobic environment. The protons are transferred to the cathode through the separator (membrane), while the generated electrons are transferred via an external circuit to the cathode. In the cathode, the protons and electrons combine with electron acceptors (e.g. Oxygen, Ferricyanide etc.) and generate electricity and water [9].

In BES, the bacterial strains have different electron transfer mechanisms from the substrate to the anode surface. Based on the electron's transfer, the BES can be classified as mediator and mediator-less system. If the electrons are directly transferred from electrolyte to the electrode surface, this is called mediator-less BES. In the mediator system, electrons are transferred to the cathode with the help of a

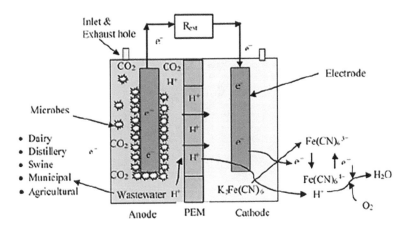

FIGURE 2.1 Schematic representation of bioelectrochemical system.

mediator (e.g. neutral red, potassium ferricyanide, viologen etc.) to the electrode [10]. But the mediators are quite expensive and cannot be recycled once added into the system. Hence, most researchers have focused on the mediator-less system for enhancing the performance of BES.

2.2.1 ELECTRON TRANSFER MECHANISM

Electrochemically active bacteria has great important in the BES since it generates not only electricity but also removal heavy metals from the wastewater. Generally, electrochemically active bacteria referred as "exo-electrogens", in which exo - stands for exocellular and electrogens are based on the ability to directly transfer electrons to the electrode surface [11]. Bacteria oxidizes the organic matter and electrons are transferred to respiratory enzymes by NADH (reduced from of NAD+). These electrons flow down a respiratory chain moving protons across an inner membrane. The protons flow back into the cell through the enzyme ATPase, creating ATP. The electrons are finally released to a soluble terminal electron acceptor and reach the surface of the electrode. When the electrons exit the respiratory chain at some reduction potential less than this of the oxygen then the bacteria obtain less energy (Figure 2.2a). The microorganism transfers the electrodes via two mechanisms such as direct electron transfer and indirect electron transfer (Figure 2.2b). Direct electron transfer is referred to as direct contact between the bacterial enzyme from the active cell membrane (conductive pili or nanowires) and the electrode surface. These microorganisms (e.g. *Geobacter* sp. and *Shewanella* sp.) have a membrane-bound electron transport protein to relay the electrons transferred from inside the bacterial cell to its outer surface or vice versa. In the indirect mechanism, electrons transfer requires a mediator to shuttle electrons between the microorganism and electrode surface [12]. Hence, mediators (e.g. methyl viologen, anthraquinone -2, 6-disulfonate (AQDS), neutral red etc.) are required to shuttle electrons between electrodes and microorganisms, and so on [13].

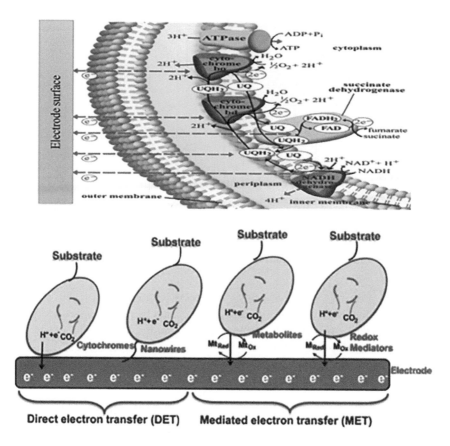

FIGURE 2.2 (a) Electron transport chain, NAD+ to NADH, ADP into ATP. (b). Electrons transfer mechanism between exoelectrogens and anode.

2.2.2 VOLTAGE GENERATION IN BES

In BES, the microorganisms act as a biocatalyst for the degradation of inorganic/organic matter in the anode and platinum (Pt) and other noble metal catalysts (Pt, Au, Ag, SS etc.) are used in the cathode side for the reduction reaction. The potential difference between the anode (low redox potential) and cathode (high redox potential) is the driving force which generates the flow of current [14]. The potential generated in the anode and cathode side can be calculated theoretically based on biochemical reactions as shown below:

Anode Reaction:

If acetate is present in the wastewater,

$$CH_3COO^- + H^+ + 2H_2O \rightarrow 2CO_2 + 8H^+ + 8e^-; \quad E^{o'} = -0.296 \text{ V} \quad (2.1)$$

Cathode Reaction:

If oxygen acts as electron acceptors,

$$O_2 + 8e^- + 8H^+ \rightarrow 4H_2O \quad E^{o'} = 0.816 \text{ V} \tag{2.2}$$

Using the Nernst equation, the Gibbs free energy value corresponds to a theoretical cell voltage of

$$E_{cell} = \frac{-\Delta G_{cell}}{nF} = \frac{-\left(-842000\,\dfrac{J}{mol}\right)}{8mole^-\,x96485\,\dfrac{c}{mole^-}} \quad E_{cell} = 0.109 \text{ V} \tag{2.3}$$

The overall cell electromotive force (E_{cell}) is the difference between these two values:

$$E_{cell} = E_{ca} - E_{an} \tag{2.4}$$

Based on the types of electron donors and acceptors, the typical reactions take place in both chamber of BES and its standard potential with reference electrodes as listed in Table 2.1.

The BES produces a maximum open circuit voltage of 1.21 V as based on the NADH and oxygen reduction potential [15]. However, in experiments, the open circuit potential is in the range between 0.6–0.8 V depending upon the experimental conditions with various losses, namely activation losses, bacterial metabolic losses, ohmic (internal resistance) losses, and mass transfer (concentration) losses, respectively [16]. These losses should be minimized to obtain the maximum performance of the BES system.

TABLE 2.1

Anode and Cathode Reaction with Standard Electrode Potential under Ambient Conditions

Electrolyte Chamber	Electron Donors/ Acceptors	Chemical Reaction	Standard Potential $E^{o'}$ (V)
Anode	Glucose	$C_6H_{12}O_6 + H_2O \rightarrow 6CO_2 + 24\,e^- + 24\,H^+$	−0.428
	Acetate	$C_2H_3O_2 + 4H_2O \rightarrow 2HCO^{3-} + 9H^+ + 8e^-$	−0.296
	Butyrate	$C_4H_8O_2 + 2H_2O \rightarrow 2C_2H_4O_2 + 4H^+ + 4e^-$	−0.28
Cathode	Oxygen	$O_2 + 4e^- + 4H^+ \rightarrow 2H_2O$	0.816
		$2O_2 + 4H^+ + 4e^- \rightarrow 2H_2O_2$	0.295
	Ferricyanide	$Fe(CN)_6^{3-} + e^- \rightarrow Fe(CN)_6^{4-}$	0.361
	Permanganate	$MnO^{4-} + 4H^+ + 3e^- \rightarrow MnO_2 + 2H_2O$	1.68
	Chromate	$Cr_2O_7^{2-} + 14H^+ + 6e^- \rightarrow 2Cr_3^- + 7H_2O$	1.33
	Persulfate	$S_2O_8^{2-} + 2e^- \rightarrow 2SO_4^{2-}$	2.01

2.2.3 PERFORMANCE OF BES

The open circuit voltage (OCV) is the maximum voltage that is produced by BES, which is lower than the maximum theoretical voltage. The OCV is determined by connecting the BES with an indefinite resistance (i.e. no resistance) or measuring the voltage between anode and cathode in the absence of current. The current and power are calculated using ohm's law equation as shown below:

$$V = I / R_{ext}, \quad I = V \times R, \tag{2.5}$$

where I is the current (mA) and R_{ext} is the external resistance across the circuit.

Columbic efficiency(η_{CE}) is another important parameter to understand the performance of BES. It is defined as the ratio of total coulombs transferred to the anode from the substrate, to the maximum possible coulombs if all substrate removals produced current. The coulombic efficiency is calculated using the equation:

$$\eta_{CE} = \frac{M \int_0^t I dt}{F b V_{an} \Delta COD}, \tag{2.6}$$

where F is Faradays constant ($F = 9.64870 \times 10^4$, C/mol), $b = 4$ is the number of electrons (e⁻) exchanged per mole of oxygen, V_{an} is the anode volume (mL), and ΔCOD is the change in COD (mg/L) with respect to time.

COD removal efficiency (η_{COD}) is another important measurement in BES. The wastewater COD concentration is measured using the potassium dichromate (closed reflux) method. The COD removal efficiency is calculated using the equation:

$$\eta_{(COD)}\% = \frac{COD_{in} - COD_{out}}{COD_{in}}\% \tag{2.7}$$

Similarly, other parameters such as BOD, TDS, and color removal efficiency can also be calculated as per the APHA Standard.

2.3 COMPONENTS AND MATERIALS USED IN BES

The main challenge in constructing the BES system is to identify the suitable anode, cathode, membrane and system architecture that maximize the power generation, coulombic efficiency, and treatment efficiency. The material must be low cost, easily available, and a simple architecture that is inherently scalable. The selection of electrodes for both anode and cathode must have high conductivity, biocompatible, non-corrosive and good mechanical strength [17]. A wide variety of materials such as platinum, gold, silver, and palladium, and so on, can be used as anode electrodes for electricity generation. Due to the cost of electrodes and weak adhesions of noble metals, recently carbonaceous electrodes such as carbon paper, cloth, foam, mesh, flake, graphite plate, rod, and graphite brush have been preferred in the BES system. These remarkably satisfy most of the requirements of MFC electrode which is highly helpful in replacing the precious metal electrodes [17, 18]. The materials used in the anode side can also be used in the cathode side. The cathode materials are carbon paper, carbon cloth, carbon brush, graphite, and so on. In order to improve energy generation, the cathode electrode surface is modified with a highly active catalyst. For example, platinum rod or platinum powder is commonly used as a catalyst to improve the reaction rate at the cathode [19]. But it is too expensive and has a tendency for biofouling and the surface poisoning of microorganisms. Due to this, non-precious/inexpensive, biofouling-resistant materials have been employed in the cathode to enhance power production [20].

Membrane is another important material which is used to separate the anode and cathode chamber. Generally, cation exchange membrane (CEM), referred to as proton exchange membrane (PEM), is employed to transfer protons from the anode to the cathode side. The most commonly used CEM is the Nafion (Nafion117, Nafion115 and Nafion112) membrane. Nafion contains sulfonated tetrafluoroethylene copolymer, and consists of a hydrophobic fluorocarbon backbone ($-CF_2-CF_2-$) to which hydrophilic sulfonate groups (SO^-_3) are attached. The first two digits indicate a material with 1100 g Equivalent Weight (EW) and last digit is membrane thickness 0.178 (Nafion 117), 0.127 (Nafion 115) and 0.051 mm (Nafion 112) mm respectively. Generally, the Nafion membrane is used in the BES due to its high proton conductivity resulting from the presence of sulfonate groups in the structure [21]. Similarly, anion exchange membrane (AEM) provides better performance compared to CEM, as protons are consumed in the anode by OH^- transferred from the cathode, without further pH reduction in the anode chamber [22].

2.4 FACTORS INFLUENCING THE BES PERFORMANCE

To improve the performance of BES and reduce the construction, operation, and maintenance costs, various factors such as types of wastewater, pH, electrode materials, and reactor configurations need to be clearly addressed.

2.4.1 TYPES OF WASTEWATER AND ITS CONCENTRATION

In BES, the substrate is one of the most important biological factors which affects electricity generation. The system efficiency and economic viability of converting organic matter into electricity mainly depends on the components present in the wastewater. It influences not only the integral composition of the bacterial community and also the power generation of the system. Besides power generation, the removal of pollutants such as COD, TDS, nitrates, sulfide, sulfates, color, and so on, were also observed in the system. Generally, acetate is preferred as a substrate due to its simple nature and is also the end product of several metabolic pathways for higher order carbon sources [7]. Glucose is also used as a substrate next to acetate in the BES. But the glucose is a fermentable substrate, implying its consumption by either fermentation or methanogenesis that suppresses the production of electricity [23].

Several researchers have used synthetic wastewater as a substrate due to its well-defined composition and easy to control concentration, pH, its conductivity, and so on. Venkata Mohan et al. (2008) have tested MFC with synthetic wastewater at different loading rates to achieve variable performances [24]. Several researchers have

TABLE 2.2

Selective Types of Substrates Used in MFC and its Peak Power Density

S.No	Substrate/Wastewater	Microorganism	PD (mW/m²)	η_{COD}(%)	Refs.
1	Artificial	Mixed culture	0.56	-	[28]
2	VFA from Food waste	Ostrich dung	240	-	[29]
3	Urban	Activated sludge	25	50	[30]
4	Paper recycling	Mixed culture	672	76 ± 4	[31]
5	Brewery	Mixed culture	170	87	[32]
6	Starch processing	Starch process waste	239.4	98	[33]
7	Cheese whey	Anaerobic sludge	46	94	[34]
8	Food processing	Activated sludge	230	86	[35]
9	Bakery	anaerobic sludge	60	80	[26]
10	Dairy	Mixed consortia	1.1 W/m³	95.5	[36]
11	Distillery wastewater	Mixed culture	348	57	[37]
12	Distillery wastewater	Mixed culture	202 ± 6	63.5	[38]

evaluated MFC performance with real (industrial) wastewater as a substrate. Nimje et al. (2012) compared bioelectricity production from various (domestic, agricultural, paper and dairy) wastewaters using mixed cultures and pure strains of *S. Oneidensis* in MFC [25]. Velasquez-Orta et al. (2011) investigated the different types of industrial wastewaters such as brewery, bakery, dairy, and paper wastewater on the performance of identical systems and the microbial composition and electrochemistry of anodes. The microbial composition of anodic biofilms differed with respect to type of wastewater used [26]. A high strength of molasses wastewater can provide a good source of organic matter for electricity production in MFC [27]. Gude has reported the wide variety of substrates ranging from pure compound (e.g. glucose, acetate and fructose etc.) to complex mixture (wastewater, lignocellulose, algal biomass etc.) with the different inoculum used in BES [4]. A wide variety of wastewater has been used in BES as given in Table 2.2. The wastewater contains different ranges of organic matter content which have dissimilar degradability that affects the rate of organic carbon removal in BES. Hence, the selection of wastewater should contain high organic matter and be easily biodegradable for maximum electricity generation in BES.

Substrate concentration is the most significant factor which limits power production in BES. When the substrate concentration and its loading rate increased, the performance of the system significantly improved to certain range during operation [39, 40]. However, some of the studies reported that a higher concentration of substrates reduces the performance of BES. This is due to a higher concentration inhibiting bacterial growth at the electrode surface which suppresses electricity generation in the system [41]. The substrate inhibition mainly depends on the types of bacteria and degradation capability which needs to be identified.

2.4.2 Types of Bacteria

The anodic electron transfer mechanism is a key issue in understanding the principle of BES operation. The electrons transfer to the electrode surface following two mechanisms such as direct and indirect electron transfer, as already described in section 2.1. Many microorganisms have the ability to transfer the electrons from the metabolism of organic matter to the anode. For example, exoelectrogenic species such as *Geobacter sulferreducens, Bacillus subtillus, Cornebacterium, Shewanella Oneidensis MR-1, Bacteroidetes, Geobacter metallireducens, Rhodopseudomonas palustris* and *Lysinibacillus sphaericus* SN-1, SN-2 have been employed in the MFC for wastewater treatment and power generation [42–46]. The microorganism either in a pure or mixed state should have higher exoelectrogenic activity than the methanogenic activity which leads enhancing power production. Recently, the MFC populated by a mixed culture have gained much attention owing to their stability, robustness due to nutrient adaptability, stress resistance, and general tendency to produce higher current densities than those obtained using pure cultures [47]. Several researchers have investigated the MFC performance by inoculating with the mixed microbial communities. But the mechanism and role of the individual microorganism contribution to power generation becomes difficult to understand when a mixed microbial community is used as a biocatalyst [46].

Besides exoelectrogens, there are also metal-reducing bacteria, denitrifying bacteria, hydrogen-scavenging microorganisms, methanogens, and so on [48]. Though a variety of microorganisms, such as isolated or mixed cultures, are used in MFCs, it requires a different pH environment for its exoelectrogenic activity and growth.

The bacteria require the pH to be close to neutral conditions for optimal growth. According to Nernst's equation, organic matter oxidation potential would shift -59 mV at each pH unit in an alkaline direction in the anolyte chamber and $+59$ mV at each pH unit in the acidic direction in the catholyte chamber in the BES [49]. Since, the electrolyte pH either in anode and cathode chamber can remarkably influence the cell voltage in MFC. Literature reported that the best performance is obtained at a pH level between 4 and 10 with various wastewater and inoculum used in the anode chamber [50]. The addition of buffering salt in the anodic chamber becomes important to maintain the constant pH for enhancing the power output in the BES. Buffering is not only used for maintaining the suitable pH conditions for the bacterial growth in the MFC, but it also increases the conductivity of the electrolyte and reduces the internal resistance of the cell [51, 52].

2.4.3 Types of Electron Acceptors in the Cathodic Chamber

Electricity generation mainly depends on reduction kinetics at the cathode. The maximum potential produced from the air-cathode MFC using acetate as a substrate is 1.105 V. However, the slow rate of oxygen reduction in the absence of catalyst on the electrode leads to a higher reduction over potential which reduces the performance of BES. In a two-chamber system, the most commonly used electrolyte next to oxygen is potassium ferricyanide ($K_3Fe(CN)^6$). It has faster reduction kinetics and relatively higher redox potential as compared to oxygen [51]. Potassium permanganate ($KMnO_4$) is another more commonly used electron acceptor by virtue of its high oxidation capacity as well as its environmental safety [53]. However, the permanganate requires continuous replacements to compensate its depletion of ions in the system. Therefore, this technique may be only applied to lab-scale level. Potassium per sulfate is also used by one of the researchers and achieved enhanced performance as compared to potassium permanganate. But it could not sustain the voltage for the longer duration. A phosphate buffer is also used for enhancing the performance of MFC. The power output increased with increase the phosphate concentration due to increase the solution conductivity. However, phosphate buffering is high cost and higher concentration of the salt caused a eutrophication condition of water bodies without removing the phosphate from the wastewater [52].

2.4.4 ELECTRODE MATERIALS

The anode or cathode electrode are the most important factors affecting the power production in BES. The anode material provides a surface for the biocatalyst to attach and react with the substrate. Consequently, the anode material has to be biocompatible, conductive, and allow extensive contact between the biocatalyst and substrate [54]. Some noble metals such as platinum, gold, silver, palladium, rhodium, and so on, have been exploited for specific applications. However, all the electrodes are quite expensive and weak adhesion of the inoculated bacteria limited the utilization of electrodes in the MFCs [55]. Recently, carbonaceous materials such as carbon paper, carbon cloth, carbon brush, carbon foam, carbon flake and graphite plate, foam, flake, and felt are used as electrodes in the MFC due to their good conductivity and surface area, biocompatibility, versatility, and low cost [56, 57]. Currently, modification of nanostructured materials (e.g. Nanocomposites such as FeO, CuO, MoO4 etc.) on carbonaceous electrodes such as carbon nanotube, porous carbon, polyaniline (PANI), 2D and 3D structured graphene nanocomposites are employed in BES for reducing ohmic losses and enhancing bacterial adhesions which helped improve performance [12, 58]. The development of effective and low-cost catalyst production plays a potential role in expanding the futuristic role of BES technology.

Cathode electrodes are most important since their reduction reaction occurs at the cathode. Platinum electrodes are commonly employed in the air-cathode system due to their high reduction kinetics [59]. But, the price of platinum makes it uneconomical for use in MFC treating wastewater. Non-precious metals (Ni, Stainless Steel, Copper, Silver etc.) have also been studied as catalysts for improving the kinetics of oxygen reduction in the cathode [60]. Currently, carbonaceous materials such as carbon paper, carbon cloth, carbon brush, graphite plate/sheet, foam, flake, felt, activated carbon, and so on, are used in BES systems due to their reduction kinetics, high active surface area, and low cost for practical applications [12]. Biocathodes using microalgae are used as a biocatalyst at the cathode to reduce the cost of BES and, moreover, utilize the generated carbon dioxide in the anode [61].

2.4.5 MEMBRANE

Membrane is another important material which separated the anode and cathode chamber due to that transfer of electrolyte (e.g. Ferricyanide or water) between each chamber is restricted. The membrane should be permeable so that the protons can be easily transferred from the anode to the cathode side. Hence, substrate losses, biofouling, oxygen diffusion, and internal resistance should be considered while selecting the membrane for BES applications [62]. Generally, Nafion is most commonly used in BES due to its high proton conductivity. This gives better proton conductivity due to better higher performance as compared to salt bridge. Another frequently used CEM is Ultrex CMI 7000 which is a strong acid polymer membrane with gel polystyrene and divinylbenzene cross-link structure, that also contains large amounts of sulphonic acid groups. It has excellent cation conductivity and mechanical durability, but has high ohmic resistance as compared to Nafion membrane. The oxygen permeability, biofouling, and high cost of CEMs has limited real-time applications [63]. AEM has better performance as compared with cation exchange membranes, as protons in the anode are consumed by OH⁻ transferred from the cathode, without further pH decreases in the anode chamber. This property of AEMs results in lowering ion transport resistance and reducing membrane fouling and the cathode resistance caused by the precipitation of transported cations. Several low-cost separators have been developed, including bipolar membrane, ultrafiltration membranes, porous fabrics, polystyrene, glass fibers, and J-cloth, and so on [64, 65]. Apart from that, various types of membrane such as PVDF, Polyvinyl alcohol -Nafion- borosilicate, sulfonated polystyrene-ethylene-butylene-polystyrene (SPSEBS), SSEBS, SPEEK, and ultrafiltration membrane are used as a separator in the MFC [12].

2.4.6 Reactor Configuration

The reactor design is one of the important factors that directly affects current production and wastewater treatment efficiency in BES. The reactor designs have ultimately proved they can meet the power requirements, efficiency, stability, and durability being developed by various researchers. Liu et al. (2004) reported that a single chamber BES consists of a single cylindrical chamber with eight graphite rods in a concentric arrangement surrounding a single cathode [14]. A two compartment BES is the anode system which consists of anode and cathode that is separated by membrane/salt bridge. This type of design restricts the contact of oxygen or other electron acceptors to the anode [66]. However, one of the researchers reported that the presence of membrane led to creating the pH gradient across the membrane [67]. Various types of design including rectangular, cylindrical, bottle type, fluidized bed, and up flow design have been developed for enhancing the performance of MFC in terms of wastewater treatment as well as energy generation [68]. However, in the various designs proposed by researchers, the individual cell potential cannot exceed greater than 1.21 V due to thermodynamic limitations. Further, to increase the system voltage or current, the MFCs can be connected in the series or parallel, depending on the requirement. Wang and Han (2009) constructed a single chamber stackable MFC which comprised four MFC units and evaluated the performance in individual mode as well as in stack mode. The parallel connection of four individual units produced electricity is four times higher than anyone of the single unit [69]. Rahimnejad et al. (2012) developed a new stack design which was composed of four anodes and three cathode compartments [70]. Zhang et al. (2017) investigated the stacked MFC with serpentine flow field and response to cell number, connection type, variable loads, and electrolyte flow rates [71]. But the unbalanced distribution of substrate in each cell usually leads to voltage reversal which may result in low electricity generation [72].

2.5 VARIOUS TYPES OF BIOELECTROCHEMICAL SYSTEM

2.5.1 MICROBIAL ELECTROLYSIS CELL

Microbial Electrolysis Cell (MEC) is another technology which comes under the category of BES that is used to produce hydrogen and other value-added compounds along with wastewater treatment. The MEC is similar to MFC in which the both anode and cathode are maintained as anaerobic environment whereas in the MFC the cathode is open to the atmosphere. In MEC, electrons and protons are produced and transferred from anode compartment during the degradation of organic matter by exoelectrogens and combined in the cathode to produce hydrogen. But this reaction does not occur spontaneously due to the thermodynamic limitations of the system [73]. Hence, an additional voltage of 0.114 to 1.21 V must be supplied to drive the reaction to form hydrogen at the cathode in the MEC [74]. The MEC has several advantages, such as producing hydrogen at low energy input, high conversion efficiency, can produce value-added products, and does not require any highly expensive catalyst (Pt) [75]. The MEC performance is influenced by operational (microorganism, wastewater, pH etc.) and design factors (applied voltage, reactor configuration, electrode, and catalyst), and so on. More research should be carried out to further improve hydrogen production using MEC technology for real application [76–78].

2.5.2 MICROBIAL DESALINATION CELL

Microbial Desalination Cell (MDC) is another system that comes under BES, shown as having a significantly potential approach for water desalination along with electricity generation and wastewater treatment [79]. This technology is based on the transfer of ionic species from saline water in proportion to current generated by bacteria. This system consists of three chambers, namely anode, cathode, and desalination chamber separated by AEM and CEM. In the anode chamber, an exoelectrogenic bacteria oxidized the organic matter and releases the protons, electrons, and CO_2. In the cathode chamber, the electron acceptors undergo reduction reaction and produce water. The anions (Cl^-, SO_4^{2-} etc.) migrate from the saltwater to anode through AEM while the cations (Na^+, Ca^{2+} etc) move to the cathode chamber through CEM. This loss of ionic species in the middle chamber is called water desalination. This does not require water pressure or electrical energy [80]. The MDC can be used with other water desalination technology, such as reverse osmosis (RO) and electrodialysis (ED). One of the most important parameters of MDC performance is the desalination rate (DR), which is greatly dependent on the salt concentration of the seawater. Much research work has been done on the improvement of MDC performance by varying the operational and design parameters [80, 81]. However, significant research is needed for potential developments in advancing MDC technology.

2.5.3 MICROBIAL REMEDIATION CELL

Microbial remediation cell (MRC) is another type of BES that provides a new approach for the effective removal of metal ions from the wastewater, since it offers a flexible platform for both oxidation and reduction reaction-oriented processes

[82, 83]. Numerous studies have been reported on the removal of metal ions (eg. Au (III), V (V), Cr (VI), Ag (I), Cu (II), Fe (III), and Hg (II), etc.) from contaminated water either in the anode side or the cathode side along with energy generation [84]. A mechanism proposed is metal recovery using abiotic and biotic cathodes with and without external power supply [85]. Bio-cathodes have shown good performance in removing and recovering metals, but high concentration metal solutions generally inhibit microbial activities and reduce system efficacy. More effective and tolerant biofilms need to be investigated. Apart from that, more than 50 types of BES system have been identified to date based on various applications [8].

2.6 CONCLUSION

BES is a novel biotechnology that targets the two significant problems: environmental sustainability and energy demand. It is the only technology that can generate electricity or other value-added compounds such as hydrogen, hydroxide, ethanol, acetic acid, and so on, with or without the external energy from different sources. This chapter mainly discussed the basic concepts of BES, electrons transferring mechanisms, and components used for the fabrication of BES. It also discussed in detail the various factors involved in improving the performance and application of BES. This technology has a lot of potential but it still under development including looking at low-cost electrodes and catalysts in the anode and cathode sides, biocathode development, low-cost membrane, complex system design, and so on. More in-depth research should be carried out looking at the development and execution of this technology for real-time applications.

REFERENCES

[1] Du Z, Li H, Gu T. A state of the art review on microbial fuel cells: A promising technology for wastewater treatment and bioenergy. *Biotechnol Adv* 2007;25:464–482. https://doi.org/10.1016/j.biotechadv.2007.05.004.

[2] Pant D, Adholeya A. Biological approaches for treatment of distillery wastewater: A review. *Bioresour Technol* 2007;98:2321–2334. https://doi.org/10.1016/j.biortech.2006.09.027.

[3] Logan BE, Hamelers B, Rozendal R, Schröder U, Keller J, Freguia S, et al. Microbial fuel cells: Methodology and technology. *Environ Sci Technol* 2006;40:5181–5192. https://doi.org/10.1021/es0605016.

[4] Gude VG. Wastewater treatment in microbial fuel cells - An overview. *J Clean Prod* 2016;122:287–307. https://doi.org/10.1016/j.jclepro.2016.02.022.

[5] Santoro C, Arbizzani C, Erable B, Ieropoulos I. Microbial fuel cells: From fundamentals to applications: A review. *J Power Sources* 2017. https://doi.org/10.1016/j.jpowsour.2017.03.109.

[6] Rahimnejad M, Adhami A, Darvari S, Zirepour A, Oh S-E. Microbial fuel cell as new technology for bioelectricity generation: A review. *Alexandria Eng J* 2015;54:745–756. https://doi.org/10.1016/j.aej.2015.03.031.

[7] Liu H, Cheng S, Logan BE. Production of electricity from acetate or butyrate using a single-chamber microbial fuel cell. *Environ Sci Technol* 2005;39:658–662. https://doi.org/10.1021/es048927c.

[8] Wang H, Ren ZJ. A comprehensive review of microbial electrochemical systems as a platform technology. *Biotechnol Adv* 2013;31:1796–1807. https://doi.org/10.1016/j.biotechadv.2013.10.001.

[9] Muthukumar H, Mohammed SN, Chandrasekaran NI, Sekar AD, Pugazhendhi A, Matheswaran M. Effect of iron doped Zinc oxide nanoparticles coating in the anode on current generation in microbial electrochemical cells. *Int J Hydrogen Energy* 2019. https://doi.org/10.1016/j.ijhydene.2018.06.046.

[10] Bond DR, Lovley DR. Electricity production by Geobacter sulfurreducens attached to electrodes. *Appl Environ Microbiol* 2003;69:1548–1555. https://doi.org/10.1128/AEM.69.3.1548-1555.2003.

[11] Logan BE. *Microbial Fuel Cell.* Wiley-Interscience A John Wiley & Sons, Inc., Publication, 2007.

[12] Palanisamy G, Jung H-Y, Sadhasivam T, Kurkuri MD, Kim SC, Roh S-H. A comprehensive review on microbial fuel cell technologies: Processes, utilization, and advanced developments in electrodes and membranes. *J Clean Prod* 2019;221:598–621. https://doi.org/10.1016/J.JCLEPRO.2019.02.172.

[13] Huang L, Regan JM, Quan X. Electron transfer mechanisms, new applications, and performance of biocathode microbial fuel cells. *Bioresour Technol* 2011;102:316–323. https://doi.org/10.1016/j.biortech.2010.06.096.

[14] Liu H, Ramnarayanan R, Logan BE. Production of electricity during wastewater treatment using a single chamber microbial fuel cell. *Environ Sci Technol* 2004;38:2281–2285. https://doi.org/10.1021/es034923g.

[15] Zhuang L, Zheng Y, Zhou S, Yuan Y, Yuan H, Chen Y. Scalable microbial fuel cell (MFC) stack for continuous real wastewater treatment. *Bioresour Technol* 2012;106:82–88. https://doi.org/10.1016/j.biortech.2011.11.019.

[16] Kadier A, Kalil MS, Abdeshahian P, Chandrasekhar K, Mohamed A, Azman NF, et al. Recent advances and emerging challenges in microbial electrolysis cells (MECs) for microbial production of hydrogen and value-added chemicals. *Renew Sustain Energy Rev* 2016;61:501–525. https://doi.org/10.1016/j.rser.2016.04.017.

[17] Wei J, Liang P, Huang X. Recent progress in electrodes for microbial fuel cells. *Bioresour Technol* 2011;102:9335–9344. https://doi.org/10.1016/j.biortech.2011.07.019.

[18] Hindatu Y, Annuar MSM, Gumel AM. Mini-review: Anode modification for improved performance of microbial fuel cell. *Renew Sustain Energy Rev* 2017. https://doi.org/10.1016/j.rser.2017.01.138.

[19] Feng Y, Shi X, Wang X, Lee H, Liu J, Qu Y, et al. Effects of sulfide on microbial fuel cells with platinum and nitrogen-doped carbon powder cathodes. *Biosens Bioelectron* 2012. https://doi.org/10.1016/j.bios.2011.08.030.

[20] Tremouli A, Martinos M, Bebelis S, Lyberatos G. Performance assessment of a four-air cathode single-chamber microbial fuel cell under conditions of synthetic and municipal wastewater treatments. *J Appl Electrochem* 2016. https://doi.org/10.1007/s10800-016-0935-3.

[21] Shahgaldi S, Ghasemi M, Wan Daud WR, Yaakob Z, Sedighi M, Alam J, et al. Performance enhancement of microbial fuel cell by PVDF/Nafion nanofibre composite proton exchange membrane. *Fuel Process Technol* 2014. https://doi.org/10.1016/j.fuproc.2014.03.015.

[22] Kim JR, Cheng S, Oh S-E, Logan BE. Power generation using different cation, anion, and ultrafiltration membranes in microbial fuel cells. *Environ Sci Technol* 2007;41:1004–1009. https://doi.org/10.1021/es062202m.

[23] Chae KJ, Choi MJ, Lee JW, Kim KY, Kim IS. Effect of different substrates on the performance, bacterial diversity, and bacterial viability in microbial fuel cells. *Bioresour Technol* 2009. https://doi.org/10.1016/j.biortech.2009.02.065.

[24] Venkata Mohan S, Mohanakrishna G, Reddy BP, Saravanan R, Sarma PN. Bioelectricity generation from chemical wastewater treatment in mediatorless (anode) microbial fuel cell (MFC) using selectively enriched hydrogen producing mixed culture under acidophilic microenvironment. *Biochem Eng J* 2008;39:121–130. https://doi.org/10.1016/j.bej.2007.08.023.

[25] Nimje VR, Chen CY, Chen HR, Chen CC, Huang YM, Tseng MJ, et al.Comparative bioelectricity production from various wastewaters in microbial fuel cells using mixed cultures and a pure strain of Shewanella oneidensis. *Bioresour Technol* 2012. https://doi.org/10.1016/j.biortech.2011.09.129.

[26] Velasquez-Orta SB, Head IM, Curtis TP, Scott K.Factors affecting current production in microbial fuel cells using different industrial wastewaters. *Bioresour Technol* 2011;102:5105–5112. https://doi.org/10.1016/j.biortech.2011.01.059.

[27] Mohanakrishna G, Venkata Mohan S, Sarma PN. Bio-electrochemical treatment of distillery wastewater in microbial fuel cell facilitating decolorization and desalination along with power generation. *J Hazard Mater* 2010;177:487–494. https://doi.org/10.1016/j.jhazmat.2009.12.059.

[28] Moon H, Chang IS, Kim BH. Continuous electricity production from artificial wastewater using a mediator-less microbial fuel cell. *Bioresour Technol* 2006;97:621–627. https://doi.org/10.1016/j.biortech.2005.03.027.

[29] Choi JDR, Chang HN, Han JI. Performance of microbial fuel cell with volatile fatty acids from food wastes. *Biotechnol Lett* 2011. https://doi.org/10.1007/s10529-010-0507-2.

[30] Rodrigo MA, Cañizares P, Lobato J, Paz R, Sáez C, Linares JJ. Production of electricity from the treatment of urban waste water using a microbial fuel cell. *J Power Sources* 2007;169:198–204. https://doi.org/10.1016/j.jpowsour.2007.01.054.

[31] Huang L, Logan BE. Electricity generation and treatment of paper recycling wastewater using a microbial fuel cell. *Appl Microbiol Biotechnol* 2008;80:349–355. https://doi.org/10.1007/s00253-008-1546-7.

[32] Feng Y, Wang X, Logan BE, Lee H. Brewery wastewater treatment using air-cathode microbial fuel cells. *Appl Microbiol Biotechnol* 2008. https://doi.org/10.1007/s00253-008-1360-2.

[33] Lu N, Zhou S, Zhuang L, Zhang J, Ni J. Electricity generation from starch processing wastewater using microbial fuel cell technology. *Biochem Eng J* 2009;43:246–251. https://doi.org/10.1016/j.bej.2008.10.005.

[34] Tremouli A, Antonopoulou G, Bebelis S, Lyberatos G. Operation and characterization of a microbial fuel cell fed with pretreated cheese whey at different organic loads. *Bioresour Technol* 2013. https://doi.org/10.1016/j.biortech.2012.12.173.

[35] Mansoorian HJ, Mahvi AH, Jafari AJ, Amin MM, Rajabizadeh A, Khanjani N. Bioelectricity generation using two chamber microbial fuel cell treating wastewater from food processing. *Enzyme Microb Technol* 2013;52:352–357. https://doi.org/10.1016/j.enzmictec.2013.03.004.

[36] Venkata Mohan S, Mohanakrishna G, Velvizhi G, Babu VL, Sarma PN. Bio-catalyzed electrochemical treatment of real field dairy wastewater with simultaneous power generation. *Biochem Eng J* 2010. https://doi.org/10.1016/j.bej.2010.04.012.

[37] Sonawane JM, Marsili E, Chandra Ghosh P. Treatment of domestic and distillery wastewater in high surface microbial fuel cells. *Int J Hydrogen Energy* 2014. https://doi.org/10.1016/j.ijhydene.2014.07.085.

[38] Samsudeen N, Radhakrishnan TK, Matheswaran M. Bioelectricity production from microbial fuel cell using mixed bacterial culture isolated from distillery wastewater. *Bioresour Technol* 2015;195:242–247. https://doi.org/10.1016/j.biortech.2015.07.023.

[39] Juang DF, Yang PC, Chou HY, Chiu LJ. Effects of microbial species, organic loading and substrate degradation rate on the power generation capability of microbial fuel cells. *Biotechnol Lett* 2011. https://doi.org/10.1007/s10529-011-0690-9.

[40] Zhang B, Zhao H, Zhou S, Shi C, Wang C, Ni J. A novel UASB-MFC-BAF integrated system for high strength molasses wastewater treatment and bioelectricity generation. *Bioresour Technol*2009. https://doi.org/10.1016/j.biortech.2009.06.045.

[41] Mathuriya AS, Sharma VN. Bioelectricity production from various wastewaters through microbial fuel cell technology. *J Biochem Technol* 2010;2:133–137.

[42] Kim MS, Cha J, Kim DH. Enhancing factors of electricity generation in a microbial fuel cell using Geobacter sulfurreducens. *J Microbiol Biotechnol* 2012. https://doi.org/10.4014/jmb.1204.04010.

[43] Hassan SHA, Kim YS, Oh SE. Power generation from cellulose using mixed and pure cultures of cellulose-degrading bacteria in a microbial fuel cell. *Enzyme Microb Technol* 2012. https://doi.org/10.1016/j.enzmictec.2012.07.008.

[44] Nimje VR, Chen CY, Chen CC, Jean JS, Reddy AS, FanCW, et al.Stable and high energy generation by a strain of Bacillus subtilis in a microbial fuel cell. *J Power Sources* 2009. https://doi.org/10.1016/j.jpowsour.2009.01.019.

[45] Ha PT, Lee TK, Rittmann BE, Park J, Chang IS. Treatment of alcohol distillery wastewater using a bacteroidetes-dominant thermophilic microbial fuel cell. *Environ Sci Technol* 2012. https://doi.org/10.1021/es203861v.

[46] Samsudeen N, Radhakrishnan TK, Matheswaran M. Effect of isolated bacterial strains from distillery wastewater on power generation in microbial fuel cell. *Process Biochem* 2016;51:1876–1884. https://doi.org/10.1016/j.procbio.2016.06.007.

[47] Watson VJ, Logan BE. Power production in MFCs inoculated with *Shewanella oneidensis* MR-1 or mixed cultures. *Biotechnol Bioeng* 2010. https://doi.org/10.1002/bit.22556.

[48] Zuo Y, Xing D, Regan JM, Logan BE. Isolation of the exoelectrogenic bacterium *Ochrobactrum anthropi* YZ-1 by using a U-tube microbial fuel cell. *Appl Environ Microbiol* 2008;74:3130–3137. https://doi.org/10.1128/AEM.02732-07.

[49] Zhuang L, Zhou S, Li Y, Yuan Y. Enhanced performance of air-cathode two-chamber microbial fuel cells with high-pH anode and low-pH cathode. *Bioresour Technol* 2010;101:3514–3519. https://doi.org/10.1016/j.biortech.2009.12.105.

[50] He Z, Huang Y, Manohar AK, Mansfeld F. Effect of electrolyte pH on the rate of the anodic and cathodic reactions in an air-cathode microbial fuel cell. *Bioelectrochemistry* 2008;74:78–82. https://doi.org/10.1016/j.bioelechem.2008.07.007.

[51] Nam JY, Kim HW, Lim KH, Shin HS, Logan BE. Variation of power generation at different buffer types and conductivities in single chamber microbial fuel cells. *Biosens Bioelectron* 2010. https://doi.org/10.1016/j.bios.2009.10.005.

[52] Samsudeen NM, Thota Karunakaran R, Manickam M. Enhancement of bioelectricity generation from treatment of distillery wastewater using microbial fuel cell. *Environ Prog Sustain Energy* 2018;37:663–668. https://doi.org/10.1002/ep.12734.

[53] Asefi B, Li SL, Moreno HA, Sanchez-Torres V, Hu A, Li J, et al. Characterization of electricity production and microbial community of food waste-fed microbial fuel cells. *Process Saf Environ Prot* 2019. https://doi.org/10.1016/j.psep.2019.03.016.

[54] Zhou M, Chi M, Luo J, He H, Jin T. An overview of electrode materials in microbial fuel cells. *J Power Sources* 2011;196:4427–4435. https://doi.org/10.1016/j.jpowsour.2011.01.012.

[55] Wei J, Liang P, Huang X. Recent progress in electrodes for microbial fuel cells. *Bioresour Technol*2011;102:9335–9344. https://doi.org/10.1016/j.biortech.2011.07.019.

[56] Kumar GG, Sarathi VGS, Nahm KS. Recent advances and challenges in the anode architecture and their modifications for the applications of microbial fuel cells. *Biosens Bioelectron* 2013;43:461–475. https://doi.org/10.1016/j.bios.2012.12.048.

[57] Cai T, Meng L, Chen G, Xi Y, Jiang N, Song J, et al. Application of advanced anodes in microbial fuel cells for power generation: A review. *Chemosphere* 2020;248:125985. https://doi.org/10.1016/J.CHEMOSPHERE.2020.125985.

[58] Kaur R, Marwah A, Chhabra VA, Kim K-H, Tripathi SK. Recent developments on functional nanomaterial-based electrodes for microbial fuel cells. *Renew Sustain Energy Rev* 2019:109551. https://doi.org/10.1016/J.RSER.2019.109551.

[59] Deng L, Yuan Y, Zhang Y, Wang Y, Chen Y, Yuan H, et al. Alfalfa leaf-derived porous heteroatom-doped carbon materials as efficient cathodic catalysts in microbial fuel cells. *ACS Sustain Chem Eng* 2017. https://doi.org/10.1021/acssuschemeng.7b01585.

[60] Ben Liew K, Daud WRW, Ghasemi M, Leong JX, Su Lim S, Ismail M. Non-Pt catalyst as oxygen reduction reaction in microbial fuel cells: A review. *Int J Hydrogen Energy* 2014;39:4870–4883. https://doi.org/10.1016/j.ijhydene.2014.01.062.

[61] Chen Y, Shen J, Huang L, Pan Y, Quan X. Enhanced Cd(II) removal with simultaneous hydrogen production in biocathode microbial electrolysis cells in the presence of acetate or NaHCO3. *Int J Hydrogen Energy* 2016. https://doi.org/10.1016/j.ijhydene.2016.06.200.

[62] Ghasemi M, Shahgaldi S, Ismail M, Yaakob Z, Daud WRW. New generation of carbon nanocomposite proton exchange membranes in microbial fuel cell systems. *Chem Eng J* 2012;184:82–89. https://doi.org/10.1016/j.cej.2012.01.001.

[63] Angioni S, Millia L, Bruni G, Ravelli D, Mustarelli P, Quartarone E. Novel composite polybenzimidazole-based proton exchange membranes as efficient and sustainable separators for microbial fuel cells. *J Power Sources* 2017. https://doi.org/10.1016/j.jpowsour.2017.02.084.

[64] Li WW, Sheng GP, Liu XW, Yu HQ. Recent advances in the separators for microbial fuel cells. *Bioresour Technol* 2011. https://doi.org/10.1016/j.biortech.2010.03.090.

[65] Rahimnejad M, Adhami A, Darvari S, Zirepour A, Oh SE. Microbial fuel cell as new technol ogy for bioelectricity generation: A review. *Alexandria Eng J* 2015;54:745–756. https://doi.org/10.1016/j.aej.2015.03.031.

[66] Ringeisen BR, Henderson E, Wu PK, Pietron J, Ray R, Little B, et al. High power density from a miniature microbial fuel cell using *Shewanella oneidensis* DSP10. *Environ Sci Technol* 2006. https://doi.org/10.1021/es052254w.

[67] Hu H, Fan Y, Liu H. Hydrogen production using single-chamber membrane-free microbial electrolysis cells. *Water Res* 2008. https://doi.org/S0043-1354(08)00262-5[pii]10.1016/j.watres.2008.06.015.

[68] He L, Du P, Chen Y, Lu H, Cheng X, Chang B, et al. Advances in microbial fuel cells for wastewater treatment. *Renew Sustain Energy Rev* 2017. https://doi.org/10.1016/j.rser.2016.12.069.

[69] Wang B, Han JI. A single chamber stackable microbial fuel cell with air cathode. *Biotechnol Lett* 2009;31:387–393. https://doi.org/10.1007/s10529-008-9877-0.

[70] Rahimnejad M, Ghoreyshi AA, Najafpour GD, Younesi H, Shakeri M. A novel microbial fuel cell stack for continuous production of clean energy. *Int J Hydrogen Energy* 2012. https://doi.org/10.1016/j.ijhydene.2011.12.154.

[71] Zhang L, Li J, Zhu X, Ye D Ding, Fu Q, Liao Q. Response of stacked microbial fuel cells with serpentine flow fields to variable operating conditions. *Int J Hydrogen Energy* 2017. https://doi.org/10.1016/j.ijhydene.2017.04.205.

[72] Kuchi S, Sarkar O, Butti SK, Velvizhi G, Venkata Mohan S. Stacking of microbial fuel cells with continuous mode operation for higher bioelectrogenic activity. *Bioresour Technol* 2018. https://doi.org/10.1016/j.biortech.2018.02.057.

[73] Jayabalan T, Matheswaran M, Naina Mohammed S. Biohydrogen production from sugar industry effluents using nickel based electrode materials in microbial electrolysis cell. *Int J Hydrogen Energy* 2019:17381–17388. https://doi.org/10.1016/j.ijhydene.2018.09.219.

[74] Naina Mohamed S, Matheswaran M, Jayabalan T. Chapter 15 - Microbial electrolysis cells for converting wastes to biohydrogen. In: Krishnaraj Rathinam N, Sani B, editors. *Biovalorisation Wastes to Renew. Chem. Biofuels*, Elsevier; 2020, 287–301. https://doi. org/https://doi.org/10.1016/B978-0-12-817951-2.00015-8.

[75] Zhang Y, Angelidaki I. Microbial electrolysis cells turning to be versatile technology: Recent advances and future challenges. *Water Res* 2014;56:11–25. https://doi. org/10.1016/J.WATRES.2014.02.031.

[76] Kadier A, Simayi Y, Kalil MS, Abdeshahian P, Hamid AA. A review of the substrates used in microbial electrolysis cells (MECs) for producing sustainable and clean hydrogen gas. *Renew Energy* 2014. https://doi.org/10.1016/j.renene.2014.05.052.

[77] Zhou M, Wang H, Hassett DJ, Gu T. Recent advances in microbial fuel cells (MFCs) and microbial electrolysis cells (MECs) for wastewater treatment, bioenergy and bioproducts. *J Chem Technol Biotechnol* 2013;88:508–518. https://doi.org/10.1002/jctb.4004.

[78] Kadier A, Simayi Y, Abdeshahian P, Azman NF, Chandrasekhar K, Kalil MS. A comprehensive review of microbial electrolysis cells (MEC) reactor designs and configurations for sustainable hydrogen gas production. *Alexandria Eng J* 2016;55:427–443. https:// doi.org/10.1016/J.AEJ.2015.10.008.

[79] Kim Y, Logan BE. Microbial desalination cells for energy production and desalination. *Desalination* 2013;308:122–130. https://doi.org/10.1016/j.desal.2012.07.022.

[80] Al-Mamun A, Ahmad W, Baawain MS, Khadem M, Dhar BR. A review of microbial desalination cell technology: Configurations, optimization and applications. *J Clean Prod* 2018;183:458–480. https://doi.org/10.1016/J.JCLEPRO.2018.02.054.

[81] Saeed HM, Husseini GA, Yousef S, Saif J, Al-Asheh S, Abu Fara A, et al.Microbial desalination cell technology: A review and a case study. *Desalination* 2015;359:1–13. https://doi.org/10.1016/j.desal.2014.12.024.

[82] Nancharaiah YV, Venkata Mohan S, Lens PNL. Metals removal and recovery in bioelectrochemical systems: A review. *Bioresour Technol* 2015. https://doi.org/10.1016/j. biortech.2015.06.058.

[83] Fu F, Wang Q. Removal of heavy metal ions from wastewaters: A review. *J Environ Manage* 2011;92:407–418. https://doi.org/10.1016/j.jenvman.2010.11.011.

[84] Mathuriya AS, Yakhmi JV. Microbial fuel cells to recover heavy metals. *Environ Chem Lett* 2014;12:483–494. https://doi.org/10.1007/s10311-014-0474-2.

[85] Feng P, Yang K, Xu Z, Wang Z, Fan L, Qin L, et al.Growth and lipid accumulation characteristics of Scenedesmus obliquus in semi-continuous cultivation outdoors for biodiesel feedstock production. *Bioresour Technol* 2014;173:406–414. https://doi. org/10.1016/j.biortech.2014.09.123.

3 Bioelectrochemical Reactors

Factors Governing Power Production and Its Applications

Subramaniapillai Niju, Elangovan Elakkiya and
Erudayadhas Lavanya
PSG College of Technology, India

CONTENTS

3.1   INTRODUCTION

Bioelectrochemical systems (BES) are novel electrochemical reactors capable of converting chemical energy in organic matter to electrical energy (Logan, 2009). Microbes capable of extracellular electron transfer are utilized as biocatalysts for the energy conversion process. The microbial metabolism is exploited for donating and accepting electrons through artificially introduced electrodes (Venkata Mohan et al., 2014). A simple BES system consists of anode and cathode chambers separated by a membrane. Exoelectrogens in the anode feed on the organics present in the anode chamber, resulting in production of protons, electrons, and carbon dioxide. The electrons and protons are transported to the cathode chamber via an external circuit and separator, respectively. In cathodes, oxygen is reduced by electrons and protons to water. BESs are gaining profound interest due to their innate potential to generate electricity directly, low environmental foot-print, and sustainable nature (Liu and Logan, 2004). BES can operate with a wide range of substrates ranging from simple defined media to complex wastewaters. Reports on BES were few in number in the 1990s but have gained considerable interest since 2003 with an increase in the volume of papers published and number of citations. This trend emphasizes the scope and importance of research on BES (Venkata Mohan et al., 2014) (Figure 3.1).

BES confers various advantages over existing energy sources but the low power production is the limiting factor in utilizing these reactors for practical application. Researchers across the globe work consistently to improve the energy recovery efficiency from these reactors. Various factors contribute to the power production from these reactors. Biological factors like the nature of substrate, inoculum used, and the rate of electron transfer from bacteria to anode affect the power production. Similarly, the physicochemical factors such as pH of electrolytes, electrode and separator materials, along with the design configuration play a vital role in determining the energy produced from these units. Wastewater treatment is the widely studied application of BES though it is capable of producing various products including hydrogen, methane, ethanol, formic acid, and various other industrially important chemicals with minor modifications (Zhang and Angelidaki, 2014). These units could be exploited as desalination units by incorporating a three-chambered system known as microbial desalinization cells (Pandit and Das, 2018). In this chapter we initially briefly glance through the basic working of BES and the principle of bioelectrogenesis. Later in the chapter we provide an elaborate review of the factors affecting BES operation and performance. In the final section a small attempt has been made to summarize the application of BES in biosensors, hydrogen production, and the green synthesis of chemicals, along with the modifications made in simple BES to attain the desired reactions.

3.2   BIOELECTROGENESIS AND BIOELECTROCHEMICAL REACTORS

The term bioelectrogenesis refers to electricity generation by living organisms. Microbial fuel cells exploit the principle of bioelectrogenesis for power production from organic matter. Metabolism in microbes involves the breakdown of carbon substrate to reducing equivalents such as electrons (e^-) and protons (H^+) (Rabaey and

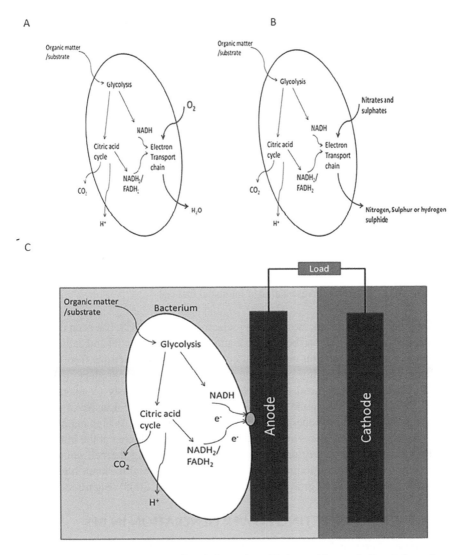

FIGURE 3.1 (A) Aerobic metabolism in bacterium (B) Anaerobic metabolism in bacterium (C) Bioelectrogenesis in MFC.

Verstraete, 2005). Generated H^+ and e^- pass through cascade of redox components to reach the terminal electron acceptor. The redox components aid in the generation of energy-rich phosphate bonds in ATP which is used later in the cellular functions. Oxygen acts as the terminal electron acceptor in oxidative phosphorylation (aerobic metabolism) and is the most preferred electron acceptor by microbes due to high electro-negativity. In the absence of oxygen, other electron-accepting molecules such as sulfates and nitrates drive the flow of electrons. In bioelectrochemical reactors (BES) an external electron acceptor (anode) is introduced and bacteria transfers electrons to the anode. Hence a negative potential is developed in the anode. The aerobic

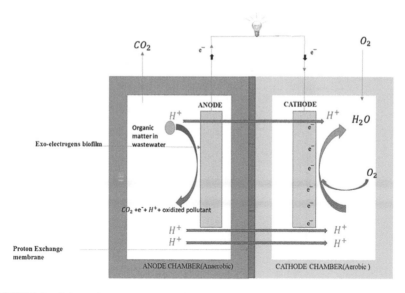

FIGURE 3.2 Schematic representation of microbial fuel cell.

and anaerobic metabolism, along with bioelectrogenesis in BES, has been illustrated in Figure 3.2. The difference between the anode negative potential and cathode positive potential is the electron motive force which drives the electrons from anode to cathode. The continuous flow of electron from anode to cathode produces a deficit in anode hence there is a need to generate electrons continuously (Venkata Mohan et al., 2014). When the anode potential is low there is higher transfer of electrons from microbes and lower energy transfer toward cell growth. At high anodic potential, the cell growth enhances and hence anodic potential is higher at the initial phase of BES start-up and slowly decreases with time, increasing the cell voltage. It is important to maintain optimum anode and cathode potential to attain high energy recovery and optimal cell growth (Rabaey and Verstraete, 2005) (Figure 3.2).

3.3   FACTORS AFFECTING POWER GENERATION IN BES

Electron discharge and power generation in BES is governed by physical, chemical, and biological components (Venkata Mohan et al., 2014). Physicochemical factors include electrode materials, spacing, surface area, pH of anolyte, and catholyte and conductivity (Oliveira et al., 2013). Biological factors influencing BES power generation are inoculum used, organic loading rate, and complexity of substrates available for bacterial metabolism.

3.3.1   BIOLOGICAL PARAMETERS

3.3.1.1   Biological Components

Microbial inoculum used in BES influence substrate degradation rate. As the rate of substrate degradation increases, a higher number of electrons are produced and transferred by the microbes to anode resulting in higher power production. Hence it is important to choose inoculum according to the substrate utilized in the anode chamber of the MFC. Several electrochemically active bacteria have been analyzed in BES studies and, in particular, Geobacter and Shewanella sp. have exhibited high electron transfer efficiency (Logan, 2009). When wastewater was used as substrate it was profitable to use a mixed culture from anaerobic or aerobic treatment plants than monoculture. Mixed cultures were efficient in degrading complex organics in wastewater and hence improved substrate utilization and wastewater treatment efficiency (Venkata Mohan et al., 2014). The overall columbic efficiency is reduced as a result of electron loss due to undesirable biochemical reactions and other mass transfer losses. Methanogens or methane production in the anode chamber of BES is one of the major losses contributing to low power recovery. Exoelectrogens and methanogens share a similar substrate and growth condition requirements making it difficult to avoid diversion of electrons for methane production. Heat treatment of the inoculum and addition of external methanogen suppressor could enhance the columbic efficiency or energy recovery of the system.

3.3.1.2 Electron Transfer to Anode

The microbes capable of transferring electrons outside the cell without the aid of artificial mediators are known as exoelectrogens. These microbes colonize on the anode surface to form a microbial biofilm. Biofilm may be comprised of single/multiple strains of exoelectrogens and is known as electroactive biofilm. The electrons are transferred to the anode via two mechanisms: (i) direct electron transfer (DET) by c-type cytochromes on membrane and conductive nano pili and (ii) mediated electron transfer (MET) with the aid of redox mediators shuttling electrons from bacteria to anode (Logan, 2009).

In single-layer biofilm the membrane-bound cytochromes play a prominent role in electron transfer whereas in thicker and multilayer biofilm the conductive pili aids in long-distance electron transfer. Mediated electron transfer is carried out by redox mediators which shuttle from oxidized and reduced state. These redox molecules are reduced by bacteria and move to the anode where it delivers the electron and gets oxidized. Mediators can be artificially added or can be naturally produced as primary or secondary bacterial metabolites. Exogenous mediators include methylene blue, neutral red, thionine, and quinones. They are toxic, costly, and unstable hence naturally produced redox mediators are preferable such as phenazines, phenoxazines, flavins, and quinones (Kumar et al., 2015). The mediators produced by one organism can be used by other bacteria in the anode chamber. This synergetic relationship aids in improved BES power performance (Logan, 2009). Along with these other electron-carrying co-factors such as heme, iron, and copper, ions play a significant role in electron transfer (Kumar et al., 2016).

The rate of exogenous electron transfer to anode by microbes plays a very important role in determining the net energy conversion efficiency. Consistent efforts have been made by researchers to enhance electron transfer by electrode modification with nanoparticles to increase the total surface area available for microbial attachment and electron transfer. These modifications are not only influenced the current density produced from BES but also the diversity of microbes in electroactive biofilm. Genetic modifications of exoelectrogens for the increased rate of electron transfer have been explored lately and have resulted in improved current density. Electron transfer from bacteria/microbes to anode is crucial in determining the power output from BES. Though the majority of the existing literature is based on the formation of single-species biofilms and its interaction with the anode, it is essential to study the relationship between microbes in electroactive biofilms to enable its practical application in wastewater treatment plants (Kumar et al., 2016).

3.3.2 Physicochemical Parameters

3.3.2.1 Electrode Materials

The synergistic association of bacteria with anode material plays an important role in electron transfer and power generation in BES, as it provides the surface for bacterial attachment and electron transfer. The basic requirements of a material to qualify as an anode in BES include bio-compatibility, electrical conductivity, chemical and mechanical stability, high surface area, and cost-efficiency. Carbonaceous materials are most widely used due to its good bio-compatibility, high conductivity, and stability (Li et al., 2017). Initial work on BES widely used plain configurations such as carbon paper, graphite/carbon rod, and plates. The limited surface available for microbial/bacterial attachment resulted in low current density, discouraging its future use. Though carbon cloth is more flexible and provided a higher surface area, it was not cost-effective when used in large-scale applications (Hernández-Fernández et al., 2015). Modification and improvements in anode materials have aimed at improving the available surface area for bacterial attachment and bio-compatibility without compromising electrical conductivity. Coating the anodes with metal/metal oxide nano particles resulted in increased surface area and bacterial attachment. Graphite and carbon foams, felts, and reticulated vitreous carbon having a porous structure provide a higher surface area for anodic biofilm attachment, but very few studies are available making it difficult to compare the variations in power production (Wei et al., 2011). A packed structure like granular graphite with improved specific surface area has been commonly used. A current collector is needed or else the bed has to be packed tightly to make it conductive. Granulated activated carbon anodes could generate higher power than carbon cloth. The porosity of these anodes was low and hence clogging was a major issue when BES has to be operated for a long time. Carbon electrodes in the brush form have higher power generation efficiency as they provide a high surface area for bacterial growth and enable reduction in electrode spacing. Metal-based anodes are much more conductive but their usage is ruled out due to their non-corrosive requirements. Stainless steel and titanium have been used as common base material anodes. The smooth surface of metal anodes restricts bacterial adhesion and no current was observed when plain titanium without coating was used. Various modifications of anode material for improving surface characteristics have been made which includes surface coatings with nano material, polymers, and metal oxides (Li et al., 2017). Further research into the mechanism of bacterial-anode interaction and work into new anode design for cost reduction are required. A reduction of charge transfer resistance, improving bio-compatibility, and electrical conductivity has been the major thrust area in anode-based studies.

Cathodes are a major performance determining factor in BES, as the oxygen reduction reaction occurring on the cathode is the slowest and rate-limiting step in BES. Materials used in BES cathode include graphite, carbon cloth, carbon paper, platinum, titanium, graphite foils, and graphite plate (Li et al., 2017). Air-cathode BES use a Pt catalyst to improve oxygen reduction reaction (ORR), but large-scale application is difficult due to the cost of cathode fabrication. Long-term fouling of Pt loaded catalyst resulted in poor performance of BES. Activated carbon has been used as a substitute for Pt catalyst and has proved to be efficient at increasing ORR. Finding a low-cost alternative to noble metals such as platinum with considerable performance (Ben Liew et al., 2014) is the challenge lying on the cathode side of BES reactors (Deng et al., 2010; Zhang et al., 2014). Though usage of granulated activated carbon as a cathode catalyst has shown a comparative performance, although studies on designing BES units with a higher volume for practical application are not available. Electrode materials are one of the important areas extensively studied in BES after reactor design. Numerous reviews are available for electrode materials and their influence on power generation. Placement of electrodes in BES also plays a crucial role in proton-electron mobility. A shorter diffusion length influences faster electrochemical reactions. Reduced electrode spacing reduces the internal resistance of the cell hence improving the power generation, but spacing electrodes in the least possible distance yielded least power. This was attributed due to oxygen diffusion into the anode chamber. Therefore optimized spacing could enhance power generation in BES (Venkata Mohan et al., 2014) and (Kim et al., 2015).

3.3.3 Operating Parameters

3.3.3.1 pH of the System

In BES, microbial metabolism continuously produces protons and continuous operation causes acidification of the anode chamber due to the slow migration of protons to the cathode chamber. In the cathode chamber, an oxygen reduction reaction contributes to alkalization. The changes in pH have a drastic effect on BES performance. This phenomenon causes a membrane pH gradient, resulting in thermodynamic/electrochemical limitation to BES performance. In the anode chamber, reduced pH causes slower microbial metabolism, reducing substrate utilization and lowering the protons' and electrons' generating capacity. In the cathode chamber the oxidation reduction reaction increases with decreasing pH. Hence it is important to maintain proper pH conditions in the anode and cathode chamber to improve BES performance (Oliveira et al., 2013). In dual-chambered BES the anolyte and catholyte pH have played a crucial role and various studies on optimizing pH for enhanced power generation are available. In single-chambered air-cathode fuel cells there is only one electrolyte and its pH had great effect on BES power output and it has been found that pH 8 to 10 is favorable for improved anodic and cathodic reactions (He et al., 2008).

A phosphate buffer has been extensively used in BES to counteract the pH effect. However, the addition of a phosphate buffer is expensive and it is important to remove phosphate from wastewater before release into the environment as it is a major cause for eutrophication of water bodies (Oliveira et al., 2013). Various other buffer systems such as bicarbonate (Fan et al., 2015), borax, and synthetic zwitterions (Nam et al., 2010) are used in BES studies. Despite the advantages of using buffers in BES systems they are not feasible to be used in practical or for real-scale application due to the increase in expenditure to BES. To overcome this, the addition of CO_2 to the cathode chamber has been studied which generates a CO_2/carbonate or bicarbonate buffer system. This idea seems promising as CO_2 is produced as waste gas in industries and hence can be readily available without adding to the cost of BES operation (Fornero et al., 2010).

3.3.3.2   Organic Loading Rate (OLR)

OLR is important factor that defines the substrate characteristics and concentration available for bacterial growth. OLR is the deciding factor of the capacity of BES reactor per unit volume (Oliveira et al., 2013). It is important to provide anodic feed at optimized OLR to improve BES performance. As the organic loading rate is increased, there is increase in power generation. The increase in OLR reduced the internal resistance of the cell which was attributed to the increased ionic strength and increased microbial activity. However, a decrease in power generation was observed when OLR was too high which may be due to substrate mediated inhibition in anodic microbial community. The CE increased with reduced OLR and substrate oxidation rate and this might be due to the competition of methanogenic bacteria with exoelectrogens at higher substrate concentrations. Hence it is very important to optimize the substrate loading rate to extract maximum power from BES (Mohan et al., 2009; Martin et al., 2010; Velvizhi and Mohan, 2012).

3.3.3.3   Hydraulic Retention Time and Shear Stress

In continuous systems the hydraulic retention time determines the substrate concentration. Higher flow rates provide higher power generation (Oliveira et al., 2013). The microbial community has to be provided with optimal time for substrate capture and its hydrolysis. Though increase in flow rate increased power generation, the COD removal and columbic efficiency was low so it is important to optimize hydraulic retention time to improve BES performance (Juang et al., 2011). Shear rates determine the current generation and biofilm thickness. At higher shear rates the biofilm thickness is lower and current generation is low. It has been observed that at high shear rates the biofilm formed was homogenous, which improved direct electron transfer and, further, more electron shuttles were produced, contributing to the better performance of anodic microbial consortia (Rickard et al., 2004).

3.3.3.4   Effect of Temperature

Temperature plays important role in determining BES performance as it affects the system kinetics, mass transfer, thermodynamic factors, and nature of the microbial community formed. Temperature determines the microbial growth and hence substrate utilization, and so it is important to maintain optimum temperature for maximum power generation (Oliveira et al., 2013).

Minor modification in the BES could extend its application to hydrogen production, methane generation, biosensing application, and desalination. Later we will review the modification made to suit BES for various applications.

3.4 APPLICATIONS OF BIOELECTROCHEMICAL SYSTEMS

3.4.1 WASTEWATER TREATMENT

When the anode chamber is fed with wastewater, the pollutants are used as substrates for microbial metabolism. Utilization of BES for wastewater treatment with simultaneous power generation is regarded as a major advantage over existing waste management technologies. The composition and complexity of substrates have influenced the bacterial metabolism rate, power production, and treatment efficiency. Wastewater rich in organic biodegradable fraction is a potential substrate for anodic bacteria. A wide spectrum of wastewater has been experimented on as substrates for microbes in BES (Pant et al., 2010). Food and food processing wastewater which is high in carbohydrate content are easily biodegradable and are favorable substrates for bacteria metabolism. Wastewater used in BES is food processing wastewater (protein, tomato pomace, starch, beverages manufacturing, beer brewery, winery, fermented apple juice, cheese whey, dairy, chocolate, and yogurt waste water), and agro processing industries such as rice mill, cassava mill, palm oil mill, mustard tuber wastewater, and livestock processing wastewater, including the meat industry, carcass wastewater, and the swine industry have been studied. Other complex wastewater like biorefinery, pharmaceutical, the paper and pulp industry, mining, petrochemical, domestic, and municipal have been studied and the power generation from BES employing wastewater as substrates have relatively increased (Pandey et al., 2016). Wastewater characteristics are one of the important factors influencing power generation in BES and it is important to analyze its characteristics and inoculum selection based on the components to attain maximum conversion of chemical energy in wastewater to electrical energy (Venkata Mohan et al., 2014). A considerable number of review articles are available which highlight power generation from BES using different substrates.

3.4.2 MICROBIAL ELECTRMOLYMSIS CELLS

The standard method for hydrogen production is electrolysis of water, but high electricity requirements discourage its application on a practical scale. This can be tackled using a microbial electrolysis cell system (MEC), which uses waste streams for biohydrogen production using minimum external potential (0.2–0.8 V) compared to the standard method (>2.1 V). BES are modified to have anaerobic anode and cathode compartments. Substrates such as glucose, glycerol, sewage sludge, other waste streams have been evaluated as the source for potential generation. In the case where acetate is used as a substrate, the chemical reaction in anodes and cathodes are as follows:

Anode: $CH_3COOH + 2H_2O \rightarrow 2CO_2 + 8H^+ + 8e^-$

Cathode: $8H^+ + 8e^- \rightarrow 4H_2$

In a traditional MEC system, the anode compartment contains organic substrates and production of bio-hydrogen takes place in the cathode compartment and is connected to an external power supply. These chambers are usually separated by an exchange unit or separators, usually anion exchange membrane (AEM), cation exchange membrane (CEM), bipolar membrane, or charged-mosaic membrane. Previously, it was assumed that presence of membrane or exchange unit avoided the loss of hydrogen but Call and Logan in 2008 designed a membrane-less single-chamber system and found an increase in bio-hydrogen production. Recently in 2019, Jafary et al. created a fully bio-based MEC (FB-MEC) consisting of bio-anodes and bio-cathodes. Bio-anodes were enriched using half biological microbial fuel cell (HB-BES, bioanode-abiocathode) inoculated with palm oil mill effluent and bio-cathode was enriched using half biological microbial electrolysis cell (HB-MEC, abioanode-biocathode) inoculated with sulfate-reducing bacteria from palm oil mill effluent. Compared to the abiotic system ($0.2 A/m^2$), current production increased in FB-MEC ($1.5–2.5 A/m^2$). But the rate of hydrogen production decreased in FB-MEC compared to the HB-MEC system. However, cathodic hydrogen recovery has increased in a fully bio-based system (56%–65%) thus showing a more efficient redox rate than HB-MEC. Pre-enrichment of anode and cathode thus have proved to have more efficiency than simultaneous enrichment in a single system (Jafary et al., 2019) (Figure 3.3).

Microbial analysis using pyro-sequencing of waste-activated sludge and the biofilm on anodes showed significant difference in microbial population. Acid-producing bacteria and exoelectrogens (mostly *Geobacter* species) were found to be predominant and associated in hydrogen production. Acid-producing bacterial oxidized sugars into simple organic acid molecules which were used along with other sugars by

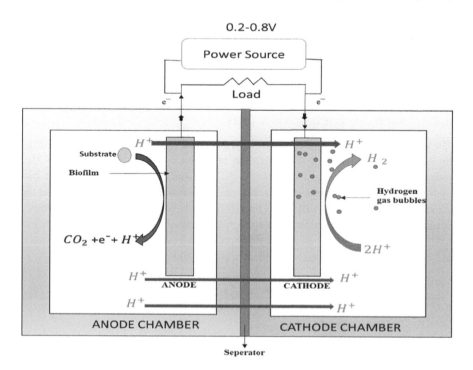

FIGURE 3.3 Schematic representation of microbial electrolysis cell.

the exoelectrogen to produce hydrogen gas. The mutual dependence between these two bacterial groups has caused the usage of various organic matter to enhance hydrogen production (Lu et al., 2012).

3.4.3 Electro-synthesis of Products

Chemicals such as methane, ethanol, hydrogen peroxides, formic acid and acetate can be produced in this system with minor modifications. Methane production is usually enhanced by applying an external voltage of 1.2 V, but this cannot replace traditional anaerobic treatment. Only low strength waste streams can be used. Hydrogen peroxide is an important industrial chemical and can be produced by providing an aerated cathode with an external voltage of 0.5 V which has an efficiency of 83%, thus requiring lower energy than normal chemical synthesis. Ethanol could be produced by using an electron mediator such as methyl viologen, and its concentration plays an important role in ethanol production but the exact mechanism of ethanol production is unknown. Methyl viologen increased ethanol production six-fold, suppressing methanogens (methane producers). A cathode potential of −550 mV was required for ethanol production. Formic acid is mostly used as a preservative and anti-bacterial agent in feed stock, in the pharmaceutical industry and in paper manufacturing. Carbon dioxide, a greenhouse gas which is produced during wastewater treatment, was electro reduced to produce formic acid. Electricity produced by a 5 series BES was used to power MEC, which in turn uses carbon dioxide to produce formic acid. The carbon dioxide was captured from two cycles of BES by sparging nitrogen into the anode chamber. The open-circuit voltage of these BES was about 2.73 V (Zhao et al., 2012).

Nevin et al. proposed that acetate can be used as an energy storing molecule via microbial electro-synthesis as the electrons are directly converted to acetate by *S. ovata* biofilm in the cathode (electron recovery-86% ± 21%) and energy is stored in a chemical bond. So the energy produced via solar cells can be stored as covalent bonds which can be transported with the available infrastructure (Nevin et al., 2010; Gopalakrishnan et al., 2012; Zhang and Angelidaki, 2014) (Table 3.1).

3.4.4 Removal of Pollutants

Since cathode potential can be controlled in MEC, the removal of organic and inorganic pollutant is possible. An increase in cathode potential enhanced the efficiency of this process multiple times higher than simple BES. Organic components such as nitrobenzene, acid orange 7 and 4-chlorophenol removal have been studied and this process provided a great alternative for pharmaceutical de-halogenation. Some examples of inorganics which can be removed are nitrates, sulfate, perchlorate, heavy metals, and so on. Metal with a high reduction potential can be easily recovered using BES, but metals with lower potential require extra energy which can be provided by MEC. Since each metal has different potential requirement such as for copper −0 V, lead −0.34 V and so on, pure metal can be isolated thus decreasing the energy required in comparison to the traditional method. Other components such as phosphate, ammonium, and so on can also be recovered.

TABLE 3.1

List of Electro-synthesized Product

Chemical	Reaction	Cathode Potential	Inference
Methane	Anode: $CH_3COOH + 2H_2O \rightarrow 2CO_2 + 8H^+ + 8e^-$ Cathode: $CO_2 + 8H^+ + 8e^- \rightarrow CH_4 + 2H_2O$	1.2 V	Not economical as compared to anaerobic digestion
Hydrogen peroxide	Anode: $CH_3COO^- + 4H_2O \rightarrow 2HCO_3^- + 9H^+ + 8e^-$ Cathode: $4O_2 + 8H^+ + 8e^- \rightarrow 4H_2O_2$	0.5 V	Low concentration achieved (0.13wt%), not enough for practical use
Ethanol	$CH_3COO^- + 5H^+ + 4e^- \rightarrow 2CH_3CH_2OH + H_2O$	−550 mV	Continuous replenishment of methyl viologen required
Formic Acid	Anode: $CH_3COOH + 2H_2O \rightarrow 2CO_2 + 8H^+ + 8e^-$ Cathode: $6H^+ + 2e^- \rightarrow 3H_2$ $CO_2 + 2H^+ + 2e^- \rightarrow HCOOH$	2.73 V	Mass transfer and cathode electrode modification required for higher conversion rate
Acetate	Anode: $4H_2O \rightarrow O_2 + 8H^+$ Cathode: $2CO_2 + 8H^+ \rightarrow CH_3CHOOH + 2H_2O$	−400 mV	Genetically modified microbe is required for full conversion of carbon dioxide to acetate

Reference: Nevin et al., 2010; Zhang and Angelidaki, 2014

3.4.5 MICROBIAL DESALINIZATION CELL

Microbial desalination cell (MDC) is a bioelectrochemical cell that oxidizes organic matter present in wastewater to fuel desalination process. Unlike other BES, MDC consist of three chambers: anode chamber, center desalination chamber, and cathode chamber. The anode chamber is fed with wastewater containing organic matter which is oxidized by bacteria on the anode. The electrons produced in the anode chamber are passed to the anode and then to the cathode via an external circuit and in the cathode, it combines with oxygen to form water. The anode and cathode chambers are separated from the desalination chamber by anion exchange membrane and cation exchange membrane respectively. The middle chamber contains saline water with a high concentration of ions. A potential gradient is created due to a reduction and oxidation in the electrode attracts anions and cations present in saline water toward the anode and cathode respectively. This causes desalination of water in the middle chamber(Pandit and Das, 2018) (Figure 3.4).

Recently, research based on combining MEC with other bioelectrochemical systems has become popular, especially MDC. Microbial electrodialysis cell (MEDC) is a combination of MEC and MDC. Hydrogen produced in MEC is used as an energy source for the desalination process. With higher external potential about 0.8 V it was found to have higher hydrogen production and about 98.8% salt removal. But several drawbacks such as pH variation, membrane fouling, and low conductivity are needed to be addressed before real seawater desalination.

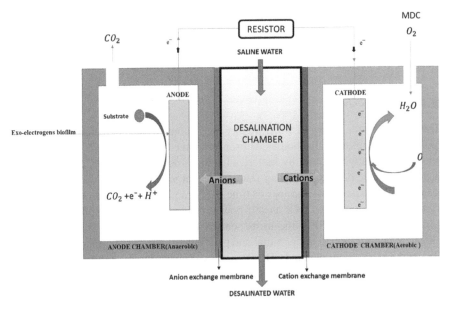

FIGURE 3.4 Schematic representation of microbial desalination cell.

Another integrated system, microbial saline-water electrolysis cell (MSC), is a modified MEDC where the position of cation exchange membrane (CEM) and anion exchange membrane (AEM) were changed. Here the anode biofilm will break down the organic compounds in the saline water and hydrogen is produced at the cathode. The energy produced is used to move anion and cations to the chamber in the middle where that salt concentration is very high, thus simultaneously removing organic (94%), saline content (84%), and hydrogen production. The major disadvantage was at high saline conditions, exo-electron transferability of the bacteria was affected permanently. So exoelectrogens with a high saline tolerance are to be employed.

To overcome the hindrances, especially lower pH in anode chamber and high pH cathode chamber in the above system, a new four-chamber system was proposed known as microbial electrolysis desalination and chemical production cell (MEDCC). Here an extra bipolar membrane was added between AEM and the anode, and then an electric field was applied to dissociate water creating H^+ and OH^- ions. The OH^- ions navigate to the anode chamber to maintain pH, and the H^+ ions migrate to the acid chamber to produce acid. The cations from the desalination chamber (cathode) bind with OH- to form alkali from oxygen reduction. But the major limitation was that only low saline content can be used otherwise production will diminish.

To create a self-sustainable system to produce hydrogen even in rural regions with limited electricity, a system called microbial reverse-electrodialysis electrolysis cell was created. A small electrodialysis stack was integrated into MEC. By using seawater and river water, they were able to produce hydrogen about 1.6 m³/m³-anolyte/day.

But this is possible only near coastal regions, and researchers have used ammonium carbonate which can be easily regenerated and was successful. To date only 10% of energy has been recovered, and further modification is required to improve the efficiency of the system (Zhang and Angelidaki, 2014).

3.4.6 BIOSENSORS

The real-time monitoring of water quality has become a prerequisite due to the continuous pollution of water bodies. Biosensors have been employed in this concern which are usually powered by batteries having a definite life time. Biosensors are battery-powered devices requiring frequent recharge or change in the power source leading to increased cost and loss of time. BES can be used both as a power supply system for biosensors as well as self-powered biosensors. In self-powered biosensors, we directly use BES due to its sensitivity to various biological and physiochemical factors such as temperature and pH. Some of the major application includes Biochemical oxygen demand (BOD) analysis, detection of toxins, detection of microbial biofilm that causes bio-corrosion, monitoring microbial activity, and other applications. Other advantages include high sensitivity, stability, and suitability for remote application without electricity requirement (Ivars-Barceló et al., 2018).

3.4.6.1 Biochemical Oxygen Demand

BOD is the amount of dissolved oxygen required to break down organic matter in a given water sample by aerobic microorganism under a given temperature and time period. Real-time monitoring is a major limitation of the Standard BOD method due to its five-day process. To overcome this, numerous techniques have been designed. One such technique was based on the measurement of dissolved oxygen consumption by immobilized microbes like yeast, *Bacillus subtitles*, *Serratiamarcescens*, or mixed culture, but limitation such as membrane fouling, frequent maintenance, and incompetency of a single organism to use a variety of substrate decreases its suitability. Luminous bacterium *Photobacteriumphosphoreum* was used to correlate luminescent intensity with BOD using and so on, but the interaction of other components in photometric or fluorescence technique decreased its performance (Chang et al., 2004). In 1977, Karube et al. constructed a biofuel cell with microbial electrodes immobilized with *Clostridium butyricum* coupled with platinum (cathode) and used glucose-glutamic acid solution as model water. A linear relationship between steady state current and BOD (concentration of model water) was detected. Relative error of ±10% with a time period of about 30–40 min was noticed while testing industrial waste water(Karube et al., 1977). Later in 2003, Kim et al. created a reliable system using a mediator-less BES having operational stability for over five years without any service and their association between BOD value and the coulomb generated was stable (Kim et al.,2003). To detect the BOD content of ground water, Zhang and Angelidaki innovated a submersible BES without the anode compartment having a standard deviation of 6%–16% compared to the standard method (Zhang and Angelidaki, 2011). For online in-situ monitoring, L. Peixoto et al. in 2013 proposed that variables such as temperature, pH, and conductivity affect the sensitivity of the sensor so calibration before analysis is required (Peixoto et al., 2011).

3.4.6.2 Toxicity Sensor

The presence of toxic compounds hinders the performance and activity of exo-electrogenic biofilm by inhibiting its capability to oxidize organic matter in effluent thus decreasing the current output. This mechanism can be used to detect toxic components in a given sample (Kim et al., 2007; Yang et al., 2015; Zhou et al., 2017). N.E. Stein et al. in 2012 conducted experiments to analyze the influence of membrane type, current, and potential with the response to toxic compounds like nickel in drinking water. No delay in response with good sensitivity was observed. Variation in membrane least affected the system, but as the nickel concentration decreased, higher potential was required for higher sensitivity thus increasing the current density (Stein et al., 2012). Later in 2017, D. Yu et al. analyzed the influence of six metals (such as Cu^{2+}, Hg^{2+}, Zn^{2+}, Cd^{2+}, Pb^{2+} and Cr^{2+}) in microbial activity usually found in water to calculate inhibitory ratio and discovered mercury had the highest value. While testing tap water, an inhibitory ratio of 28.13% was calculated due to the presence of a mixture of heavy metals (Yu et al., 2017). Addressing the sensitivity limitation, Y. Jiang et al. in 2015 optimized the flow configuration and control modes. They discovered that flow through anode had increased sensitivity by 15–41 folds and controlled anode potential was better than constant external resistance. To avoid electro deposition of copper which can cause bias in toxicity measurement, higher potential about −0.15 V was required(Jiang et al., 2015) (Figure 3.5).

3.4.7 MICROBIAL ACTIVITY MONITORING

Microbial activity monitoring has various applications such as checking for contaminants in food industries (Patchett et al., 1988), to keep in check bioremediation and natural attenuation of ground water (Tront et al., 2008) and to check the microbial activity of re-vegetated soil (Jiang et al., 2018). The microbial activity and number of

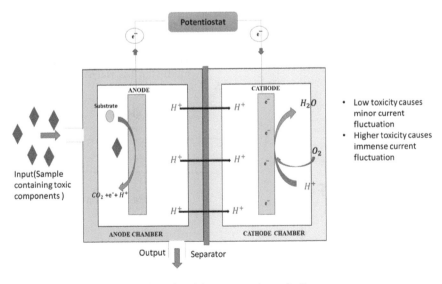

FIGURE 3.5 Microbial fuel-based toxicity sensor schematic diagram.

microbes can be measured by correlating current and microorganism. Two approaches can be used: measurement of microbial respiration or using other parameters to replace biomass concentration as an expression of active microorganism concentration (ATP concentration). Respiration by microbe is by transferring electrons to the anode in a BES system which produce current, thus providing a correlation between microbial respiration and current (Yang et al., 2015). This was analyzed by J.M. Tront et al. in 2008 by using *G.sulfurreducens* to detect and to determine electron donor availability through biological current generation. To check sensitivity and capacity, the microbes where subjected to shock load (pulse of oxygen) which proved electricity generation resumed instantly when the load was removed (Tront et al., 2008). Another approach was proposed by Zhang in 2011, where ATP concentration was used as an indicator of microbial concentration. ATP concentration analysis was less time-consuming and measured only living cells rather than biomass concentration which doesn't distinguish living and non-living cells. Thus ATP concentration was correlated with the current produced. They also proposed the usage of fresh anodes while checking for microbial activity in a given sample (Zhang and Angelidaki, 2011).

3.4.8 MONITORING OF CORROSIVE BIOFILM

In the oil and gas industry, oil pipelines can provide a suitable anaerobic environment for the formation of biofilm, especially by sulfate-reducing bacteria. These cause microbiologically-influenced corrosion as during starvation it switches to iron

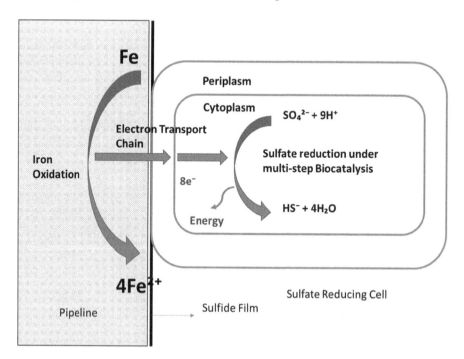

FIGURE 3.6 Schematic diagram of microbial influenced corrosion by sulfate-reducing bacteria showing transport of extracellular electrons into the cell.

causing iron reduction. Thus, in forming an electrogenic biofilm, measuring the electrogenecity of these biofilms can be used to detect its presence. These corrosive bacteria are also found in other industries such as water utilities and so on, which cause huge loss for the industries (Yang et al., 2015). A fresh inert electrode can be inserted into these pipelines, allowing biofilm formation. If there is reduction of sulfates, nitrates and so on when this electrode is used as a cathode with open-circuit potential indicates the presence of these bacteria (Xu and Gu, 2014; Yang et al., 2015).

Anode: $4Fe \rightarrow 4Fe^{2+} + 8e^-$ (iron oxidation)

Cathode: $SO_4^{2-} + 9H^+ + 8e^- \rightarrow HS^- + 4H_2O$ (biocatalytic cathodic sulfate reduction)

This can provide better and more accurate results than already existing equipment as they can't differentiate between oil layer and biofilm layer (Figure 3.6).

3.5 CONCLUSION

Bioelectrochemical systems are robust devices allowing electrochemical and biological reaction to occur in a single reactor. Biosensors based on BES are superior to their counterparts as they are self-sustainable systems with high sustainability and reliability. Wastewater treatment in BES has been widely studied and the combination of electrochemical and biological reactions occurring in the same cell contributes to superior treatment efficiency. Hydrogen production in MEC and BES has opened up an opportunity for the conversion of waste to a usable form of energy. Though BES confers numerous advantages, the cost of material used in its fabrication questions its suitability for its practical-scale development. Further stability studies should be carried out for novel material synthesized to check its applicability in BES.

REFERENCES

Ben Liew, K., Daud, W.R.W., Ghasemi, M., Leong, J.X., Su Lim, S., Ismail, M., 2014. Non-Pt catalyst as oxygen reduction reaction in microbial fuel cells: A review. *Int. J. Hydrogen Energy* 39, 4870–4883. https://doi.org/10.1016/j.ijhydene.2014.01.062

Chang, I.S., Jang, J.K., Gil, G.C., Kim, M., Kim, H.J., Cho, B.W., Kim, B.H., 2004. Continuous determination of biochemical oxygen demand using microbial fuel cell type biosensor. *Biosens. Bioelectron.* 19, 607–613. https://doi.org/10.1016/S0956-5663(03)00272-0

Deng, Q., Li, X., Zuo, J., Ling, A., Logan, B.E., 2010. Power generation using an activated carbon fiber felt cathode in an upflow microbial fuel cell. *J. Power Sources* 195, 1130–1135. https://doi.org/10.1016/j.jpowsour.2009.08.092

Fan, Y., Hu, H., Liu, H., Qian, F., Morse, D.E., Mink, J.E., Qaisi, R.M., Logan, B.E., Hussain, M.M., Choi, S., Florida State University, 2015. Sustainable power generation in microbial fuel cells using bicarbonate buffer and proton transfer mechanisms. *Biosens. Bioelectron.* 69, 8154–8158. https://doi.org/10.1021/es071739c

Fornero, J.J., Rosenbaum, M., Cotta, M.A., Angenent, L.T., 2010. Carbon dioxide addition to microbial fuel cell cathodes maintains sustainable catholyte ph and improves anolyte ph, alkalinity, and conductivity. *Environ. Sci. Technol.* 44, 2728–2734. https://doi.org/10.1021/es9031985

Gopalakrishnan, K., van Leeuwen, J., Brown, R.C., 2012. Sustainable Bioenergy and Bioproducts: Value Added Engineering Applications 001, Green Energy and Technology. https://doi.org/10.1007/978-1-4471-2324-8

He, Z., Huang, Y., Manohar, A.K., Mansfeld, F., 2008. Effect of electrolyte pH on the rate of the anodic and cathodic reactions in an air-cathode microbial fuel cell. *Bioelectrochemistry* 74, 78–82. https://doi.org/10.1016/j.bioelechem.2008.07.007

Hernández-Fernández, F.J., Pérez De Los Ríos, A., Salar-García, M.J., Ortiz-Martínez, V.M., Lozano-Blanco, L.J., Godínez, C., Tomás-Alonso, F., Quesada-Medina, J., 2015. Recent progress and perspectives in microbial fuel cells for bioenergy generation and wastewater treatment. *Fuel Process. Technol.* 138, 284–297. https://doi.org/10.1016/j.fuproc.2015.05.022

Ivars-Barceló, F., Zuliani, A., Fallah, M., Mashkour, M., Rahimnejad, M., Luque, R., 2018. Novel applications of microbial fuel cells in sensors and biosensors. *Appl. Sci.* 8. https://doi.org/10.3390/app8071184

Jafary, T., Wan Daud, W.R., Ghasemi, M., Abu Bakar, M.H., Sedighi, M., Kim, B.H., Carmona-Martínez, A.A., Jahim, J.M., Ismail, M., 2019. Clean hydrogen production in a full biological microbial electrolysis cell. *Int. J. Hydrogen Energy* 44, 30524–30531. https://doi.org/10.1016/j.ijhydene.2018.01.010

Jiang, Y., Liang, P., Zhang, C., Bian, Y., Yang, X., Huang, X., Girguis, P.R., 2015. Enhancing the response of microbial fuel cell based toxicity sensors to Cu(II) with the applying of flow-through electrodes and controlled anode potentials. *Bioresour. Technol.* 190, 367–372. https://doi.org/10.1016/j.biortech.2015.04.127

Jiang, Y., Deng, H., Qin, H., Han, C., Zhong, W., 2018. Indication of soil microbial activity by electrical signals of microbial fuel cells with re-vegetated red soils. *Pedosphere* 28, 269–276. https://doi.org/10.1016/S1002-0160(18)60009-8

Juang, D.F., Yang, P.C., Chou, H.Y., Chiu, L.J., 2011. Effects of microbial species, organic loading and substrate degradation rate on the power generation capability of microbial fuel cells. *Biotechnol. Lett.* 33, 2147–2160. https://doi.org/10.1007/s10529-011-0690-9

Karube, I., Matsunaga, T., Mitsuda, S., Suzuki, S., 1977. Microbial electrode BOD sensors. *Biotechnol. Bioeng.* 19, 1535–1547. https://doi.org/10.1002/bit.260191010

Kim, B.H., Chang, I.S., Gil, G.C., Park, H.S., Kim, H.J., 2003. Novel BOD (biological oxygen demand) sensor using mediator-less microbial fuel cell. *Biotechnol. Lett.* 25, 541–545. https://doi.org/10.1023/A:1022891231369

Kim, M., Sik Hyun, M., Gadd, G.M., Joo Kim, H., 2007. A novel biomonitoring system using microbial fuel cells. *J. Environ. Monit.* 9, 1323–1328. https://doi.org/10.1039/b713114c

Kim, K.Y., He, W., Feng, Y., Saikaly, P.E., 2015. Assessment of microbial fuel cell configurations and power densities. *Environ. Sci. Technol. Lett.* 2, 206–214. https://doi.org/10.1021/acs.estlett.5b00180

Kumar, R., Singh, L., Wahid, Z.A., Din, M.F., 2015. Exoelectrogens in microbial fuel cells toward bioelectricity generation: a review. *Int. J. Energy Res.* 39, 1048–1067. https://doi.org/10.1002/er.3305

Kumar, R., Singh, L., Zularisam, A.W., 2016. Exoelectrogens : Recent advances in molecular drivers involved in extracellular electron transfer and strategies used to improve it for microbial fuel cell applications. *Renew. Sustain. Energy Rev.* 56, 1322–1336. https://doi.org/10.1016/j.rser.2015.12.029

Li, S., Cheng, C., Thomas, A., 2017. Carbon-based microbial-fuel-cell electrodes: From conductive supports to active catalysts. *Adv. Mater.* 29, 1–30. https://doi.org/10.1002/adma.201602547

Liu, H., Logan, B.E., 2004. Electricity Generation Using an Air-Cathode Single Chamber Microbial Fuel Cell in the Presence and Absence of a Proton Exchange Membrane 38, 4040–4046.

Logan, B.E., 2009. Exoelectrogenic bacteria that power microbial fuel cells. *Nat. Rev. Microbiol.* 7, 375–381. https://doi.org/10.1038/nrmicro2113

Lu, L., Xing, D., Ren, N., 2012. Pyrosequencing reveals highly diverse microbial communities in microbial electrolysis cells involved in enhanced H 2 production from waste activated sludge. *Water Res.* 46, 2425–2434. https://doi.org/10.1016/j.watres.2012.02.005

Martin, E., Savadogo, O., Guiot, S.R., Tartakovsky, B., 2010. The influence of operational conditions on the performance of a microbial fuel cell seeded with mesophilic anaerobic sludge. *Biochem. Eng. J.* 51, 132–139. https://doi.org/10.1016/j.bej.2010.06.006

Mohan, S.V., Raghavulu, S.V., Peri, D., Sarma, P.N., 2009. Integrated function of microbial fuel cell (MFC) as bio-electrochemical treatment system associated with bioelectricity generation under higher substrate load. *Biosens. Bioelectron.* 24, 2021–2027. https://doi.org/10.1016/j.bios.2008.10.011

Nam, J.Y., Kim, H.W., Lim, K.H., Shin, H.S., Logan, B.E., 2010. Variation of power generation at different buffer types and conductivities in single chamber microbial fuel cells. *Biosens. Bioelectron.* 25, 1155–1159. https://doi.org/10.1016/j.bios.2009.10.005

Nevin, K.P., Woodard, T.L., Franks, A.E., 2010. Microbial Electrosynthesis : Feeding Microbial Electrosynthesis : Feeding Microbes Electricity To Convert Carbon Dioxide and Water to Multicarbon Extracellular Organic. https://doi.org/10.1128/mBio.00103-10.Editor

Oliveira, V.B., Simões, M., Melo, L.F., Pinto, A.M.F.R., 2013. Overview on the developments of microbial fuel cells. *Biochem. Eng. J.* 73, 53–64. https://doi.org/10.1016/j.bej.2013.01.012

Pandey, P., Shinde, V.N., Deopurkar, R.L., Kale, S.P., Patil, S.A., Pant, D., 2016. Recent advances in the use of different substrates in microbial fuel cells toward wastewater treatment and simultaneous energy recovery. *Appl. Energy* 168, 706–723. https://doi.org/10.1016/j.apenergy.2016.01.056

Pandit, S., Das, D., 2018. Fundamentals of Microbial Desalination Cell Chapter 18 Fundamentals of Microbial Desalination Cell. https://doi.org/10.1007/978-3-319-66793-5

Pant, D., Van Bogaert, G., Diels, L., Vanbroekhoven, K., 2010. A review of the substrates used in microbial fuel cells (MFCs) for sustainable energy production. *Bioresour. Technol.* 101, 1533–1543. https://doi.org/10.1016/j.biortech.2009.10.017

Patchett, R.A., Kelly, A.F., Kroll, R.G., 1988. Use of a microbial fuel cell for the rapid enumeration of bacteria. *Appl. Microbiol. Biotechnol.* 28, 26–31. https://doi.org/10.1007/BF00250492

Peixoto, L., Min, B., Martins, G., Brito, A.G., Kroff, P., Parpot, P., Angelidaki, I., Nogueira, R., 2011. In situ microbial fuel cell-based biosensor for organic carbon. *Bioelectrochemistry* 81, 99–103. https://doi.org/10.1016/j.bioelechem.2011.02.002

Rabaey, K., Verstraete, W., 2005. Microbial fuel cells : novel biotechnology for energy generation. *Trends Biotechnol.* 23, 291–298. https://doi.org/10.1016/j.tibtech.2005.04.008

Rickard, A.H., McBain, A.J., Stead, A.T., Gilbert, P., 2004. Shear rate moderates community diversity in freshwater biofilmsshear rate moderates community diversity in freshwater biofilms.*Appl. Environ. Microbiol.* 70, 7426–7435. https://doi.org/10.1128/AEM.70.12.7426-7435.2004

Stein, N.E., Hamelers, H.M.V., Van Straten, G., Keesman, K.J., 2012. On-line detection of toxic components using a microbial fuel cell-based biosensor. *J. Process Control* 22, 1755–1761. https://doi.org/10.1016/j.jprocont.2012.07.009

Tront, J.M., Fortner, J.D., Plötze, M., Hughes, J.B., Puzrin, A.M., 2008. Microbial fuel cell biosensor for in situ assessment of microbial activity. *Biosens. Bioelectron.* 24, 586–590. https://doi.org/10.1016/j.bios.2008.06.006

Velvizhi, G., Mohan, S.V., 2012. Electrogenic activity and electron losses under increasing organic load of recalcitrant pharmaceutical wastewater. *Int. J. Hydrogen Energy* 37, 5969–5978. https://doi.org/10.1016/j.ijhydene.2011.12.112

Venkata Mohan, S., Velvizhi, G., Annie Modestra, J., Srikanth, S., 2014. Microbial fuel cell: Critical factors regulating bio-catalyzed electrochemical process and recent advancements. *Renew. Sustain. Energy Rev.* 40, 779–797. https://doi.org/10.1016/j.rser.2014.07.109

Wei, J., Liang, P., Huang, X., 2011. Recent progress in electrodes for microbial fuel cells. *Bioresour. Technol.* 102, 9335–9344. https://doi.org/10.1016/j.biortech.2011.07.019

Xu, D., Gu, T., 2014. Carbon source starvation triggered more aggressive corrosion against carbon steel by the Desulfovibrio vulgaris biofilm. *Int. Biodeterior. Biodegrad.* 91, 74–81. https://doi.org/10.1016/j.ibiod.2014.03.014

Yang, H., Zhou, M., Liu, M., Yang, W., Gu, T., 2015. Microbial fuel cells for biosensor applications. *Biotechnol. Lett.* 37, 2357–2364. https://doi.org/10.1007/s10529-015-1929-7

Yu, D., Bai, L., Zhai, J., Wang, Y., Dong, S., 2017. Toxicity detection in water containing heavy metal ions with a self-powered microbial fuel cell-based biosensor. *Talanta* 168, 210–216. https://doi.org/10.1016/j.talanta.2017.03.048

Zhang, Y., Angelidaki, I., 2011. Submersible microbial fuel cell sensor for monitoring microbial activity and BOD in groundwater: Focusing on impact of anodic biofilm on sensor applicability. *Biotechnol. Bioeng.* 108, 2339–2347. https://doi.org/10.1002/bit.23204

Zhang, Y., Angelidaki, I., 2014. Microbial electrolysis cells turning to be versatile technology: Recent advances and future challenges. *Water Res.* 56, 11–25. https://doi.org/10.1016/j.watres.2014.02.031

Zhang, X., Xia, X., Ivanov, I., Huang, X., Logan, B.E., 2014. Enhanced activated carbon cathode performance for microbial fuel cell by blending carbon black. *Environ. Sci. Technol.* 48, 2075–2081. https://doi.org/10.1021/es405029y

Zhao, H., Zhang, Y., Chang, Y., Li, Z., 2012. Conversion of a substrate carbon source to formic acid for carbon dioxide emission reduction utilizing series-stacked microbial fuel cells. *J. Power Sources* 217, 59–64. https://doi.org/10.1016/j.jpowsour.2012.06.014

Zhou, T., Han, H., Liu, P., Xiong, J., Tian, F., Li, X., 2017. Microbial fuels cell-based biosensor for toxicity detection: A review. *Sensors (Switzerland)* 17, 1–21. https://doi.org/10.3390/s17102230

4 Photosynthetic Microbial Fuel Cells

Advances, Challenges and Applications

Subramaniapillai Niju, Karuppusamy Priyadharshini and Elangovan Elakkiya
PSG College of Technology, India

CONTENTS

4.1   INTRODUCTION

The energy needs of the world have skyrocketed dramatically over the last two centuries and there is an urge to shift from non-renewable fossil fuels to renewable, sustainable, clean energy forms because of the rapid global warming and climate change [1]. A recent report from the International Energy Agency (IEA) reveals that the energy produced from biofuels and waste have the highest potential among other sustainable energy forms [2]. Bioelectrochemical systems (BESs) have substantially advanced over the past decade for their contribution as an emerging sustainable technology [3]. Their applications include waste remediation, bioelectricity generation (microbial fuel cells), hydrogen production (microbial electrolysis cells) and bioelectrosynthesis of various valuable by-products (microbial electrosynthesis cells) [4–6]. Microbial fuel cells exploit the microbial interaction with solid electron acceptors/donors to convert organic compounds into electricity [7]. However, the low efficiency is a great challenge for MFC development, and it is recognized that it will be beneficial to couple MFCs with other technologies to improve the efficiency [8]. In this regard, photosynthetic microbial fuel cell technology (PMFC) is a newly developing technology that utilizes solar energy to produce electricity [9]. The solar energy is the primary source of energy for life on earth. This energy is inserted into biosphere through photosynthesis. Photosynthesis is a physico-chemical process by which plants, algae, and certain bacteria convert solar energy into chemical energy from organic matter [10]. Integrating phototrophic microorganisms into MFCs occurred in the past ten years with increasing interests in MFC technology, and there has been active research in microbiology and system development [11]. Biomass of phototrophic organisms can serve as a substrate to feed bacteria at the anode [12], assist the anode to consume and extract electrons from organic compounds [13], and assist the cathode by providing oxygen and remove nutrients as a second stage treatment process [14]. This chapter will discuss in detail about the application of microalgae and photosynthetic bacteria in MFC systems and their influence on the system performance.

4.2 GENERAL CONCEPTS

4.2.1 MICROBIAL FUEL CELL TECHNOLOGY

Microbial fuel cell (MFC) is a type of bioelectrochemical system, in which electricity is generated by degradation of organic matter with the help of a biocatalyst. In the sense, MFC converts chemical energy from an organic substrate directly into electrical energy through a cascade of redox reactions [15]. MFCs use exoelectrogenic microorganisms that biologically oxidize organic matter and transfer electrons to the anode [16]. These electrons flow through an external circuit to the cathode, where they combine with an electron acceptor through electrocatalytic or biocatalytic reductions [17]. This flow of electrons through an external circuit is measured as electric current. Almost any biodegradable organic matter can be used as substrate in MFCs, including simple molecules such as carbohydrates or proteins as well as a complex mixture of organic matter such as wastewater. Thus MFC can generate bioelectricity from any renewable biomass [18, 19].

In addition to electricity generation, MFCs have some other important applications such as waste remediation; bioelectrosynthesis of various valuable by-products and hydrogen production at lower applied potential. MFCs are constructed with a variety of materials in many different configurations, of which single-chambered

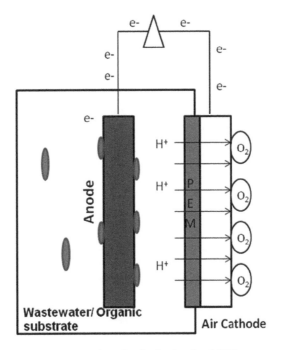

FIGURE 4.1   Schematic representation of a single-chamber MFC.

(Figure 4.1) and dual-chambered (Figure 4.2) are the most common ones. The performances of MFCs are governed by various physical, physio-chemical, chemical, biological and electro-chemical parameters [8]. Electron losses that are encountered during fuel cell operation are a major hindrance that limits the power output. Also the scale-up of this technology still remains as a major issue. Oxygen is the most common electron acceptor in MFCs due to its ample availability and high redox potential. But slow rate of oxygen reduction causes higher cathodic overpotential resulting in diminished power production. The cathodic over-potential for oxygen can be lowered by the use of chemical mediators, or expensive metal-based catalysts such as platinum, or biocatalysts [20, 21]. Chemical mediators such as potassium ferricyanide which is commonly used in lab-scale MFCs cannot be used for largescale because of their inability to be regenerated and highly polluting nature [22]. Though the platinum catalyst is very effective in improving the performance of MFCs, its cost is an obstacle to its application in large-scale MFCs. Hence the use of biocatholytes is given importance in MFC research nowadays. Accumulation of carbon dioxide in MFCs could also be reduced by the use of sustainable biocatholytes [23].

4.2.2 PHOTOTROPHIC MICROORGANISMS

Phototrophs are organisms that use light as their source of energy to produce ATP and carry out various cellular processes. Not all phototrophs are photosynthetic but they all constitute a food source for heterotrophic organisms. All phototrophs either use an electron transport chain or direct proton pumping to establish an electro-chemical

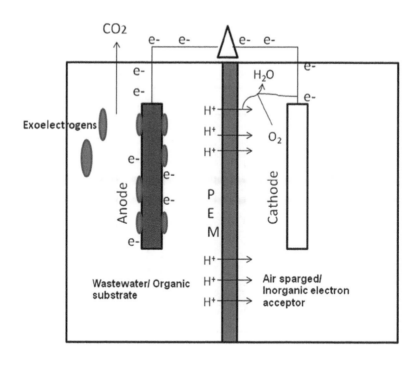

Anode chamber (Anaerobic)   Cathode chamber (Aerobic)

FIGURE 4.2 Schematic representation of a dual-chambered MFC.

gradient utilized by ATP synthase to provide molecular energy for the cell. Phototrophs can be of two types based on their metabolism –photoautotrophs and photoheterrotrophs.

Photoautotrophs – These are organisms that carry out photosynthesis. Using energy from sunlight, carbon dioxide and water are converted into organic materials to be used in cellular functions such as biosynthesis and respiration. In photosynthetic bacteria and cyanobacteria that build up carbon dioxide and water into organic cell materials using energy from sunlight, starch is produced as final product. This process is an essential storage form of carbon, which can be used when light conditions are too poor to satisfy the immediate needs of the organism.

Photoheterotrophs – These obtain their energy from sunlight and carbon from organic material and not carbon dioxide. Photoheterotrophs produce ATP through photophosphorylation but use environmentally obtained organic compounds to build structures and other bio-molecules.

4.2.2.1 Microalgae

Microalgae, one of the oldest living organisms, are microscopic, unicellular photosynthetic plants, capable of photosynthetic conversion of solar energy to biomass, which can further be converted into renewable energy such as hydrogen, bioelectricity, and other biofuels [24]. They can be found in marine and freshwater environments. It accounts for more than half of the Earth's primary photosynthetic throughput. Microalgae have better growth rates compared to forest-derived biomass, agricultural residues, and aquatic species with high conversion efficiency of solar irradiance [25]. The rapid growth rates, ability to survive stringent environmental conditions, round the year availability, cultivation on non-arable land, and non-competitiveness with food and feed make it a potential alternative renewable fuel. Algae can be categorized into seven major types, each with distinct sizes, functions, and color. The different divisions include: Euglenophyta (euglenoids), Chrysophyta (golden-brown algae and diatoms), Pyrrophyta (fire algae), Chlorophyta (green algae), Rhodophyta (red algae), Paeophyta (brown algae), Xanthophyta (yellow-green algae). They can also be classified into three types based on their metabolism viz (i) autotrophic – can be sub-categorized into photoautotrophic and chemoautotrophic; (ii) heterotrophic – can be sub-categorized into photoheterotrophic and chemo-heterotrophic; (iii) mixotrophic – can be phototrophic, heterotrophic or both. Mixotrophic microalgae have many advantages over those with another type of metabolism since they do not present dark respiration-related problems, which have a negative effect on biomass production, and besides, the consumption of organic elements is lower. In a mixotrophic mode of nutrition, respiratory and photosynthetic metabolism pathways are carried out simultaneously (Figure 4.3).

More than 100,000 different species of microalgae exist around the world, but no more than 30,000 have been investigated and classified, according to their color, size, pigments, cell wall constituencies, or metabolism, as suitable for human needs. Microalgae also include the unicellular organisms (phytoplankton) existing in natural water, which are an essential source of carbon for aquatic fauna.

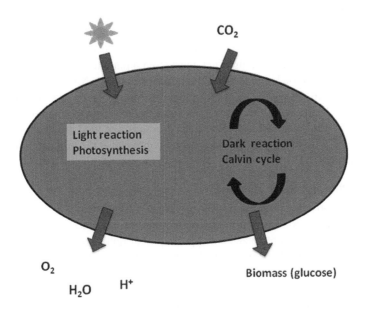

FIGURE 4.3   Schematic representation of mixotrophic growth mode in microalgae.

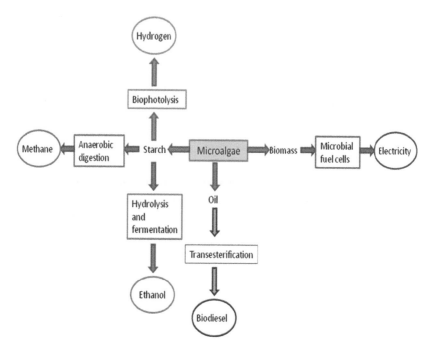

FIGURE 4.4   Different options for bioenergy production from microalgae.

Microalgal biomass is a promising substrate for bioenergy production. As algae biomass was capable of producing much more oil yield per unit cultivation area than other biomass such as corn and soybean, production of algal biodiesel has drawn the interest of scientists because of its potential prospect for practical application. As carbon dioxide (CO_2) sequestration via algae was one to two orders of magnitude higher than terrestrial plants, algae are considered a potential carbon sink in removing overloaded carbon dioxide from the atmosphere [26]. Microalgae has several applications: mitigating CO_2 to reduce global warming, wastewater treatment [27], production of biofuels such as bioethanol, biodiesel, biohydrogen and methane, bio-electricity, biofertilizer, and other important products like food products, antibiotics, and pigments. Figure 4.4 displays the different options for the production of biofuel from microalgae.

4.2.2.2 Photosynthetic Bacteria

They are phototrophic prokaryotes that can convert light energy to chemical energy by photosynthesis process. Currently, phototropic prokaryotes are mainly divided into two categories, oxygenic and anoxygenic phototrophic bacteria (OPB and APB) (Figure 4.5). Cyanobacteria are deemed as the major member of OPB, which are also the typical photoautotrophs that obtain carbon from atmospheric carbon dioxide. The metabolic process does not produce oxygen as a byproduct of the reaction in APB, which chiefly contain purple bacteria (BChl a and b), green bacteria (BChl c and d), and heliobacteria (BChl g) based on the different bacteriochlorophylls (BChl).

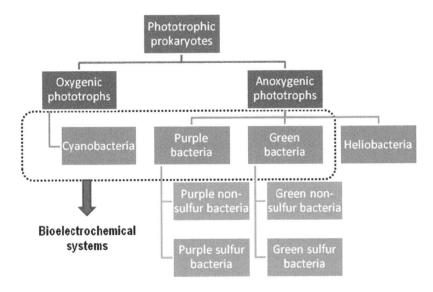

FIGURE 4.5 Classification of phototrophic prokaryotes.

4.2.2.2.1 Cyanobacteria

Cyanobacteria are one of the largest and most important groups of bacteria on earth. They are aquatic and photosynthetic. Many Proterozoic oil deposits are attributed to the activity of cyanobacteria. They are also important providers of nitrogen fertilizer in the cultivation of rice and beans. The cyanobacteria have also been tremendously important in shaping the course of evolution and ecological change throughout Earth's history. The oxygen atmosphere that we depend on was generated by numerous cyanobacteria during the Archaean and Proterozoic eras. Before that time, the atmosphere had a very different chemistry, unsuitable for life as we know it today.

The other great contribution of the cyanobacteria is the origin of plants. The chloroplast with which plants make food for themselves is actually a cyanobacterium living within the plant's cells. Sometime in the late Proterozoic, or in the early Cambrian, cyanobacteria began to take up residence within certain eukaryote cells, making food for the eukaryote host in return for a home. This event is known as endosymbiosis, and is also the origin of the eukaryotic mitochondrion.

Because they are photosynthetic and aquatic, cyanobacteria are often called "blue-green algae". This name is convenient for talking about organisms in the water that make their own food, but does not reflect any relationship between the cyanobacteria and other organisms called algae. Cyanobacteria are relatives of the bacteria, not eukaryotes, and it is only the chloroplast in eukaryotic algae to which the cyanobacteria are related.

Some species of cyanobacteria include *Microcystis aeruginosa*, *Cylindrospermopsis raciborskii*, *Anabaena circinalis*, *Cyanophora paradoxa*, and *Nostoc commune*.

The bioelectrochemical systems based on cyanobacteria are called as biophotovoltaics (BPVs), which generate bioelectricity through the light-driven oxidation of water that occurred in an oxygenic photosynthetic reaction center. These systems harvest light energy without the intervention of heterotrophic organisms.

4.2.2.2.2   Purple Bacteria

Anoxygenic phototrophic purple bacteria are a major group of phototrophic microorganisms that inhabit aquatic and terrestrial environments. Nearly 50 genera of these organisms are known and some have become prime model systems for the experimental dissection of photosynthesis. Purple sulfur bacteria differ from purple nonsulfur bacteria on both metabolic and phylogenetic grounds, but species of the two major groups often coexist in illuminated anoxic habitats in nature. Purple sulfur bacteria are strong photoautotrophs and capable of limited photoheterotrophy, but they are poorly equipped for metabolism and growth in the dark. By contrast, purple nonsulfur bacteria, nature's preeminent photoheterotrophs, are capable of photoautotrophy, and possess diverse capacities for dark metabolism and growth. Several purple bacteria inhabit extreme environments, including extremes of temperature, pH, and salinity.

Purple bacteria contain photosynthetic pigments–bacteriochlorophylls and carotenoids – and can grow autotrophically with CO_2 as the sole carbon source. Many genera of purple bacteria are known and the organisms share many basic properties with their nonphototrophic relatives.

Some species of purple bacteria include *Allochromatium vinosum* and *Thiocapsa roseopersicina* (purple sulfur bacteria); *Rhodobacter capsulatus*, *Rhodobacter sphaeroides*, *Rhodospirillum rubrum*, and *Rhodopseudomonas palustris* (purple nonsulfur bacteria).

4.2.2.2.3   Green Bacteria

Green sulphur bacteria – The green sulfur bacteria are obligate photosynthetic bacteria and depend on light as an energy source for growth. They are generally nonmotile. The chlorosomes are very sensitive light receptors and enable the green sulfur bacteria to grow at minute amounts of light. This has important ecological consequences, because the efficient light-harvesting determines the ecological niche of these bacteria at the lowermost part of stratified environments, where the least light is available. In contrast to plants, green sulfur bacteria mainly use sulfide ions as electron donors. These autotrophs fix carbon dioxide using the reverse tricarboxylic acid (RTCA) cycle.

Some species of green sulphur bacteria include *Chlorobium tepidum*, *Chlorobium chlorochromatii*, *Rhodovulum sulfidophilum*, and *Pelodictylon phaeum*.

Green non-sulphur bacteria – The green non-sulphur bacteria (Chloroflexia) are filamentous bacteria that belong to one of the six classes of phylum Chloroflexi. They are typical photoorganoheterotrophs and well adapted to their changing environment by their gliding motility, tactic responses, and versatile physiology. All characterized members of the green nonsulfur bacteria (phylum Chloroflexi) comprise Gram-negative, filamentous bacteria which exhibit gliding motility. Cells of *Oscillochloris chrysea* are an exception since they stain Grampositive. Most strains are facultatively aerobic; when living aerobically they are not photosynthetic.

Some species of green non-sulphur bacteria are *Chloroflexus auranticus*, *Chloroflexus aggregans*, *Oscillochloris trichoides*,and *Roseiflexus castenholzii*.

4.3 CLASSIFICATION OF PHOTOSYNTHETIC MFCS

The studies involving light-harvesting BESs has led to the generation of a large variety of different system designs. Systems where photosynthetic organisms are used for current generation can be classified into two major groups viz *photo-MFCs* (photosynthetic microbial fuel cells) [28] and BPVs (biophotovoltaics) [29]. One important distinction between these systems is whether an external fuel source other than light is used to drive current production. In BPVs, oxygenic photosynthetic organisms or their counterparts assist light-driven oxidation of water [30] (water splitting). Here there is no need for external fuel source supply. In case of Photo-MFCs, the system may or may not split water, but it needs an external supply of fuel source [31]. BPVs can also be differentiated from systems that utilize oxygenic photosynthetic organisms (such as algae or vascular plants) that harvest light energy but are subsequently used only as a feed stock [32] or a source of organic fuels (e.g. plant MFCs).

Thus a photosynthetic microbial fuel cell (PMFC) can be defined as a bioelectrochemical system capable of converting sunlight into electricity based on the exploitation of biocatalytic reactions within active microbial cells. The current classification of photo-MFC mainly contains three devices: (1) sub-cellular photo-MFC; (2) cellular photo-MFC; (3) complex cellular photo-MFC.

4.3.1 SUB-CELLULAR PHOTO-MFC

A sub-cellular photo-MFC system is where the purified anoxygenic photo-components of the photosynthetic bacteria are directly attached to the electrode surface. In the first reported sub-cellular photo-MFC, the bacterial photosynthetic RC from *Rhodopseudomonas sphaeroides* was isolated and dried as a thin film onto the surface of an SnO_2 electrode for bioelectricity generation [33]. The photosynthetic RC complexes based on a light-induced charge separation process can generate 70 mV of photovoltages and 0.3 $\mu A/cm^2$ of photocurrents in an external circuit. Studies show that individual photosystem complexes can act as light-driven, electron pumps and may be useful as current generators in nanoscale electric circuits, which is an exciting future prospect. However, given the energy cost in producing the complexes, it seems unlikely they will have a role in large-scale power generation.

4.3.2 CELLULAR PHOTO-MFC

These photo-MFCs use living chemoautotrophic microbes (typically anoxygenic photosynthetic microbial species) to generate electricity under anaerobic conditions in a light-dependent manner. The microbes in these systems do not contain a PSII-type reaction center and thus require an exogenous supply of reducing equivalents to grow and function. They use living cells and are therefore significantly more robust than systems that use sub-cellular photosynthetic MFCs. Whole cells are relatively far more resilient, capable of self-repair and reproduction, and can produce current under both light and dark conditions. These photo-MFCs presently hold the record for the highest light-driven current outputs achieved using an intact photosynthetic organism.

The first example of light-dependent electrical interactions between intact, living photosynthetic microbes and an electrode was reported in 1964 using the PNS species *Rhodospirillum rubrum* fed with malate [34]. Almost 40 years later, Rosenbaum et al. [35] took this further by examining the effects of different feedstock compositions on H_2 production and light conversion, using another PNS species *R. sphaeroides*. When fed with a mixture of *E. coli* fermentative by-products under anaerobic conditions and continuous illumination, H_2 produced by *R. sphaeroides* resulted in a maximum power output of 183 mW m^{-2}(at 800 mA m^{-2}). for photobiological H_2 production. Further optimizations have led to some of the highest BES currents measured using axenic *Rhodospirillum. sphaeroides* cultures, with maximum outputs of 790 mW m^{-2} under light conditions, but only 0.5 mW m^{-2} in the dark [36].

There are examples of systems that do not rely on anoxygenic photosynthetic species for generating current that can still be defined as photo-MFCs. A system using a green algal species (*Chlamydomonas reinhardtii*) maintained in an acetate-supplemented feedstock was developed [37]. By inhibiting the oxygen evolving complex (OEC) activity of PSII in *C. reinhardtii* through sulphur deprivation, O_2 can be depleted from the culture by reducing the photosynthetic O_2 production rate below the rate of mitochondrial respiration. Under anaerobic conditions the native hydrogenase activities are not inhibited and *C. reinhardtii* is able to perform light-dependent H_2 evolution over a period of several days. This phenomenon was exploited to produce electricity using a conductive polymer-coated platinum resulting in a maximum power output of 7 mW L^{-1} (at 30 mA L^{-1} (i.e. per liter of liquid culture)).

More recently, the metabolically versatile anoxygenic purple non-sulphur species *Rhodopseudomonas palustris* was shown to be able to metabolize a feedstock consisting of intact filamentous cyanobacteria (*Arthrospira maxima*) while producing current outputs of 5.9 mW m^{-3} (27.9 mA m^{-3}) [38]. Notably, both cyanobacterial growth and thus ultimately *Rh. palustris* power outputs were driven only by light. Such systems exist at the borders of what might be considered a complex photo-MFC.

4.3.3 Complex Cellular Photo-MFCs

Complex Photo MFCs comprise a broad variety of different kinds of light-harvesting BESs that contain both living heterotrophic and autotrophic species. It contains different form: algae and heterotrophic bacteria, anoxygenic photosynthetic bacteria and heterotrophic bacteria, algae and anoxygenic photosynthetic bacteria. BESs that do not contain living autotrophic species within the cell setup, for example MFCs fed with algal-based substrates, are excluded from this definition of complex photo-MFCs. These include devices (i) based on soil sediments, (ii) with anodic liquid-culture consortia, (iii) with phototrophic biocathodes, and (iv) that utilize rhizosphere-based heterotrophic microbes nourished with higher plant root exudates (plant MFCs).

Although they are typically mediator-less and require moderately low maintenance, complex photo-MFCs are highly difficult to characterize in terms of the molecular biological factors contributing to power outputs. This stems from (i) the use of a microbial consortium at the anode and/or (ii) the use of effluent feedstocks that are likely to be variable or poorly characterized. Complex photo-MFCs are challenging to optimize and replicate experimentally, and often it takes weeks for the exoelectrogenic microbial populations to develop. Nevertheless, due to ease of setup, these systems can readily be integrated into other renewable bio-processes, including anaerobic digestion, biomass production, and plant agriculture. Although good progress has been made towards increasing the long-term sustainability of these systems, overall percentage conversion of light into electrical energy remains low [39].

4.4 INTEGRATING PHOTOSYNTHETIC ORGANISMS WITH MFC

One of the most promising technologies to emerge is BESs fueled by light energy [40]. This development is of particular importance because of the nearly limitless supply of energy offered by solar radiation. Thus there is an increasing interest to integrate phototrophic microorganisms into microbial fuel cells (MFCs) to assist electricity generation [41]. In general, this integration can be accomplished in three ways: (1) phototrophic microorganisms function as or provide a substrate for supplying electrons; (2) photo-heterotrophic microorganisms catalyze the anode reaction; and (3) photoautotrophic microorganisms provide oxygen as an electron acceptor to the cathode reaction.

4.4.1 PHOTOTROPHIC MICROORGANISMS AT ANODE

4.4.1.1 Microalgal Biomass as Substrate

The significance of microalgae as a biomass feedstock have been recognized and their potential in biofuel production, which appears to be a promising alternative to fossil fuels, have been examined extensively. Microbial fuel cells (MFCs) offer a possibility to directly generate electrical power from various kinds of organic-rich wastes and biomasses while treating them [19]. Compared to the conventional biogas or bioethanol production from microalgae, producing bioelectricity from microalgal biomass has the advantage of avoiding energy conservation and transportation cost [42]. MFCs also have the advantage of producing fewer soluble microbial products compared to conventional aerobic treatment of algal biomass. Algal biomass can be used as a substrate for electricity generation in MFCs, either as living cells [32] (cultivated or naturally occurred) or dry mass [43]. Algal biomass is mainly composed of proteins, carbohydrates, and lipids, which are complex substrates for microorganisms, so no great amount of energy has been produced at this moment. It has been pointed out that energy production from microalgae biomass should be achieved through different types of pre-treatment [44], although the cost of pre-treatment arises as a challenge for the application of these technologies [45].

The production of electrons and protons in the anodic chamber is carried out by algae mainly during the light phase. CO_2 and light allow algae to perform photosynthesis. During this process, organic substrates, algal biomass, and oxygen are generated. Anodic microorganisms degrade algae wastes and excreted solutes, producing electrons and protons (Eq. 1 in Figure 4.6). The oxygen produced during the light

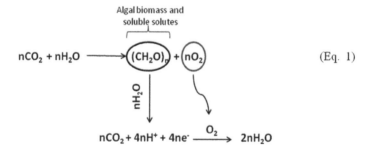

FIGURE 4.6   Reactions involved in anode and cathode of MFC where microalgae is employed as substrate.

period is consumed by algae in the dark phase. In this stage, organic substrates generated during the photosynthesis process are oxidized to obtain energy (Eq. 2 in Figure 4.6). As can be observed, the oxygen produced is consumed by algae during the whole respiration process. It is not expected that BES can completely degrade algal biomass, and thus metabolic products will be generated during electricity production. It was found that algal degradation in an MFC produced several byproducts, among which acetate and lactate were believed to be the major intermediate compounds responsible for electricity generation [46] (Figure 4.6).

In this case of photosynthetic microbial fuel cells, some of the major factors influencing the electricity production are source and intensity of light, the algal species used as feedstock, the type of electrode material used, type of biomass (fresh or dry biomass), and the type of pretreatment method employed. Living cells are usually obtained from cultivation in photobioreactors connected to MFCs [46] or from natural sources like an eutrophicated water body. Because algal cell walls can be resistant to hydrolysis, pretreatment will be necessary to improve the conversion efficiency of algal biomass to electricity. Some of the major pretreatment methods employed are biological pretreatment using an anaerobic digestor, heat treatment [47], microwave treatment [47], ultrasonic treatment, acidic [48] or alkaline pretreatment [49], and extraction of algal organic matter (AOM) [50]. In this supply of algal biomass as feedstock in MFC, there exists a syntrophic activity, in which phototrophic microorganisms are alive and provide energy-rich compounds to electricigens via their phototrophic activities (Figure 4.7).

4.4.1.2   Phototrophic Microorganisms Assisting the Anode Process

The biochemical reaction that takes place during illumination of PMFC in anode can be demonstrated as follows:

$$C_6H_{12}O_6 + 6H_2O \rightarrow 6CO_2 + 24H^+ + 24e^-$$

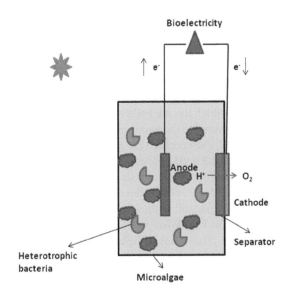

FIGURE 4.7 Schematic representation of MFC where microalgae is employed as substrate.

4.4.1.2.1 Oxygenic Photosynthetic Organisms as Biocatalyst - Without Heterotrophic Bacteria

Traditionally, heterotrophic bacteria that oxidize organic substrates are used as bio-catalysts in MFC systems. These bacteria, that are capable of exocellular electron transfer under anaerobic conditions, transfer electrons from respiratory pathways to the anode which acts as the terminal electron acceptor [16]. But all bacteria are not exoelectrogenic and there are limitations of feeding a complex organic matter to a mixed heterotrophic bacterial community in MFC due to low columbic efficiency [51]. Electrogenic microalgae or cyanobacteria can be used in the anode as electron donors like any other electrogenic bacteria [30, 52, 53]. Thus, bacteria-based MFC systems can be replaced with biomass-based ones such as algae. This will greatly reduce the cost as there will be no restriction for anaerobic condition and further, algal biomass produced can be utilized for biofuels and other value-added product formation.

This category of MFCs comes under cellular BPVs, in which living oxygenic photosynthetic microbes are used to generate current through the photolysis of water, and supply it to an electrode without the aid of heterotrophic species [29]. Notably, microbes in cellular BPVs can also generate current in the dark via the respiratory breakdown of internal carbon reserves accrued during the light, and are capable of producing power throughout a diurnal period. Cyanobacteria more commonly than algae have been employed in anode-assisting components because of their simpler physiology.

The anodic chamber should be free of or at least have controlled levels of oxygen in order to sustain the process. However, the oxygen liberated during photosynthesis (light phase) greatly influences the power output. This is the main problem associated with such MFCs. Researchers have tried to overcome this problem by purging the nitrogen gas or by adding highly concentrated salt solution [30]. In such cases, the algae should possess osmotic tolerance so as to maintain the normal working of cell without compromising its physiology. These options such as using salts in bulk are not commercially feasible and also produce very low current.

Microalgae/cyanobacteria are generally cultivated in the anode side of the PMFC in order to develop a biofilm which has the ability to assimilate a substrate generating electrons, which are then delivered to the anode either directly or via a mediator [54]. *Spirulina platensis* is one of the species that has the ability to shuttle electrons directly to the anode without the need of mediator, and based on this ability, it was investigated by several authors for current generation using a single chamber membrane-less and mediator-free PMFC with different electrode types [55–57].

Another one of the electrogenic microalgae is *Chlorella pyrenoidosa*, which generated a maximum power of 30.15 mW/m^2 in a Photo-MFC using a potassium ferricyanide solution as catholyte. The effect of light on the PMFC power density was studied in terms of light source and intensity, as they can affect greatly the chlorophyll development, stomata opening, and photosynthesis process in microalgal cells [30]. In order to be confirmed, the *Chlamydomonas reinhardtii* was used as a bioanode catalyst in PMFC and was illuminated with monochromatic red and blue lights with different intensities. The obtained PMFC power density was directly proportional to light intensity, with the superiority of red light to blue one, with maximum value of 13 mW/m^2[58]. The same microalgal strain was used in another study to optimize the electrode distance within a two-chamber PMFC with graphite electrodes and the application of dynamic light/dark regime. Maximum power density of 0.82 mW/m^2 was achieved at an electrode distance of 14.7 cm, with a significant reduction in the internal resistance [59].

It has been observed that electrons installed at the electrode in light conditions were higher compared to that installed in dark conditions under oxygenic environment thus affecting the power production. The limitations of this process can be overcome by increasing the photosynthetic efficiency of the cells for enhanced electric current. There is no need of supplying oxygen externally as the oxygen liberated by algae during photosynthesis can be utilized at cathode to produce water. Biomass-based MFCs offer some noticeable advantages over heterotrophic MFCs or photovoltaic cells. It can generate electricity by utilizing only sunlight, water, and carbon dioxide, which is ample in nature. Thus, loading of the anodic chamber with organic compounds can be totally omitted [60]. Utilization of CO_2 also contributes in the carbon sequestration resulting in a clean environment. It is capable of producing power continuously irrespective of day or night.

Using photosynthetic organisms at anode, as a substrate for bacteria needs pretreatment of cells to improve hydrolysis for better reaction, but algae as a model organism doesn't require any pre- treatment and is, thus, cost effective. There are no chances of inhibitors/acid formation at anode, if algal biomass is used. The replacement of pre-cultured algal cells at anode with the fresh one after a defined period will help in maintaining the production rate with time [61].

4.4.1.2.2 Anoxygenic Photosynthetic Organisms as Biocatalyst – Without Heterotrophic Bacteria

The different anoxygenic photosynthetic bacteria (APB) have different metabolic pathways. It can be applied either to the anode or to the cathode. APB can degrade organic matters for bioelectricity generation in anode, and it also can be used as the electron acceptor in cathode.

Bioelectricity generation of APB contain two pathways in photo-MFC anode mainly. One is that APB can produce electrons by anoxygenic photosynthesis or endogenous respiration [62]; and the other is that hydrogen from APB photosynthesis is used as medium for electron generation [41]. According to the photosynthetic process of APB, photons can stimulate cytochrome protein complex from the reaction center. The stimulated cytochrome protein complexes have a strong reducibility. Hence electrons are more easily migrated from APB to electrode. To date, the main APB from the reported papers contain *Rhodospirillum*, *Rhodobacter*, *Rhodopseudomonas*, *Rhodovulum*, and *Chlorobium*. This category of photo-MFCs comes under cellular photo-MFCs.

The most used APB is *Rhodopseudomonas* in photo-MFC anode. It is also regarded as the model bacteria in wastewater treatment. Compared to other genera, *Rhodopseudomonas palustris* DX-1, isolated from an MFC, which is a phototrophic purple non-sulfur bacteria, exhibited a very high activity in direct electron transfer to an anode electrode. This photo-MFC with *Rhodopseudomonas palustris* DX-1 generate a higher power density (2780 mW/m^2) than mixed cultures in the same photo-MFC anode [63]. This strain could use a wide range of organic compounds, including acetate, lactate, fumarate, ethanol, and glycerol, which can be found in many domestic or industrial wastewaters. Further study found that *R. palustris* could also consume the whole cells of cyanobacterium *Arthospora maxima* to generate electricity [38]. Because hydrogen is the product of organic oxidation by *R. palustris*, a hypothesis was proposed that suppressing hydrogen production might improve electricity generation by *R. palustris* in an MFC. This was examined by using gene manipulation to suppress hydrogen production, resulting in a higher power density by the mutant compared with the wild type. *Rhodopseudomonas* was identified as a dominate cluster of bacteria with *Rhodobacter* in a phototrophic consortium on the anode of an MFC; this microbial community produced soluble electron mediators to assist electricity generation, and it was observed that illumination had a positive effect on electricity generation, but this phenomenon was not explained [62].

A few APB (e.g. *Rhodovulum sulfidephilum*, *Chlorobium*) are also explored, but the power output from photo-MFC with *Chlorobium* is near zero [64]. The phenomenon possibly shows that purple bacteria are better suited to photo-MFC than green bacteria. For engineering application, the mixed cultures can consider an electron donor in photo-MFC anode due to easy operation. However, there are only a few reports of the mixed cultures [65]. Thus, the future exploration should pay more attention to the photo-MFC with mixed APB.

4.4.1.2.3   Syntrophic Relation between Phototrophic Organisms and Heterotrophic Bacteria Assisting the Anode Process

Phototrophic activities can result in production of energy-rich compounds, such as hydrogen by photoheterotrophic bacteria or organic compounds by photoautotrophic microorganisms. Those compounds can be converted into electricity in an MFC through a syntrophic relationship between phototrophic and electricigenic microorganisms [39]. It should be noted that this approach is different from the supply of biomass of phototrophic microorganisms described [66]. The performance of such systems is not only influenced by the system configuration and operation, but also by choice of taxa included, as well as conditions to promote phototrophic growth and synergistic interactions between photoautotrophs and heterotrophs in the microbial communities. A synergistic relationship between photosynthetic producers and mixed culture consumers can be established in various in situ and ex situ systems, such as microorganisms or plants with mixed cultures bacteria in sediments. Therefore, a self-sustained photosynthetic MFC can be developed to produce electricity continuously under light illumination through this synergistic interaction, without the need for external exogenous organics or nutrients input [67].

Photosynthetic organisms inside MFC set up take up CO_2 produced by bacteria in the anode, and buffer the hydroxide production. Thus, they reduce the pH gradient between anode and cathode, thereby benefiting current generation. The hydrogen produced by photoheterotrophic bacteria such as *Rhodobactercapsulatus* can be used as a substrate in a PEM fuel cell by linking with a photobioreactor. This approach is simplified through integrating phototrophic bacteria into the anode of a fuel cell. Moreover, the use of two-step biohydrogen production for electricity generation from organic compounds was accomplished through connecting dark fermentation (by *Escherichia coli*) and photo fermentation (by *Rhodobacter sphaeroides*) [68].

Using defined binary cultures in an MFC, the researchers demonstrated that a non-phototrophic electricigen *Geobacter sulfurreducens* could use formate produced by a green alga *Chlamydomonas reinhardtii* for generating electricity, which provides a proof of syntrophic relationship between two microorganisms during electricity generation. This relationship was further demonstrated in an MFC system containing either monoculture of isolated photosynthetic bacterium *Chlorobium*, electricigen *Geobacter*, or a co-culture of the two. It was found that light-responsive current generation was observed only in the co-culture MFC, which was likely from *Geobacter* oxidizing acetate produced from glycogen (via dark fermentation) that was released by *Chlorobium* during photosynthesis [69].

4.4.2   Phototrophic Microorganisms Assisting the Cathode Process

A limiting step in power generation is the rate of oxygen reduction to water in the cathode chamber. There has been a greater interest in applying phototrophic microorganisms (microalgae and cyanobacteria) in the cathode of an MFC with multiple benefits such as supplying oxygen, reducing carbon dioxide, producing valuable biomass, and/or polishing wastewater effluent [70]. These phototrophic microorganisms can serve as biocatholytes in MFCs because the oxygen produced by them can act as an electron acceptor for the electrons harvested from the anode compartment

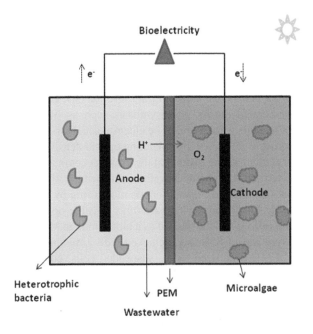

FIGURE 4.8   Schematic representation of algal biocathode MFC.

(Figure 4.8). Also, these biocathodes can cooperate in reducing the emitted CO_2 from bacterial metabolism and respiration. Results from different researches demonstrate that current generation in an MFC can be stimulated by light illumination in the presence of photosynthetic microorganisms, and concurrent variation of dissolved oxygen provides strong evidence that the cathode reaction benefits from oxygen supplied via photosynthesis [71].

The chemical equations for photosynthetic reactions in the cathodic chamber of PMFCs (phototrophic organisms employed at cathode) are as follows:

Light reaction

$$6CO_2 + 12H^+ + 12e^- \rightarrow C_6H_{12}O_6 \text{ (biomass)} + 3O_2$$

Dark reaction

$$C_6H_{12}O_6 \text{ (biomass)} + 6O_2 \rightarrow 6CO_2 + 6H_2O$$

Cathodic reaction

$$O_2 + 4H^+ + 4e^- \rightarrow 2H_2O$$

Chlorella vulgaris is one of the most common microalgal species that was investigated by several authors as a biocathode in PMFC. One of the first trials was conducted by Powell et al. to study the capture of CO_2 by *C. vulgaris* which acted as an electron acceptor in the cathode chamber of PMFC. The maximum cell growth rate

(3.6 mg/L/h) was achieved under a concentration of 10% CO_2 with obtained power density of 2.7 mW/m² [72]. A study of a cyanobacterium, *Microcystis aeruginosa* IPP in the cathode, provided the first experimental proof that photosynthetic microorganisms can also catalyze the cathode reaction. It was observed that the photosynthetic biocathode increased the current density by 245%, compared with the abiotic cathode that was aerated to maintain a certain level of DO, and the reactive oxygen species (ROS) generated during photosynthesis such as hydrogen peroxide and superoxide anion radicals were found to be electron acceptors for electricity generation [73]. Monitoring algal growth by collecting optical density measurements in the cathode chamber is challenging given that the growth occurs in three forms: as algae biofilm on the electrode and chamber walls, as suspended algal aggregates, and as suspended biomass in solution. Algae biomass monitoring by weighing dry biomass samples or assaying for protein content are other feasible options. When *Chlorella vulgaris* was immobilized and introduced as a biocathode in MFC, Chemical oxygen demand (COD) removal of 85% was achieved with a maximum power density of 2485.35 mW/m³ [74]. Compared to the suspended cells, coulombic efficiency of immobilized ones was improved by 58%. Some investigators proved that the application of immobilized cells as compared to suspended cells could increase the columbic efficiency up to 88%.

With the application of different illumination regimes to the cathode (12–24 h of light), the critical role of light in PMFCs was again confirmed by observing the decline in dissolved oxygen in dark mode with cell voltage following the same trend of dissolved oxygen profile, resulting in a power density of 13.5 mW/m² during the acclimation phase [14]. Illumination is an important condition for growth of photosynthetic microorganisms, and the researchers mainly focused on illumination intensity and length. Several studies found that increasing illumination intensity could significantly increase electricity generation, likely through improving the production of dissolved oxygen [75]. However, Juang et al. reported that their MFCs produced more electricity under lower light intensity. Those results suggest that there are optimal light intensities for the MFCs with photosynthetic cathode, which will be related to microbial species and operating conditions [76]. Adjusting light intensity can also affect the production of some valuable chemicals such as carotenoid in the cathode of an MFC. A long illumination period may benefit oxygen production and thus electricity generation when illumination is a limiting factor. But an extended illumination was found to decrease electricity production, indicating that dark period is necessary to maintain a healthy community of photosynthetic microorganisms. Increasing the frequency of light/dark cycle decreased the production of both electricity and algal biomass, confirming the importance of a proper dark period for a photosynthetic MFC. The effect of light intensity on the power production of PMFC was also inspected by using a microalgal co-culture of *Chlorella* and *Phormidium*. Covered anodic PMFC exposed to lower power of light (6 and 12 W) exhibited higher coulombic efficiency and power than the one with higher powers (18 and 26 W), which means that high light power should be avoided if algal photosynthesis is the source of oxygen in the cathodic chamber of PMFCs [76]. This study drew the attention toward the benefits of using more than one strain of microalgae in PMFCs by employing water ponds containing mixed cultures of microalgae.

In case of dual chamber MFC, the cathode compartment is generally filled with buffer solution or a defined nutrition medium that supports the growth of photosynthetic organisms. This creates an additional demand for water, which becomes economically challenging. Alternatively, when wastewater is used for catholyte purposes, it can support the growth of photosynthetic organism thus creating biomass and simultaneously remove contaminants from wastewater. However, feeding raw wastewater into a cathode compartment will stimulate the growth of heterotrophic bacteria and organic compounds will act as an electron donor competing with the cathode electrode, which will damage the electricity-generating function of MFCs. A possible approach is to feed the wastewater treated by the anode into the cathode, and this treated wastewater will provide nutrients (e.g., nitrogen and phosphorus) for algal growth and may also be further polished by the cathode for removal of organic residues and nutrients.

As a result of photosynthesis, biomass of photosynthetic microorganisms is accumulated and may be used to produce energy or value-added compounds. The concentration of biomass in the solution varies from 100 to more than 4000 mg/L. The highest biomass concentration in the two-chamber MFCs was 2800 mg/L [71]. It should be noted that biomass concentration can also be affected by hydraulic retention time (HRT). A long HRT (e.g., 410 days) tends to accumulate more biomass than a shorter HRT. For example, several fed-batch-operated MFCs could produce 4300 mg/L algal biomass at an HRT of 410 days; while a continuously operated integrated photo-bioelectrochemical (IPB) system with an HRT of three days produced a relatively low biomass concentration of 128 mg/L [77]. Algal biomass produced in an MFC can be used to extract pigment, which contains some high value compounds such as carotenoids, and composition of pigment was affected by light intensity and nutrient supply. Micro-algal biomass production can be accelerated under light and nutrient stress. The identified carotenoids (β-carotene and secondary carotenoids and lutein) have added value as antioxidants, pharmaceuticals, and nutraceuticals. Stimulating microalgal growth under stress conditions can give rise to valuable compounds without compromising power output. Use of blue green algae in MFCs not only produces pigments but also releases potent toxins. Microcystins are the toxins produced by blue green algae in lake waters when the algal cells rupture and die. Microcystin cannot be treated with conventional treatment technologies such as coagulation, flocculation, sedimentation, and filtration. The presence of blue green algae MFCs cannot effectively harvest bioelectricity but can successfully treat toxic microcystins Microcystin-RR and -LR types (90.7% and 91%) [78]. Synthetic plastic production from algal polysaccharides is another useful product of algal biomass. Cyanobacteria is an excellent feedstock for bio-plastic production due to its high yield and ability to grow in a variety of environments.

4.5 APPLICATION ASPECTS OF PHOTOSYNTHETIC MICROBIAL FUEL CELL

4.5.1 Microbial Carbon Capture Cell

It is known that an MFC with a photosynthetic biocathode can fix carbon dioxide. Instead of externally supplying carbon dioxide, an approach was developed to have photosynthetic microorganisms in the cathode of an MFC growing on carbon dioxide produced from its anode (oxidation of organic compounds), which is named microbial carbon capture cells (MCC). Thus, it is a carbon-neutral technology with strong sustainability.

Wang et al. developed an MCC by introducing gas generated through bacterial respiration and metabolism at the anode into a cathode in which a photosynthetic microorganism (*Chlorella vulgaris*) was growing [79]. They demonstrated the proof-of-concept of organic waste removal from the anode chamber with simultaneous electricity generation and carbon sequestration in the cathode chamber without the need for external energy inputs. MCCs were constructed, by another group of scientists, with cyanobacteria growing in a photo biocathode in dual-chambered flat plate mediator-less MFCs separated by an anion exchange membrane from the anode compartment containing *Shewanella putrefaciens*[80]. The performance of the MCC with *Anabaena* sparged with CO_2-air mixture was compared with that of a conventional cathode sparged with air only. The power outputs of MCCs were better than those of dual-chambered MFCs. Flue gas which contains large amounts of CO_2 might provide a suitable buffering capacity and high NOx-containing flue gas could be used for power augmentation since nitrate can act as potential electron acceptor. Power generation in an MCC depends on both light and bicarbonate utilization. Factors such as light intensity, superficial gas velocity, and reactor design need to be optimized for maximizing power generation in MCCs.

4.5.2 MICROBIAL DESALINATION CELL

A microbial desalination cell is a biological electrochemical system that implements the use of electro-active bacteria to power desalination of water in situ, resourcing the natural anode and cathode gradient of the electro-active bacteria and thus creating an internal supercapacitor. Using biocathode in MDC systems improved the performance of the systems many fold [81]. Phototrophic biocathode was also studied in an MDC for driving desalination, and the results show that the MDC with microalgae *Chlorella vulgaris* in the cathode removed 40.1% of salt, almost twice that of an abiotic cathode MDC. In a photo- MDC system, it was estimated that the produced algal biomass represented an energy content of 0.21 kW h m^{-3}, which further improved the energy benefits of this system to 2.01 kW h m^{-3} (without desalination energy credit). The algal bio-cathode showed a desalination rate of 0.161 g/L/d, which is about 2.1 times higher than the corresponding air cathode operation (0.076 g/L/d) [82]. The higher potential difference between electrodes might be the probable reason to stimulate the ions transfer from middle chamber to anode and cathode based on charge.

4.5.3 PHOTOSYNTHETIC SEDIMENT MFCs

Sediment MFC is a type of MFC in which the anode is buried in the sediment, while the cathode is placed on top of the sediment. This configuration is also referred to as benthic MFC. The biocatalyst present in the sediment oxidizes the organic material while the cathodic reactions include reduction of dissolved oxygen in the water. The naturally existing differences in electropotential of the anode and cathode help in generating bioelectricity [83]. A new approach of photosynthetic sediment MFC was proposed wherein microalgae is incorporated in the cathode compartment. The CO_2 produced by anodic bacterial activity is consumed by algal cells, and the O_2 produced by the algae is consumed by the PSMFC cathode compartment for current production. This PSMFC is generally consisting of an anode placed in the middle of a sediment layer which is covered with sand and a cathode compartment which is filled with microalgal culture medium. The PSMFC is normally operated in the presence of a light source in order for the photosynthesis to take place [84]. He et al. have constructed a sediment-type photosynthetic MFC which produced a maximum current of 0.054 ± 0.002 mA at a resistance of 1 kΩ in a system which had been operated for over 145 days [39]. Another group of scientists developed a membrane-less sediment-type MFC consisting of a photosynthetic biocathode containing a complex microbial community along with microalgae and cyanobacteria which were able to produce a maximum power density $11 mWm^{-2}$ over 180 days with no feeding [85].

4.6 CONCLUSIONS AND PERSPECTIVES

Applying phototrophic microorganisms for electricity generation in MFCs makes those systems sustainable with no requirement for replacement. The challenges associated with integrating photosynthetic microorganisms into MFCs should be clearly understood for focusing future research and development. Though PMFCs has the advantage of producing biomass in addition to electricity generation and wastewater treatment, in some cases one needs to choose between improving electricity generation and harvesting biomass. Most studies have focused on a pure strain of phototrophic microorganism, while in practical application (especially for wastewater treatment) a mixed culture will exist in the bioreactor. More work will be needed to understand the activities of phototrophic microorganisms in a mixed microbial community and their interaction with other microorganisms (competition and/or syntrophic relationships). The role of illumination in electrochemical catalysis of the anode reaction by phototrophic microorganisms should be further investigated.

Use of algae in the anode chamber or as substrate, lowers level of output while in the case of microalgae-assisted cathodes, the results are promising. The overall recovered energy from the PMFC could be magnified by increasing the electrode surface area, which positively results in enriched biofilm and overcomes the competition of non-electrochemically active microorganisms. To elucidate the process as economically viable and environmentally benign, the PMFCs should be scaled up for a successful application.

REFERENCES

[1] A. R. Osmani, "Conventional energy to renewable energy: Perspectives for India," *NEHU J.*, vol. XII, no. 2, pp. 41–60, 2014.
[2] International Energy Agency, "2018 World Energy Outlook: Executive Summary," Int. Energy Agency, p. 11, 2018.
[3] S. Bajracharya, M. Sharma, G. Mohanakrishna, X. Dominguez, D.P.B. Strik, and P. M. Sarma, "An overview on emerging bioelectrochemical systems (BESs): Technology for sustainable electricity , waste remediation , resource recovery , chemical production and beyond," *Renew. Energy*, vol. 98, pp. 153–170, 2016.

[4] C. K. Pallavi and T. H. Udayashankara, "A review on microbial fuel cells employing wastewaters as substrates for sustainable energy recovery and wastewater treatment," *IOSR J. Environ. Sci. Toxicol. Food Technol.*, vol. 10, no. 12, pp. 31–36, 2016.

[5] S. V. Mohan, "Harnessing of biohydrogen from wastewater treatment using mixed fermentative consortia : Process evaluation towards optimization," *Int. J. Hydrogen Energy*, vol. 34, no. 17, pp. 7460–7474, 2009.

[6] R. Karthikeyan, R. Singh, and A. Bose, "Microbial electron uptake in microbial electrosynthesis: a mini-review," *J. Ind. Microbiol. Biotechnol.*, no. 0123456789, 2019.

[7] A. Hirose, T. Kasai, R. Koga, Y. Suzuki, A. Kouzuma, and K. Watanabe, "Understanding and engineering electrochemically active bacteria for sustainable biotechnology," *Bioresour. Bioprocess.*, vol. 6, no. 1, pp. 1–15, 2019.

[8] S. V. Mohan, G. Velvizhi, J. A. Modestra, and S. Srikanth, "Microbial fuel cell : Critical factors regulating bio-catalyzed electrochemical process and recent advancements," *Renew. Sustain. Energy Rev.*, vol. 40, pp. 779–797, 2014.

[9] J. Greenman, I. Gajda, and I. Ieropoulos, "Microbial fuel cells (MFC) and microalgae; Photo microbial fuel cell (PMFC) as complete recycling machines," *Sustain. Energy Fuels*, vol. 3, no. 10, pp. 2546–2560, 2019.

[10] B. Demmig-Adams and W. W. Adams, "Harvesting sunlight safely," *Nature*, vol. 403, no. 6768, pp. 371–374, 2000.

[11] D. P. B. Strik, R. A. Timmers, M. Helder, K. J. J. Steinbusch, H. V. M. Hamelers, and C. J. N. Buisman, "Microbial solar cells: Applying photosynthetic and electrochemically active organisms," *Trends Biotechnol.*, vol. 29, no. 1, pp. 41–49, 2011.

[12] S. B. Velasquez-Orta, T. P. Curtis, and B. E. Logan, "Energy from algae using microbial fuel cells," *Biotechnol. Bioeng.*, vol. 103, no. 6, pp. 1068–1076, 2009.

[13] S. Mateo, A. G. Campo, P. Cañizares, J. Lobato, M. A. Rodrigo, and F. J. Fernandez, "Bioelectricity generation in a self-sustainable Microbial Solar Cell," *Bioresour. Technol.*, vol. 159, pp. 451–454, 2014.

[14] A. González Del Campo, P. Cañizares, M. A. Rodrigo, F. J. Fernández, and J. Lobato, "Microbial fuel cell with an algae-assisted cathode: A preliminary assessment," *J. Power Sources*, vol. 242, pp. 638–645, 2013.

[15] B. E. Logan et al., "Microbial fuel cells: Methodology and technology," *Environ. Sci. Technol.*, vol. 40, no. 17, pp. 5181–5192, 2006.

[16] B. E. Logan, "Exoelectrogenic bacteria that power microbial fuel cells," *Nat. Rev. Microbiol.*, vol. 7, pp. 375–381, 2009.

[17] D. Ucar, Y. Zhang, I. Angelidaki, and A. E. Franks, "An overview of electron acceptors in microbial fuel cells," *Front. Microbiol.*, vol. 8, no. April, pp. 1–14, 2017.

[18] P. Pandey, V. N. Shinde, R. L. Deopurkar, S. P. Kale, S. A. Patil, and D. Pant, "Recent advances in the use of different substrates in microbial fuel cells toward wastewater treatment and simultaneous energy recovery," *Appl. Energy*, vol. 168, pp. 706–723, 2016.

[19] D. Pant, G. Van Bogaert, L. Diels, and K. Vanbroekhoven, "A review of the substrates used in microbial fuel cells (MFCs) for sustainable energy production.," *Bioresour. Technol.*, vol. 101, no. 6, pp. 1533–1543, 2010.

[20] S. Srikanth, T. Pavani, P. N. Sarma, and S. Venkata Mohan, "Synergistic interaction of biocatalyst with bio-anode as a function of electrode materials," *Int. J. Hydrogen Energy*, vol. 36, no. 3, pp. 2271–2280, 2011.

[21] H. Liu, R. Ramnarayanan, and B. E. Logan, "Production of electricity during wastewater treatment using a single chamber microbial fuel cell," *Environ. Sci. Technol.*, vol. 38, no. 7, pp. 2281–2285, 2004.

[22] A. E. Franks and K. P. Nevin, "Microbial fuel cells, a current review," *Energies*, vol. 3, no. 5, pp. 899–919, 2010.

[23] S. Srikanth, M. Kumar, D. Singh, M. P. Singh, and B. P. Das, "Electro-biocatalytic treatment of petroleum refinery wastewater using microbial fuel cell (MFC) in continuous mode operation," *Bioresour. Technol.*, vol. 221, pp. 70–77, 2016.

[24] X. Liu, A. F. Clarens, and L. M. Colosi, "Algae biodiesel has potential despite inconclusive results to date," *Bioresour. Technol.*, vol. 104, pp. 803–806, 2012.

[25] J. D. E. Noue and N. D. E. Pauw, "The Potential of Microalgal Biotechnology : a Review of Production and Uses of Mocroalgae," *Biotechnol. Adv.*, vol. 6, pp. 725–770, 1988.

[26] S. H. Ho, C. Y. Chen, D. J. Lee, and J. S. Chang, "Perspectives on microalgal CO2-emission mitigation systems - A review," *Biotechnol. Adv.*, vol. 29, no. 2, pp. 189–198, 2011.

[27] N. Abdel-Raouf, A. A. Al-Homaidan, and I. B. M. Ibraheem, "Microalgae and wastewater treatment," *Saudi J. Biol. Sci.*, vol. 19, no. 3, pp. 257–275, 2012.

[28] F. Fischer, "Photoelectrode, photovoltaic and photosynthetic microbial fuel cells," *Renew. Sustain. Energy Rev.*, vol. 90, no. June 2017, pp. 16–27, 2018.

[29] T. Wenzel and C. J. Howe, "Biophotovoltaics: oxygenic photosynthetic organisms in the world of bioelectrochemical systems," *Energy Environ. Sci.*, vol. 00, pp. 1–18, 2015.

[30] C. Xu, K. Poon, M. M. F. Choi, and R. Wang, "Using live algae at the anode of a microbial fuel cell to generate electricity," *Environ. Sci. Pollut. Res.*, vol. 22, no. 20, pp. 15621–15635, 2015.

[31] Z. Yang, H. Pei, Q. Hou, L. Jiang, L. Zhang, and C. Nie, "Algal biofilm-assisted microbial fuel cell to enhance domestic wastewater treatment: Nutrient, organics removal and bioenergy production," *Chem. Eng. J.*, vol. 332, pp. 277–285, 2018.

[32] F. Ndayisenga, Z. Yu, Y. Yu, C. H. Lay, and D. Zhou, "Bioelectricity generation using microalgal biomass as electron donor in a bio-anode microbial fuel cell," *Bioresour. Technol.*, vol. 270, pp. 286–293, 2018.

[33] A. F. Janzen and M. Sibert, "Photoelctrochemical conversion using reaction centre electrodes," *Nature*, vol. 286, pp. 584–585, 1980.

[34] R. S. Berk and J. H. Canfield, "Bioelectrochemical energy conversion," *Appl. Microbiol.*, vol. 12, no. 1, pp. 10–12, 1964.

[35] M. Rosenbaum and F. Scholz, "In situ electrooxidation of photobiological hydrogen in a photobioelectrochemical fuel cell based on Rhodobacter sphaeroides," *Environ. Sci. Technol.*, vol. 39, no. 16, pp. 6328–6333, 2005.

[36] J. R. Benemann, "The technology of biohydrogen," *Biohydrogen*, pp. 19–30, 1998.

[37] M. Rosenbaum, U. Schröder, and F. Scholz, "Utilizing the green alga Chlamydomonas reinhardtii for microbial electricity generation: a living solar cell," *Appl. Microbiol. Biotechnol.*, pp. 753–756, 2005.

[38] A. E. Inglesby, D. A. Beatty, and A. C. Fisher, "Rhodopseudomonas palustris purple bacteria fed Arthrospira maxima cyanobacteria : demonstration of application in microbial fuel cells," *RSC Adv.*, vol. 17001, no. 2007, pp. 4829–4838, 2012.

[39] Z. He, J. Kan, F. Mansfeld, L. T. Angenent, and K. H. Nealson, "Self-sustained phototrophic microbial fuel cells based on the synergistic cooperation between photosynthetic microorganisms and heterotrophic bacteria," *Environ. Sci. Technol.*, vol. 43, no. 5, pp. 1648–1654, 2009.

[40] A. Elmekawy, H. M. Hegab, K. Vanbroekhoven, and D. Pant, "Techno-productive potential of photosynthetic microbial fuel cells through different configurations," *Renew. Sustain. Energy Rev.*, vol. 39, pp. 617–627, 2014.

[41] Y. K. Cho, T. J. Donohue, I. Tejedor, M. A. Anderson, K. D. McMahon, and D. R. Noguera, "Development of a solar-powered microbial fuel cell," *J. Appl. Microbiol.*, vol. 104, no. 3, pp. 640–650, 2008.

[42] M. Ray, N. Kumar, V. Kumar, and S. Negi, Microalgae: A way forward approach towards wastewater treatment and bio-fuel production. *Appl. Microbiol. Bioeng.*, pp. 229–243, 2019.

[43] N. Rashid, Y. F. Cui, S. U. R. Muhammad, and J. I. Han, "Enhanced electricity generation by using algae biomass and activated sludge in microbial fuel cell," *Sci. Total Environ.*, vol. 456–457, pp. 91–94, 2013.

[44] C. M. Fernández-Marchante et al., "Thermally-treated algal suspensions as fuel for microbial fuel cells," *J. Electroanal. Chem.*, 2018.

[45] A. Khandelwal, A. Vijay, A. Dixit, and M. Chhabra, "Microbial fuel cell powered by lipid extracted algae: A promising system for algal lipids and power generation," *Bioresour. Technol.*, vol. 247, pp. 520–527, 2018.

[46] A. E. Inglesby and A. C. Fisher, *"Downstream application of a microbial fuel cell for energy recovery from an Arthrospira maxima fed anaerobic digester effluent,"* RSC Adv., 2013.

[47] V. Gadhamshetty, D. Belanger, C. J. Gardiner, A. Cummings, and A. Hynes, "Evaluation of Laminaria-based microbial fuel cells (LbMs) for electricity production," *Bioresour. Technol.*, vol. 127, pp. 378–385, 2013.

[48] S. Kondaveeti, K. S. Choi, R. Kakarla, and B. Min, *"Microalgae Scenedesmus obliquus as renewable biomass feedstock for electricity generation in microbial fuel cells (MFCs),"* 2013.

[49] H. Wang, L. Lu, F. Cui, D. Liu, Z. Zhao, and Y. Xu, "Simultaneous bioelectrochemical degradation of algae sludge and energy recovery in microbial fuel cells," *RSC Adv.*, vol. 2, pp. 7228–7234, 2012.

[50] H. Wang, D. Liu, L. Lu, Z. Zhao, Y. Xu, and F. Cui, "Degradation of algal organic matter using microbial fuel cells and its association with trihalomethane precursor removal," *Bioresour. Technol.*, vol. 116, pp. 80–85, 2012.

[51] S. Pandit, K. Chandrasekhar, and R. Kakarla, *"Basic Principles of Microbial Fuel Cell : Technical Challenges and Economic Feasibility,"* vol. 1, 2017.

[52] K. S. Madiraju, D. Lyew, R. Kok, and V. Raghavan, "Carbon neutral electricity production by Synechocystis sp . PCC6803 in a microbial fuel cell," *Bioresour. Technol.*, vol. 110, pp. 214–218, 2012.

[53] R. W. Bradley, P. Bombelli, S. J. L. Rowden, and C. J. Howe, "Biological photovoltaics : intra- and extra-cellular electron transport by cyanobacteria," *Biochem. Soc. Trans.*, vol. 40, pp. 1302–1307, 2012.

[54] Y. Zou, J. Pisciotta, R. B. Billmyre, and I. V. Baskakov, "Photosynthetic microbial fuel cells with positive light response," *Biotechnol. Bioeng.*, vol. 104, no. 5, pp. 939–946, 2009.

[55] C. C. Fu, T. C. Hung, W. T. Wu, T. C. Wen, and C. H. Su, "Current and voltage responses in instant photosynthetic microbial cells with Spirulina platensis," *Biochem. Eng. J.*, vol. 52, no. 2–3, pp. 175–180, 2010.

[56] C. C. Lin, C. H. Wei, C. I. Chen, C. J. Shieh, and Y. C. Liu, "Characteristics of the photosynthesis microbial fuel cell with a Spirulina platensis biofilm," *Bioresour. Technol.*, vol. 135, pp. 640–643, 2013.

[57] C. C. Fu, C. H. Su, T. C. Hung, C. H. Hsieh, D. Suryani, and W. T. Wu, "Effects of biomass weight and light intensity on the performance of photosynthetic microbial fuel cells with Spirulina platensis," *Bioresour. Technol.*, vol. 100, no. 18, pp. 4183–4186, 2009.

[58] J. C. Lan, K. Raman, C. Huang, and C. Chang, "The impact of monochromatic blue and red LED light upon performance of photo microbial fuel cells (PMFCs) using Chlamydomonas reinhardtii transformation F5 as biocatalyst," *Biochem. Eng. J.*, vol. 78, pp. 39–43, 2013.

[59] K. Raman and J. C. Lan, "Performance and kinetic study of photo microbial fuel cells (PMFCs) with different electrode distances," *Appl. Energy*, vol. 100, pp. 100–105, 2012.

[60] M. Otadi, S. Poormohamadian, F. Zabihi, and M. Goharrokhi, "Microbial Fuel Cell Production with Alga," *World Appl. Sci. J.*, vol. 14, no. SPL ISS 3, pp. 91–95, 2011.

[61] X. A. Walter, J. Greenman, B. Taylor, and I. A. Ieropoulos, "Microbial fuel cells continuously fuelled by untreated fresh algal biomass," *Algal Res.*, vol. 11, pp. 103–107, 2015.

[62] X. Cao, X. Huang, N. Boon, P. Liang, and M. Fan, "Electricity generation by an enriched phototrophic consortium in a microbial fuel cell," *Electrochem. Commun.*, vol. 10, no. 9, pp. 1392–1395, 2008.

[63] S. Cheng and J. M. Regan, "Electricity Generation by Rhodopseudomonas palustris," vol. 42, no. 11, pp. 4146–4151, 2008.

[64] I. Torres, R. Krajmalnik-Brown, and J. P. Badalamenti, "Light-responsive current generation by phototrophically enriched anode biofilms dominated by Green Sulfur Bacteria," *Biotechnol. Bioeng.*, vol. 110, no. 4, pp. 1020–1027, 2013.

[65] R. Chandra, G. Venkata Subhash, and S. Venkata Mohan, "Mixotrophic operation of photo-bioelectrocatalytic fuel cell under anoxygenic microenvironment enhances the light dependent bioelectrogenic activity," *Bioresour. Technol.*, vol. 109, pp. 46–56, 2012.

[66] Z. Baicha, M. J. Salar-García, V. M. Ortiz-Martínez, F. J. Hernández-Fernández, and A. P. D. Ríos, "*A critical review on microalgae as an alternative source for bioenergy production : A promising low cost substrate for microbial fuel cells,*" 2016.

[67] K. Nishio, K. Hashimoto, and K. Watanabe, "Light/electricity conversion by defined cocultures of chlamydomonas and geobacter," *J. Biosci. Bioeng.*, vol. 115, no. 4, pp. 412–417, 2013.

[68] S. Malik et al., "A self-assembling self-repairing microbial photoelectrochemical solar cell," *Energy Environ. Sci.*, vol. 2, no. 3, pp. 292–298, 2009.

[69] J. P. Badalamenti, C. I. Torres, and R. Krajmalnik-Brown, "Coupling dark metabolism to electricity generation using photosynthetic cocultures," *Biotechnol. Bioeng.*, vol. 111, no. 2, pp. 223–231, 2014.

[70] I. Gajda, J. Greenman, C. Melhuish, and I. Ieropoulos, "Photosynthetic cathodes for Microbial Fuel Cells," *Int. J. Hydrogen Energy*, pp. 1–6, 2013.

[71] L. Gouveia, C. Neves, D. Sebastião, B. P. Nobre, and C. T. Matos, "Effect of light on the production of bioelectricity and added-value microalgae biomass in a Photosynthetic Alga Microbial Fuel Cell," *Bioresour. Technol.*, vol. 154, pp. 171–177, 2014.

[72] E. E. Powell, M. L. Mapiour, R. W. Evitts, and G. A. Hill, "Growth kinetics of Chlorella vulgaris and its use as a cathodic half cell," *Bioresour. Technol.*, vol. 100, no. 1, pp. 269–274, 2009.

[73] P. Cai, X. Xiao, Y. He, W. Li, G. Zang, and G. Sheng, "Reactive oxygen species (ROS) generated by cyanobacteria act as an electron acceptor in the biocathode of a bio-electrochemical system," *Biosens. Bioelectron.*, vol. 39, no. 1, pp. 306–310, 2013.

[74] M. Zhou, H. He, T. Jin, and H. Wang, "Power generation enhancement in novel microbial carbon capture cells with immobilized Chlorella vulgaris," *J. Power Sources*, vol. 214, pp. 216–219, 2012.

[75] Y. ChengWu, Z. JieWang, Y. Zheng, Y. Xiao, Z. Hui Yang, and F. Zhao, "Light intensity affects the performance of photo microbial fuel cells with Desmodesmus sp. A8 as cathodic microorganism," *Appl. Energy*, vol. 116, pp. 86–90, 2014.

[76] D. F. Juang, C. H. Lee, and S. C. Hsueh, "Comparison of electrogenic capabilities of microbial fuel cell with different light power on algae grown cathode," *Bioresour. Technol.*, vol. 123, pp. 23–29, 2012.

[77] L. Xiao, E. B. Young, J. A. Berges, and Z. He, "Integrated Photo-Bioelectrochemical System for Contaminants Removal and Bioenergy Production," 2012.

[78] Y. Yuan, Q. Chen, S. Zhou, L. Zhuang, and P. Hu, "Bioelectricity generation and micro-cystins removal in a blue-green algae powered microbial fuel cell," *J. Hazard. Mater.*, vol. 187, no. 1–3, pp. 591–595, 2011.

[79] X. Wang et al., "Sequestration of CO 2 discharged from anode by algal cathode in microbial carbon capture cells (MCCs)," *Biosens. Bioelectron.*, vol. 25, no. 12, pp. 2639–2643, 2010.

[80] S. Pandit, B. K. Nayak, and D. Das, "Microbial carbon capture cell using cyanobacteria for simultaneous power generation, carbon dioxide sequestration and wastewater treatment," *Bioresour. Technol.*, vol. 107, pp. 97–102, 2012.

[81] H. M. Saeed et al., "Microbial desalination cell technology: A review and a case study," *Desalination*, vol. 359, pp. 1–13, 2015.

[82] B. Kokabian and V. G. Gude, "Photosynthetic microbial desalination cells (PMDCs) for clean energy, water and biomass production," *Environ. Sci. Process. Impacts*, vol. 15, no. 12, pp. 2178–2185, 2013.

[83] C. E. Reimers, L. M. Tender, S. Fertig, and W. Wang, "Harvesting energy from the marine sediment - Water interface," *Environ. Sci. Technol.*, vol. 35, no. 1, pp. 192–195, 2001.

[84] H. J. Jeon et al., "Production of algal biomass (Chlorella vulgaris) using sediment microbial fuel cells," *Bioresour. Technol.*, vol. 109, pp. 308–311, 2012.

[85] A. S. Commault, G. Lear, P. Novis, and R. J. Weld, "Photosynthetic biocathode enhances the power output of a sediment-type microbial fuel cell," *New Zeal. J. Bot.*, vol. 52, no. 1, pp. 48–59, 2014.

5 Pretreatment of Paddy Straw for Sustainable Bioethanol Production

Suchitra Rakesh and N. Arunkumar
Central University of Tamil Nadu, India

Karthikeyan Subburamu and Ramesh Desikan
Tamil Nadu Agricultural University, India

CONTENTS

5.1 INTRODUCTION

Clean energy is created through clean, harmless, and non-polluting methods and has less impact on the environment than other conventional energy sources. It establishes a negligible amount of carbon dioxide, and its use can reduce pollution that contributes to the reduction of global warming. On the other hand, most of the renewable energy sources are in the clean energies category. The major problem is the seasonal burning of crop residues, especially paddy straw, by farmers in their fields, which causes the release of enormous quantities of particulate matter, along with other harmful gases. An estimate shows that the crop residue generated in India is about 350×10^6 kg y^{-1} (Shashidhar et al., 2018) and 140 million tonnes of residue burnt in fields with paddy straw contributes about 40%. The quantity of rice production is in the range of 0.7–1.4 kg per kg of milled rice, which depends on rice varieties, the cutting height of the straw and its moisture content (IRRI, 2019). Due to increased mechanization of agriculture, the use of combine harvesters has become widespread, and has left the stubble about a foot long. Ploughing this back into the soil is difficult, and uprooting, or otherwise cutting, and collecting it is labor-intensive and

93

increasingly expensive. The easiest and cheapest method for farmers is to burn agricultural waste into the field itself. This air pollution contributes around 12%–60% of particulate concentrations depending on the generation of other pollutants in different locations, winds, temperatures, and other local factors (D'Amato et al., 2014). Farmers are more severely harmed due to the local air pollution caused and by the loss of soil nutrients such as nitrogen, potassium, phosphorus, and sulfur due to the burning (Gadde et al., 2009). There are many options for handling and utilization of paddy straw stubble, but each involves not merely available technologies, but appropriate research *via* university mechanisms.

In paddy straw, the polysaccharides, cellulose, and hemicellulose are intimately associated with lignin. The lignin component acts as a physical barrier and must be removed to make the carbohydrates available for further hydrolysis processes (Alvarez et al., 2016). Hence, the conversion of lignin into fermentable sugar is a pre-requisite for bioethanol production. Bioconversion of cellulosic biomass into fermentable sugar for the production of ethanol using microorganisms, especially cellulose-degrading fungi, makes bioethanol production economical, environmentally friendly, and also renewable. There are different pretreatment methods like milling and grinding, pyrolysis, high-energy radiation, high pressure steaming, alkaline or acid hydrolysis, hydrogen peroxide treatment, hydrothermal treatment, stream explosion, wet oxidation and biological treatments such as an enzyme or microbial conversion (Soni et al., 2010; Elsayed, 2013). Bioconversion offers a cheap and eco-friendly way of disposing of agricultural waste into a usable form like bioethanol production. The results, along with reducing environmental pollution which is caused by the burning of agricultural residues, also helps in producing bioethanol, which will accelerate the clean energy revolution.

5.2 POTENTIAL

Paddy straw is an agricultural waste that can be used for economic and sustainable biofuel production. Though theoretical, ethanol yields from sugar and starch are higher than from lignocellulose, but these conventional sources are insufficient for worldwide bioethanol production (Binod et al., 2010). In that aspect, agricultural waste, that is, paddy straw is renewable, less costly, and abundantly available in nature. Paddy straw does not demand separate land, water, and energy requirements, and does not have food value either. Hence bioethanol production from agricultural waste like paddy straw could be the route to the effective utilization of agricultural residues. Rice, as a significant agrarian waste in terms of quantity of biomass available, will drastically reduce the production cost of biofuel. Price is an essential factor for the large-scale expansion of biofuel production. The biofuels produced from lignocellulosic wastes can avoid the existing complications of food versus fuel caused by grain-based biofuel production.

Lignocellulosic paddy straw has been projected to be one of the leading resources for economically attractive biofuel production, as India is one of the second-largest producers of rice in the world with an area of ca. 434 lakh hectares. Though theoretical ethanol yields from sugar and starch are higher than from lignocellulose, these conventional sources are insufficient for worldwide bioethanol production. In that aspect, agricultural waste, that is, paddy straw is renewable, less costly, and abundantly available in nature (Sung et al., 2013). Paddy straw so not demand separate land, water, and energy requirements. They do not have food value, as well. The hindrances like feedstock conversion technology, hydrolysis process, and fermentation configuration need to be overcome for economic biofuel production.

Major feedstock obstacles are cost, supply, harvesting, and handling. As regards conversion technology, the hindrances are biomass processing, proper and cost-effective pretreatment technology to liberate cellulose and hemicellulose from their complex with lignin. The main challenge in the hydrolysis process is to recover the high concentration of fermentable sugar. In this aspect, enzymatic hydrolysis may be the most potent alternative process for the saccharification of the complex polymer. Several efforts have already been made in recent years to reduce the cost of cellulase enzymes to optimize the enzymatic hydrolysis process. Proper pretreatment and efficient technology are to be applied, so that biofuel production from paddy straw may be successfully developed and optimized soon (Li et al., 2009).

5.3 CHEMICAL COMPOSITION

Paddy straw constitutes a high amount of cellulose and hemicellulose for bioethanol production. Chemical composition of paddy straw comprises hemicellulose (19%–27%), cellulose (32%–47%), and lignin (5%–24%) (Garrote et al., 2002; Saha, 2003). Paddy straw is resistant to degradation due to heterogeneity in its composition (Moiser et al., 2005) and is connected strongly *via* hydrogen and covalent bonds. These bonds could be broken down by using various methods of pretreatment (Limayem and Ricke, 2012). Cellulose, as one of the most abundant materials on the earth, has high stability due to its crystalline structure and orderly arrangement (Tayyab et al., 2017). It is composed of a β-glucan linear polymer of glucose, which is linked by β-1,4-glycosidic bonds. Due to its crystalline character, cellulose structure is difficult to break without enzymatic hydrolysis (Ruel et al., 2012). Hemicellulose is made of different sugar units like uronic acids (Peng et al., 2012). Monosaccharides in hemicelluloses are pentoses and hexoses (Cardona and Sanchez, 2017). Lignin is challenging to degrade and composed of phenylpropanoid units, which is a heterogeneous polymer. Due to cross-linking with other biopolymers in lignin, the pretreatment stage is very crucial for efficient biodiesel production (Boerjan et al., 2003).

5.4 CONVERSION OF PADDY STRAW INTO BIOETHANOL

Thermochemical and biochemical conversion methods can be used for the production of biofuels from lignocellulosic biomass. In the case of thermochemical process, the biomass is converted into intermediate gas or liquid using a non-biological catalyst in a reactor. Then, end products are further transformed into different fuels like methanol, hydrogen, ethanol and hydrocarbons (Foust et al., 2009). The thermo chemical conversion process is not suitable for paddy straw due to a higher ash content, which lead to clinkers formations in the reactions and periodical ash removal issues in the reactor.

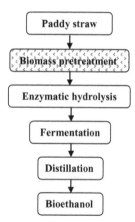

FIGURE 5.1   Bioethanol production from paddy straw. (Red color indicates process requires more attention to reduce process cost and energy).

Biochemical conversion requires microorganisms or enzymes to produce biofuels (liquid or gaseous form) and value-added products from lignocellulosic biomass feedstocks (Foust et al., 2009; Achinas and Euverink, 2016). Biochemical conversion methods are more popular to produce bioethanol based on the compositions of biomass (Sims et al., 2010). For bioethanol production from paddy straw, the process sequence is biomass pretreatment, enzymatic hydrolysis, fermentation with microbes, and distillation (Kang et al., 2014). Paddy straw contains lignin (5%–24%) (Garrote et al., 2002; Saha, 2003) and the biomass pretreatment is an essential process to break the lignin barrier to release fermentable sugars. Pretreatment may be a physical, biological, or chemical method and this step will improve the reduction of sugar yield *via* enzymatic hydrolysis from cellulose and hemicellulose. In other words, pretreatment can facilitate the production of simple sugars by making cellulose and hemicellulose more readily available for enzymatic hydrolysis. Broeur et al. (2011) have reported the importance of pretreatment like improving the sugar recovery *via* enzymatic hydrolysis. The added advantages of the pretreatment process are reduction of inhibitor formation and value-added products from lignin. The process flowchart for bioethanol production is shown in Figure 5.1. The red is marked for the biomass pretreatment process, which indicates that more research has to be focused on the biomass process to optimize the parameters for better pretreatment efficiency and reduce the cost and energy used in the process.

5.5   PHYSICAL PRETREATMENT METHODS

Physical biomass pretreatment is a size-reduction process that can be achieved via chopping, steaming, irradiation, heating, extrusion, grinding, or milling and can lead to significant improvement in the enzymatic hydrolysis of paddy straw (Al-Haj, 2012). Several size-reduction machines are commercially available on the market such as hammer mills, refiners, colloid mills, and multi-blade cutters. Among them, hammer mills are preferred due to their ability to finely grind the straw at a relatively low price with an accessible mode of operation (Lopo, 2002). Wet disk milling is another proven method for significant glucose recovery and energy saving (Hideno et al., 2009).

Pretreatment with irradiation is used to accelerate the subsequent enzymatic hydrolysis of paddy straw (DelaRosa et al., 1983). Various studies reported that the irradiation lowers the crystallinity and degree of polymerization of the lignocellulose and thereby increases the accessibility of the cellulose crystalline area (Manas et al., 2018). Gramma irradiation remarkably improves acid hydrolysis and produces more reduced sugar at higher dosages. Electron beam-irradiated paddy straw on enzymatic hydrolysis yields glucose almost doubles that of the untreated one (Jin et al., 2009). Microwave irradiation changes the ultrastructure of cellulose and improves the enzymatic hydrolysis of lignocellulosic paddy straw (Xiong et al., 2000). Extrusion is another type of physical pretreatment, in which the substrate is allowed to pass through the extruder, resulting in physical and chemical modification and thereby increasing the accessibility of sugar before hydrolysis and fermentation (Alvira et al., 2010).

5.6 CHEMICAL PRETREATMENT METHODS

Chemical pretreatment of paddy straw is the most promising pretreatment which involves the use of acids, alkalis, and oxidizing agents (Al-Haj, 2012). Alkali pretreatment is also known as peroxide, ozonolysis, and wet oxidation pretreatment, which uses alkaline solutions like potassium hydroxide or sodium hydroxide at higher concentrations and breaks the ester bonds between cellulose, hemicellulose, and lignin (Talebnia et al., 2010). It will also improve the saccharification yields of paddy straw (Gaspar et al., 2007). As per the previous studies, among the alkaline pretreatments, NaOH pretreatment is considered one of the best methods (Kim, 2018). Alkaline pretreatments perform at low temperatures and pressure by using less corrosive chemicals, which will swell the lignocellulose and, thereby, decreases the degree of polymerization and crystallinity of the material. In addition to this, solubilization of the saccharide portion is comparatively less during alkaline pretreatment, which will increase the retrieval of maximum saccharides in the succeeding steps and thereby enhance higher production (Shalley and Preeti, 2018). The high cost of alkalis is one of the significant drawbacks in this pretreatment method, which can be sorted out by using it with the combination of other pretreatments like oxidative, fungal, or wet oxidation (Al-Haj, 2012).

Ammonia pretreatment is also preferred for the conversion of lignocellulosic biomass, as ammonia is a useful swelling reagent and has high volatility with a non-polluting and non-corrosive nature. Ammonia cleaved the C-O-C bonds in lignin as well as ester and ether bonds in a carbohydrate-lignin complex (Kim and Lee, 2007). Diluted or concentrated inorganic acids can be effectively used as a pretreatment to remove hemicellulose from the paddy straw and improves the yield of fermentable sugars. Monosaccharides recovery via enzymatic hydrolysis of paddy straw by acid pretreatment was recorded to be around 46 g/L (Vlasenko et al., 1997). Concentrated acid pretreatment does not involve the use of any enzymes for saccharification and yields higher fermentable sugar (Galbe and Zacchi, 2002; Sun and Cheng, 2002; Talebnia et al., 2010). Dilute acid pretreatment involves the use of cellulose enzymes for hydrolysis.

The two-stage process to reduce the enzyme requirements was developed at the National Renewable Energy Laboratory in Golden, Colorado (Sumphanwanich et al., 2008).

The oxidative agents like per-acetic acid or hydrogen peroxide are used to pretreat the lignocellulosic paddy straw. Reactions involved in the oxidative pretreatment are displacement of the side chain, electrophilic substitution, cleavage of alkyl aryl ether linkages, or oxidative cleavage of aromatic nuclei (Hon and Shiraishi, 2001). Hydrogen peroxide pretreatment solubilizes and loosens the lignocellulosic matrix and, hence, improves the enzymatic digestibility (Martel and Gould, 1990). The organic solvent pretreatment improves the efficiency of hemicellulose removal and enzymatic reactions by leaving easily hydrolyzable more sugar yield at higher rates. It usually involves the use of hot organic solvents like ethanol at acidic pH for fractionating the lignocellulosic biomass (Mantanis et al., 1995). Removal of organic solvent after pretreatment is required as solvents might inhibit the enzymatic hydrolysis and fermentation (Xuebing et al., 2009). Maheshwari et al. (2017) used pearl millet biomass for bioethanol production via chemical pretreatment and reported more sugar yield than other pretreatment methods.

5.7 BIOLOGICAL PRETREATMENT METHODS

Biological pretreatment is safe, environment friendly, and uses less energy compared to other pretreatment methods. These methods are less preferred over other pretreatment processes due to the prolonged rate of hydrolysis and long duration, which reduces the overall yield of fermentable sugars. In this method, lignin degrading fungi or bacteria are used as whole cell or as enzymes to pretreat the lignocellulosic biomass. Laccases, manganese peroxidase, lignin peroxidase, and versatile peroxidase are the enzymes used in lignin degradation. Fungi are the best-studied microorganism for biological pretreatment due to its capability of degrading cellulose, hemicellulose, and lignin (Julie et al., 2018). White-rot fungi (*Phanerochaete chrysosporium, Ceriporiopsis subvermispora, Pleurotus ostreatus, and Termetes versicolor*) of the class Basidiomycetes are the most promising fungi for biological pretreatment (Taniguchi et al., 2005). Pretreatment of paddy straw in combination with *Pleurotus ostreatus* and alkali ammonia yields high sugar at a relatively shorter period (Balan et al., 2007). A combination of two fungi, that is, *Aspergillus awamori* and *Aspergillus niger* for paddy straw pretreatment, has shown higher sugar yield and ethanol after fermentation (Patel et al., 2007).

5.8 COMBINED PRETREATMENT METHODS

A combination of one pretreatment with another suitable one will effectively improve the yield of fermentable sugars. Zhu et al. (2006) have reported different combinations of microwave pretreatment with acids/alkalis and found that the combination of a microwave and acid can remove more lignin from paddy straw compared to that of an alkali catalyst alone. A combination of steam explosion with superfine grinding has enhanced the sugar yield (Jin and Chen, 2006). Nutwan et al. (2010) have reported that the combination of ultrasonic with acid pretreatment before enzymatic hydrolysis can enhance the sugar yield up to 44% and thereby increases bioethanol yield (Table 5.1).

TABLE 5.1
Ethanol Production from Rice Straw Using Various Pretreatment Methods

Pretreatment Method	Pretreatment Conditions	Ethanol Production (g L⁻¹)	References
Steam pretreatment	Autoclaving at 121°C for 30 min	1.1	Arora et al., 2016
Alkali pretreatment	1% NaOH at 121°C for 30 min	4.03	Sharma et al., 2019
Acid pretreatment	1% Sulfuric acid at 121°C for 15 min.	6.5	Belal, 2013
Alkali+acid pretreatment	1% HCl after 1% NaOH treatment incubation at 150 rpm for 24 h at room temperature	6.13	Hashem et al., 2013
Acid pretreatment+ ultrasound pretreatment+*Trichoderma reesei* cellulase	Ultrasound at 40 W for 10 min after 1% Sulfuric acid treatment followed by enzymatic treatment (0.495 U/g rice straw)	11.0	Belal, 2013
Popping method	rice strawat75%moisture, reactor heat 220°C, 1.96 MPa and rapidly opening	25.8	Wi et al., 2013

5.9 CONCLUSIONS

Paddy straw is a lignocellulosic agricultural waste, which can be utilized for bioethanol production by performing suitable pretreatment techniques. Bioethanol production from paddy straw is not only cost-effective but is also environmentally sustainable. The major hindrance for the conversion of lignocellulosic biomass to fermentable sugars is the strong linkage bonds between cellulose, hemicellulose, and lignin. Suitable pretreatment can break the link and make the enzymatic hydrolysis easy for conversion into fermentable sugars. There are many pretreatments like physical, chemical, and biological conversion of paddy straw into fermentable sugars, but there are several methods available for pretreating paddy straw which are cost-effective, and economically feasible pretreatment is still a significant hindrance to bioethanol production. A balanced and efficient combination of pretreatment for achieving a high yield of fermentable sugar has to be selected for maximum efficacy of the process and will become available technology in the future.

REFERENCES

Achinas, S. and Euverink, G.J.W., 2016, Consolidated briefing of biochemical ethanol production from lignocellulosic biomass. *Electrol. J. Biotechnol.*, 23: 44–53.

Al-Haj, I., 2012, Pretreatment of straw for bioethanol production. *Energy Procedia*, 14: 542–551.

Alvarez, C., Reyes-Sosa, F.M. and Diez, B., 2016, Enzymatic hydrolysis of biomass from wood. *J. Microbial. Biotechnol.*, 9(2): 149–156.

Alvira, P., Thomas-Pejo, E., Ballesteros, M. and Negro, M.J., 2010, Pretreatment technologies for an efficient bioethanol production process based on enzymatichydrolysis. *Bioresour. Technol.*,101(13): 4851–4861.

Arora, A., Priya, T., Sharma, P., Sharma, S. and Nain, L., 2016, Evaluating biological pretreatment as a feasible methodology for ethanol production from paddy straw. *Biocatal. Agric. Biotechnol.*, 8: 66–72.

Balan, V., Chundawat, S.P.S. and Dale, B.E., 2007, *Effect of fungal conditioning on ammonia fibre expansion (AFEX) pretreatment of rice straw.* Abstracts of papers: *233rd ACS National Meeting*, Chicago, IL, United States, CELL-04025.

Belal, E.B., 2013, Bioethanol production from rice straw residues. *Braz. J. Microbiol.*, 44(1), 225–234.

Binod, P., Sindhu, R., Singhania, R.R., Vikram, S., Devi, L., Nagalakshmi, S., Kurien, N., Sukumaran, R.K. and Pandey, A., 2010, Bioethanol production from rice straw: An overview. *Bioresour. Technol.*, 101(3): 4767–4774.

Boerjan, W., Ralph, J. and Baucher, M., 2003, Lignin biosynthesis. *Annu. Rev. Plant Biol.*, 54: 519–546.

Broeur, G., Yau, E., Badal, K., Collier, J., Ramachandran, K.B. and Ramakrishnan, S., 2011, Chemical and physicochemical pretreatment of lignocellulosic biomass: A review. *Enzyme Res.*, 2011: 17.

Cardona, C.A. and Sanchez, O.J., 2017, Fuel ethanol production: Process design trends and integration opportunities. *Bioresour.Technol.*, 98(2): 2415–2457.

D'Amato, G., Bergmann, K.C., Cecchi, L., Annesi-Maesano, I., Sanduzzi, A., Liccardi, G., Vitale, C., Stanziola, A. and D'Amato, M., 2014, Climate change and air pollution – Effects on pollen allergy and other allergic respiratory diseases. *Allerg. J. Int.*, 23(1): 17–23.

DelaRosa, A.M., DelaMines, A.S., Banzon, R.B. and Simbul-Nuguid, Z.F., 1983, Radiation pretreatment of cellulose for energy production. *Radia. Physic. and Chem.*, 22(3–5): 861–867.

Elsayed, B.B., 2013, Bioethanol production from rice straw residues. *Braz. J. Microbiol.*, 44(1): 225–234.

Foust, T.D., Aden, A., Dutta, A. and Phillips, S., 2009, An economic and environmental comparison of a biochemical and a thermo chemical lignocellulosic ethanol conversion processes. *Cellulose*, 16(4): 547–565.

Gadde, B., Bonnet, S., Christoph, M. and Garivait, S., 2009, Air pollutant emissions from rice straw open field burning in India, Thailand and the Philippines. *Environ. Pollut.*, 157(5): 1554–1558.

Galbe, M. and Zacchi, G., 2002, A review of the production of ethanol from softwood. *Appl. Microbiol. Biotechnol.*, 59 (6): 109–628.

Garrote, G., Dominguez, H. and Parajo, J.C., 2002, Autohydrolysis of corncob: study of non-isothermal operation for xylooligosaccharides production. *J. Food Eng.*, 52(3): 211–218.

Gaspar, M., Kalman, G. and Reczey, K., 2007, Cornfibre as a raw material for hemicellulose and ethanol production. *Process Biochem.*, 42 (7): 1135–1139.

Hashem, M., Ali, E.H. and Abdel-Basset, R., 2013, Recycling rice straw into biofuel. *J. Agric. Sci. Technol.*, 15(4): 709–721.

Hideno, A., Inoue, H., Tsukahara, K., Fujimoto, S., Minowa, T., Inoue, S., Endo, T. and Sawayama, S., 2009, Wet disk milling pretreatment without sulphuric acid for enzymatic hydrolysis of rice straw. *Bioresour.Technol.*, 10(100): 2706–2711.

Hon, D.N.S. and Shiraishi, N., 2001, *Wood and Cellulosic Chemistry*, 2nd Edn. (Marcel Dekker, New York and Basel), 914–8881.

IRRI (International Rice Research Institute), 2019, *Rice Straw.* http://ricestraw.irri.org/ (November 19, 2019).

Jin, S. and Chen, H., 2006, Superfine grinding of stream-exploded rice straw and its enzymatic hydrolysis. *Biochem.Eng. J.*, 30(3): 225–230.

Jin, S.B., Ja, K.K., Young, H.H., Byung, C.L., In-Geol, C. and Heon, K.K., 2009, Improved enzymatic hydrolysis yield of rice straw using electron beam irradiation pretreatment. *Bioresour. Technol.*, 100(3): 1285–1290.

Julie, B., Bikash, K.N., Ritika, S., Sachin, K., Ramesh, C.D., Deben, C.B. and Eeshan, K., 2018, Recent trends in pretreatment of lignocellulosic biomass for value added products. *Fronti. Energy Res.*, 6: 1–19.

Kang, Q., Appels, L., Tan, T. and Dewi, R., 2014, Bioethanol from lignocellulosic biomass: current findings determine research priorities. *Sci.Wor. J.*, 298153.

Kim, D., 2018, Physio-chemical conversion of lignocelluloses: Inhibitor effects and detoxification strategies. *Molecules*, 23(2): 309.

Kim, T.H. and Lee, Y.Y., 2007, Pretreatment of corn stover by soaking in aqueous ammonia at moderate temperatures. *Appl.Biochem.Biotechnol.*, 137(1–12): 81–92.

Li, Q., He, Y.C., Xian, M., Jun, G., Xu, X., Yang, J.M. and Li, L., 2009, Improving enzymatic hydrolysis of wheat straw using ionic liquid 1-ethyl-3-methyl imidazolium diethyl phosphate pretreatment. *Bioresour.Technol.*, 100(14): 3570–3575.

Limayem, A. and Ricke, S.C., 2012, Lignocellulosic biomass for bioethanol production: Current perspectives, potential issues and future prospects. *Prog. Energy Combust. Sci.*, 38(4): 449–467.

Lopo, P., 2002, The right grinding solution for you: roll, horizontal or vertical. *Feed Manage.*, 53: 23–26.

Maheshwari, P., Karthikeyan, S., Ramesh, D., Sivakumar, U., Marimuthu, S. and Soundarapandian, K., 2017, Comparison of chemical pretreatment for recovery of fermentable sugars and enzymatic saccharification. *Madras Agric. J.*, 104: 273–278.

Manas, R.S., Ajith, S., Ajay, K.S. and Deepak, K.T., 2018,Bioethanol production from rice and wheat straw: An overview. In: *Bioethanol Production from Food Crops*. (DBT-IOC Centre for Advance Bioenergy Research Centre, Faridabad,India), pp. 213–231.

Mantanis, G.I., Young, R.A. and Rowell, R.M., 1995, Swelling of compressed cellulose fiber webs in organic liquids. *Cellul.*, 2(1): 1–22.

Martel, P. and Gould, J.M., 1990, Cellulose stability and delignification after alkaline hydrogen peroxide treatment of straw. *J. Appl.Polym.Sci.*, 39(3): 707–714.

Moiser, N., Wyman, C., Dale, B., Elander, R., Lee, Y., Holtzapple, Y. and Ladisch, M., 2005, Features of promising technologies for pretreatment of lignocellulosic biomass. *Bioresour.Technol.*, 96(6): 673–686.

Nutwan, Y., Phattayawadee, P., Pattranit, T. and Mohammed, N.E., 2010, Bioethanol production from rice straw. *Energy Res. J.*, 1(1): 26–31.

Patel, S.J., Onkarappa, R. and Shobha, K.S., 2007, Study of ethanol production from fungal pretreated wheat and rice straw. *The Internet. J. Microbiol.*, 4: 1–5.

Peng, F., Peng, P., Xu, F. and Sun, R.C., 2012, Fractional purification and bioconversion of hemicelluloses. *Biotechnol.Adv.*, 30(4): 879–903.

Ragauskas, A.J., Williams, C.K., Davison, B.H., Britovsek, G., Cairney, J. and Eckert, C.A., 2006, The path forward for biofuels and biomaterials. *Science*, 311(5760): 484–489.

Ruel, K., Nishiyama, Y. and Joseleau, J.P., 2012, Crystalline and amorphous cellulose in the secondary walls of Arabidopsis. *Plant Sci.*, 193–194: 48–61.

Saha, B.C., 2003, Hemicellulose bioconversion. *Ind.Microbiol.Biotechnol.*, 30(5): 279–291.

Shalley, S., Preeti, N. and Anju, A., 2018, Ethanol production from NaOHpretreated rice straw: a cost effective option to manage rice crop residue. *Waste and biomass Valoriz.*, 1–8.

Sharma, S., Nandal, P. and Arora, A., 2019, Ethanol production from NaOHpretreated rice straw: a cost effective option to manage rice crop residue. *Waste Biomass Valoriz.*, 10(11), 3427–3434.

Shashidhar, K.S., Bhuvaneswari, S., Premaradhya, N., Kumar Sambhavand Samuel and Jeberson. 2018. Sustainable options for rice residue managementin Manipur. *CAUFarmMagazine*, 8(4): 2–6.

Sims, R.E.H., Mabee, W., Saddler, J.N. and Taylor, M., 2010, An overview of second generation biofuel technologies. *Bioresour.Technol.*, 101(6): 1570–1580.

Soni, S.K., Batra, N., Bansal, N. and Soni, R., 2010, Bioconversion of sugarcane bagasse into second generation bioethanol after enzymatic hydrolysis with in-house produced cellulases from *Aspergillus* sp. S4B2F. *BioResour.*, 5(2): 741–758.

Sumphanwanich, J., Leepipatpiboon, N., Srinorakutara, T. and Akaracharanya, A., 2008, Evaluation of dilute-acid pretreated bagasse, corn cob and rice straw for ethanol fermentation by Saccharomyces cerevisiae. *Ann.Microbiol.*, 58(2): 219–225.

Sun, Y. and Cheng, J., 2002, Hydrolysis of lignocellulosic materials for ethanol production: a review. *Bioresour.Technol.*, 83(1): 1–11.

Sung, B.K., Sang, J.L., Ju, H., You, R.J., Laxmi, P.T., Jun, S.K., Youngsoon, U., Chulhwan, P. and Seung, W.K., 2013, Pretreatment of rice straw with combined process using dilute sulphuric acid and aqueous ammonia. *Biotechnol. Biofuels*, 6: 109–120.

Talebnia, K., Karakashev, S. and Angelidaki, I., 2010, Production of bioethanol from wheat straw: An overview on pretreatment, hydrolysis and fermentation. *Bioresour. Technol.*, 101(13): 4744–4753.

Taniguchi, M., Suzuki, H., Watanabe, D., Sakai, K., Hoshino, K. and Tanaka, T., 2005, Evaluation of pretreatment with Pleurotusostreatus for enzymatic hydrolysis of rice straw. *J. Biosci. Bioeng.*, 100(6): 637–643.

Tayyab, M., Noman, A., Islam, W., Waheed, S., Arafat, Y., Ali, F., Zaynab, M., Lin, S., Zhang, G. and Lin, W., 2017, Bioethanol production from lignocellulosic biomass by environment-friendly pretreatment method. *Appl. Ecol. Environ. Res.*, 16(1): 225–249.

Vlasenko, E.Y., Ding, H., Labavitch, J.M. and Shoemaker, S.P., 1997,Enzymatic hydrolysis of pretreated rice straw. *Bioresour.Technol.*, 59 (2–3): 109–119.

Wi, S.G., Choi, I.S., Kim, K.H., Kim, H.M. and Bae, H.J., 2013, Bioethanol production from rice straw by popping pretreatment. *Biotechnol. Biofuels*, 6(1): 166–172.

Xiong, J., Ye, J., Liang, W.Z. and Fan, P.M., 2000, Influence of microwave on the ultrastructure of cellulose. *J. South China. Univ.Technol.*, 28: 84–89.

Xuebing, Z., Keke, C. and Dehuda, L., 2009,Organosolventpretreatment of lignocellulosic biomass for enzymatic hydrolysis. *Appl.Microbiol.Biotechnol.*, 82(5): 27–815.

Zhu, S., Wu, Y., Zhao, Y., Tu, S. and Xue, Y., 2006, Fed-batch simultaneous saccharification and fermentation of microwave/acid/alkali/H_2O_2 pretreated rice straw for production of ethanol. *Chem.Eng.Commun.*, 193(5): 639–648.

6 Bio-based Coagulants for the Remediation of Environmental Pollutants

Mansi Kikani and Chanchpara Amit
CSIR-Central Salt & Marine Chemicals Research Institute, India

Doddabhimappa Ramappa Gangapur and Madhava Anil Kumar
CSIR-Central Salt & Marine Chemicals Research Institute, India; Academy of Scientific and Innovative Research, India

Muthulingam Seenuvasan
Hindusthan College of Engineering and Technology, India

CONTENTS

6.1 INTRODUCTION

Water quality, its goodness, and deterioration are governed by several factors such as natural conditions and anthropogenic effects. Water finds its use in domestic and industrial supplies, irrigation, nurturing, breeding and preservation of aquatic life, power generation, transport, and recreational activities. Everyday activities and their associated processes generate pollution at all spheres: atmosphere, lithosphere, biosphere, hydrosphere, and sphere of the environment. Regulatory agencies are concerned with protecting the environment into two categories: (a) point and (b) non-point sources.

6.1.1 Water Quality Parameters

Water quality describes the wellness of the water to be used for supporting the different aspects of a healthier life. The main characteristics determining water quality are physical, chemical, and biological parameters.

- Physical parameters: Temperature, color, odor, turbidity.
- Chemical parameters: Total solids (suspended, dissolved, and settleable), organic matter (biochemical and chemical oxygen demand, total organic carbon), total nitrogen-ammonia, nitrite, nitrate), total phosphorous (organic and inorganic), pH, alkalinity, chlorides, oils and grease, sulfates, heavy metals, priority pollutants, and hardness.
- Biological parameters: bacteria, archaea, algae, fungi, protozoa, viruses, and helminths.

6.1.2 Turbidity-Sources and Factors Their Impacts

Turbidity in water is the cloudiness or murkiness of water and it is caused by particles suspended or dissolved in water that scatter light. Turbidity is a pollution indicator and many factors attribute to the turbidity of water and it directly correlates the presence of pathogenic microbes. Turbidity is commonly used for operational monitoring of control measures included in water safety plans. The suspended particulates in water include sediment, clay, and silt. Sometimes, fine organic and inorganic matter, soluble-colored organic compounds, algae, and other microscopic organisms impart turbidity. The commonly used method for measuring turbidity is nephelometric and is expressed as Nephelometric Turbidity Unit (NTU). The effluents from paper and pulp, starch, the textile industry and dairy waste water industries are known to impart high turbidity.

The materials that impart turbidity are constituents such as clays, soils, silts, sediments, and natural organic matter. High turbid water is also an indication of hazardous materials and high microbial loads thereby contaminating the water quality significantly. Water turbidity has implications at all levels, beginning from the sources of water to water treatment, distribution systems, storage and end use.

The turbidity imparted in this level may be due to the rapid changes in water quality, environmental attributes like storms, floods, agricultural run-offs, and/or through indiscriminate anthropogenic activities. The turbidity in groundwater bodies is caused by the percolation and intrusion of pollutants. The turbidity changes over time and an escalated level of turbidity in sources of water indicates pollution levels and drastically determines the efficiency of water treatment techniques, particularly in the filtration, clarification, and disinfection. Turbidity in the storage reservoirs and distribution systems are observed as the result of faults and flaws in the pipelines, re-suspension of sediments due to turbulent flow of water, and detachment of adhered biofilms. The impact of turbidity when considering the end-user aspect is governed by aesthetic values and domestic storage for intended use. Crystal clear water has an ideal turbidity of <1 NTU, while a values of >4 NTU becomes murky. Disinfection should be practised when turbidity levels exceed 5 NTU and periodic monitoring should be ensured at the distribution networks which are part corrective and preventive actions (CAPA). Additionally, an emergency preparedness plan must be included for unpredictable events like failed operation, flow adjustments/modifications during treatment and disinfection.

Shore sediments are often considered the primary factor impacting turbidity, but there are other factors too, which may be natural or human-induced, erosion, runoffs, silts from stream channel areas, and agriculture. Algal growth, eutrophication, mixing of biodegradable materials like leaves, and dead animals all contribute to turbidity. Anthropogenic factors are heavily blamed as they are known to escalate erosion and the dumping of waste at large. Turbidity is largely driven by numerous factors such as the presence of dissolved and suspended solids, morphology and the composition of the materials and pigment content. The other factors contributing to turbidity are flow, source and composition of sedimentary particles, and their transport mechanisms.

6.1.3 IMPACTS OF TURBIDITY

The pollutants either from point sources or non-point sources imparting high turbidity may not be intrinsically harmful, but their effects can be detrimental. The high turbid water significantly reduces the aesthetic values of the water bodies and eventually has a detrimental impact. In order to avert the problem of turbidity, water treatment techniques are required to make the water fit for consumption and other utilities. High turbid waters are harmful to aquatic life systems and they hamper food chain, food supplies, deteriorate breeding water beds, and impart dysfunction of marine life's metabolism. The probable harmful and direct impacts of turbidity on freshwater fish are: (1) reduced biomass of edible fish, (2) makes fish prone to diseases, (3) disturbed development of eggs and larvae, (4) deviations in their migration pattern. The effects of turbidity are a reduction in yield of edible fish quality and reduced fish haul efficiency. High turbid water thus necessitates an increase in the cost incurred for purification. In domestic water supply, elevated turbidity levels are associated with the outbreaks of diseases (Mann et al. 2007; Tinker et al. 2010; Beaudeau et al. 2014).

6.2 TYPES OF WASTEWATER TREATMENT

Physical treatment methods are used to clarifying water such as sedimentation and aeration; screening and skimming are used to remove solids in wastewater. No chemical is involved in treatment. The first techniques of physical wastewater treatment include sedimentation, which is the process of suspending the heavy particles/insoluble from wastewater. Once the insoluble material settles down at the bottom, the supernatant is used as pure water. The second technique includes aeration, a process which consists of circulating air through the water to provide oxygen to it. Another technique is filtration which separates all contaminants from the wastewater by using a filter to pass the wastewater, and separate the contaminants and insoluble material present in it. The sand filter is most commonly used filter. Another treatment technique is chemical methods that use aluminum sulphate, ferric chloride, ferrous sulphate, and bentonite. Chlorine is an oxidizing agent used to kill microorganisms which disinfect them. Organic matter which is present in water is degraded using biological materials. Microorganisms metabolize organic matter in the wastewater in a biological treatment. This is divided into main three categories:

- *Aerobic processes:* Oxygen is used in this aerobic process. Bacteria decomposes the organic matter and converts into carbon dioxide that can be used by plant.
- *Anaerobic processes:* Oxygen is not used in this aerobic process. At a specific temperature, fermentation is used for festering the waste.
- *Composting:* Wastewater is treated by mixing it with sawdust or other carbon sources in this type of aerobic process.
- *Sludge wastewater treatment:* This is the solid-liquid separation process. The least residual moisture is required in the solid phase and the lowest residual moisture is required in the separated liquid phase. Centrifuge is used for removing solid materials from the wastewater; it is best device for solid-liquid separation.

6.3   COAGULATION

The coagulation-flocculation in water treatment is generally followed by sedimentation, filtration, and disinfection prior to distribution to the end-users and in storage plants. Coagulation is an effective tool to prevent water-borne diseases and makes water fit for human use by reducing the *Escherichia coli* count, residual chlorine, and turbidity >5 NTU. Coagulation involves the removal of colloidal suspensions from water and they are known to possess meritorious features as they do not require complicated unit processes and operations, and require minimal energy. Coagulation is widely used in treating surface water, domestic wastewater, and industrial wastewater, and is achieved by the addition of coagulants. Coagulants destabilize and neutralize the charge of particles and larger particles are formed through agglomeration of microflocs. The rapid mixing prompts the dispersion of the coagulants and promotes collision among the suspended particles to generate microflocs (Table 6.1).

The aggregation of colloidal suspensions in a solution is facilitated by double layer compression, flocs formation, adsorption, and charge neutralization. In coagulation, the coagulants with positive charge neutralize the negative charge of the

---

TABLE 6.1

Selected List of Chemical Coagulants Used in the Water Treatment

Pollutant	Chemical Coagulant	References
Grey water	MIEX chemical	Pidou et al. (2008)
Dairy industry waste water	Aluminium sulphate and ferrous sulfate	Parmar et al. (2011)
Disperse red dye from synthetic water	Aluminium sulphate and ferric chloride	Merzouk et al. (2011)
Leachate	Chitosan and ferric chloride	Ramli and Aziz (2015)

suspended particles, and agglomerates them to make larger particles which are easily removed by solid-liquid separation. Coagulants facilitate the generation and precipitation of flocs on which the pollutants are absorbed. This method is often considered as feasible for the reduction of turbidity, decolorization, attenuating toxicity, removal of suspended solids, reduction in heavy metal contents, and pathogenic counts (Dawood and Li 2012). Several factors affect the performance of coagulation such as pH, alkalinity, temperature, ionic strength, coagulant dosage, and type of coagulants.

6.4 BIO-BASED COAGULANTS

The chemical coagulants and the coagulant aids help in the removal of water pollutants but are associated with disadvantages such as they become ineffective in cold water bodies, they are high in cost, and there is the problem of secondary pollution due to their partial biodegradability, chemical costs, and there is significant sludge production and elevation of pH levels. These chemical coagulants exerts some adverse effects to human and environment. Thus, the negatives outlined open up the opportunity to explore natural coagulants with less toxicity, that are biodegradable, that generate low sludge volumes at low cost as an alternative to chemically synthesized coagulants. Many natural coagulants derived from living organisms either micro and/or macro levels have been identified.

6.5 PLANT-BASED COAGULANT MATERIALS

The plant-based coagulants in waste water treatment are preferred over chemical coagulants as they are abundant, cost-effective, and do not increase pH levels in the treated water. They are also biodegradable, so no additional costs would be required for their disposal (Vijayaraghavan et al. 2011). The plant-based coagulants are used extensively as they are unaltered by environmental factors and/or the developmental stage of the plant (Table 6.2).

TABLE 6.2
Different Coagulant and Their Turbidity Removal Efficiencies

Plant-based Coagulants	Types of Water	Turbidity Removal (%)	References
Musa paradisica	Textile effluent	83.0	Daverey et al. (2019)
Banana peel	Synthetic wastewater	88.0	Mokhtar et al. (2019)
Trigonella foenum-graecum and *Astragalus gossypinus*	Water treatment plant	63.5	Kashfi et al. (2019)
Mucilaginous seed of *Salvia hispanica*	Landfill leachate	62.0	Tawakkoly et al. (2019)
Ocimum basilicum	Paper recycling wastewater	85.0	Mosaddeghi et al. (2020)
Moringa oleifera seeds	Hospital wastewater	71–76	Nonfodji et al. (2020)

Vegetables, pulses, fruit-based trees, flower, cereals, and tree species are used for the treatment of waste water most of the plant-based coagulants are used to treat the physio-chemical properties of the wastewater. Vegetables are rich in minerals and micronutrients and these are helpful in the treatment of turbidity of water. A few vegetable-based natural coagulants are listed in the Table 6.3.

The seeds of leguminous crop plants are used as food, and on splitting they produce dal which is rich in proteins such as green gram, black gram, Bengal gram, peas, cow-pea, pigeon-pea, soybean, lentil, lathyrus. These seeds can be used as natural coagulants to treat wastewater. Most studies are conducted by using plant-based coagulants for the treatment of physio-chemical properties of wastewater. A few pulse species are mentioned in Table 6.4. The countryside mainly depends on agriculture, especially in rural areas where there is a scarcity of chemical coagulants to treat wastewater.

The availability of drinking water is reduced in the under- or less-developed countries; and even less in some regions of sub-Saharan Africa. In such cases there is a fruit-based tree species which grows naturally in the dryland ecosystem. During this, plant-based natural coagulants are the better source to treat the turbidity of water as shown in the Table 6.5. Water treatment processes exist to provide safe drinking water and most commonly the municipalities carry out this service through physical and chemical processes. Especially in the region covered with a dry belt and with a tribal population, there is a scarcity of safe drinking water. Research conducted on the usage of tree-based plant coagulants to treat the wastewater are listed in Table 6.6.

TABLE 6.3
Vegetable-based Coagulants Used for Waste Water Treatment

Plant Name	Plant Part	References
Abelmoschus esculentus	Mucilage	Freitas et al. (2015)
Coccinia indica	Fruit	Patale and Pandya (2012); Jadhav and Mahajani (2013)
Cyamopsis tetragonoloba	Gum	Omer et al. (2013)
Cuminum cyminum and *Trigonella foenum-graecum*	Seeds	Ramamurthy et al. (2012)
Lablab purpureus	Beans peel	Shilpaa et al. (2012)

TABLE 6.4
Pulse Plant-based Coagulants Used for Treatment of Waste Water

Plant Name	Part	References
Phaseolus vulgaris	Seed	Antov et al. (2010)
Dolichos lablab	Seed	Unnisa et al. (2010)
Glycine max	Seed	Graham and Vance (2003)
Vigna unguiculata	Seed	Megersa et al. (2014)
Phaseolus mungo and *Pisum sativam*	Seed	Hossain (2012)

TABLE 6.5
Fruit-based Coagulants Used for Treatment of Waste Water

Plant Name	Part	References
Aesculus hyppocastanum	Seed	Šćiban et al. (2009)
Carica papaya	Seed	Yongabi et al. (2011)
Nephelium lappaceum	Seed	Zurina et al. (2014)
Opuntia species	Plant	Miller et al. (2008)
Opuntia ficus-indica	Plant	Shilpaa et al. (2012)
Phoenix dactylifera	Seed and pollen sheath	Mukheled (2012)
Solanum incanum	Leaves	Kihampa et al. (2011)

TABLE 6.6
Tree-based Coagulants Used for Treatment of Waste Water

Plant Name	Part	References
Acacia catechu and *Guazuma ulmifolia*	Bark	Thakur and Choubey (2014)
M. oleifera	Seed	Pritchard et al. (2010)
S. potatorum	Seed	Babu and Chaudhuri (2005)
P.juliflora	Seed	Diaz et al. (1999)
Parkia biglobosa	Seed	Adie et al. (2014)
Persea americana	Seed	Yongabi et al. (2011)
Jatropha curcas	Seed	Zurina et al. (2014)
Azadirachta indica	Seed	Ramesh et al. 2015
Chess nut	Seed	Šćiban et al. (2009)

TABLE 6.7
Flower-based Coagulants Used for Treatment of Waste Water

Plant Name	Part	References
Amorpha fruticosa	Seed	Šćiban et al. (2009)
Cassia alata	Leaves	Rak and Ismail (2012)
Castanea sativa and *Ceratonia siliqua*	Seed	Šćiban et al. (2009)

Drinking water from most raw water sources, along with poor sanitation and hygiene, accounts for spreading diarrhea, especially in children, the elderly, and people suffering from chronic diseases. Some of the research work carried out on the flower plant-based coagulants for the treatment of physio-chemical properties of wastewater is given in the Table 6.7.

Slightly different to plants, trees are employed in combating the problem of environmental pollution, particularly the problem of turbidity. The efficiencies of the plant-based coagulants could be enhanced with some modifications in the preparation and process, and tailoring the parameters needs to be critically evaluated.

6.6 CONCLUSIONS

The coagulation-flocculation process presented in this chapter highlights the meritorious features of plant-based natural coagulants. The overall discussion shows the pervasive applications of different parts of plants that are employed in water treatment. The plant-based biocoagulants have shown distinctive performance but they do not deviate from being a threat to the environment due to their disposal after use. Conclusively, they provide remarkable opportunities to be explored for diverse water and wastewater treatment.

ACKNOWLEDGMENTS

The authors are thankful to Director, CSIR-CSMCRI for the support and the manuscript has been assigned CSIR-CSMCRI: 50/2020 registration.

REFERENCES

Adie DB, Sanni MI, Tafida A. 2014. Treatment of domestic waste water with activated carbon from locust bean (*Parkia biglobosa*) pod. *J Occup Env Med*. 2:193–198.

Antov MG, Šćiban MB, Petrović NJ. 2010. Proteins from common bean (*Phaseolus vulgaris*) seed as a natural coagulant for potential application in water turbidity removal. *Bioresour Technol*. 101(7):2167–2172.

Babu R, Chaudhuri M. 2005. Home water treatment by direct filtration with natural coagulant. *J Water Health*. 3(1):27–30.

Beaudeau P, Schwartz J, Levin R. 2014. Drinking water quality and hospital admissions of elderly people for gastrointestinal illness in Eastern Massachusetts, 1998--2008. *Water Res*. 52:188–198.

Daverey A, Tiwari N, Dutta K. 2019. Utilization of extracts of *Musa paradisica* (banana) peels and *Dolichos lablab* (Indian bean) seeds as low-cost natural coagulants for turbidity removal from water. *Environ Sci Pollut Res*. 26(33):34177–34183.

Dawood AS, Li YL. 2012. Response surface methodology (RSM) for wastewater flocculation by a novel (AlCl3-PAM) hybrid polymer. *Adv Mater Res*. 560:529–537.

Diaz A, Rincon N, Escorihuela A, Fernandez N, Chacin E, Forster CF. 1999. A preliminary evaluation of turbidity removal by natural coagulants indigenous to Venezuela. *Process Biochem*. 35(3–4):391–395.

Freitas T, Oliveira VM, De Souza MTF, Geraldino HCL, Almeida VC, Fávaro SL, Garcia JC. 2015. Optimization of coagulation-flocculation process for treatment of industrial textile wastewater using Okra (*A. esculentus*) mucilage as natural coagulant. *Ind Crops Prod*. 76:538–544.

Graham PH, Vance CP. 2003. Legumes: importance and constraints to greater use. *Plant Physiol*. 131(3):872–877.

Hossain MT. 2012. International Journal of Green and Herbal Chemistry. *Green Chem*. 1(3):296–301.

Jadhav MV, Mahajani YS. 2013. A comparative study of natural coagulants in flocculation of local clay suspensions of varied turbidities. *Int J Civ Environ Eng*. 35:26–39.

Kashfi H, Mousavian S, Seyedsalehi M, Sharifi P, Hodaifa G, Salehi AS, Takdastan A. 2019. Possibility of utilizing natural coagulants (*Trigonella foenum-graecum* and *Astragalus gossypinus*) along with alum for the removal of turbidity. *Int J Environ Sci Technol*. 16(7):2905–2914.

Kihampa C, Mwegoha WJS, Kaseva ME, Marobhe N. 2011. Performance of *Solanum incunum Linnaeus* as natural coagulant and disinfectant for drinking water. *African J Environ Sci Technol*. 5(10):867–872.

Mann AG, Tam CC, Higgins CD, Rodrigues LC. 2007. The association between drinking water turbidity and gastrointestinal illness: a systematic review. *BMC Public Health*. 7(1):256.

Megersa M, Beyene A, Ambelu A, Woldeab B. 2014. The use of indigenous plant species for drinking water treatment in developing countries: a review. *J Biodivers Environ Sci*. 5(3):269–281.

Merzouk B, Gourich B, Madani K, Vial C, Sekki A. 2011. Removal of a disperse red dye from synthetic wastewater by chemical coagulation and continuous electrocoagulation. A comparative study. *Desalination*. 272(1–3):246–253.

Miller SM, Fugate EJ, Craver VO, Smith JA, Zimmerman JB. 2008. Toward understanding the efficacy and mechanism of *Opuntia* spp. as a natural coagulant for potential application in water treatment. *Environ Sci Technol*. 42(12):4274–4279.

Mokhtar NM, Priyatharishini M, Kristanti RA. 2019. Study on the effectiveness of banana peel coagulant in turbidity reduction of synthetic wastewater. *Int J Eng Technol Sci*. 6(1):82–90.

Mosaddeghi MR, Pajoum Shariati F, Vaziri Yazdi SA, Nabi Bidhendi G. 2020. Application of response surface methodology (RSM) for optimizing coagulation process of paper recycling wastewater using *Ocimum basilicum*. *Environ Technol*. 41(1):100–108.

Mukheled A-S. 2012. A novel water pretreatment approach for turbidity removal using date seeds and pollen sheath. *J Water Resour Prot*. 4: 79–92.

Nonfodji OM, Fatombi JK, Ahoyo TA, Osseni SA, Aminou T. 2020. Performance of *Moringa oleifera* seeds protein and *Moringa oleifera* seeds protein-polyaluminum chloride composite coagulant in removing organic matter and antibiotic resistant bacteria from hospital wastewater. *J Water Process Eng*. 33:101103.

Omer RM, El Hassan BM, Hassan EA, Sabahelkheir MK. 2013. Effect of guar gum (*Cyamopsis tetragonolobus*) powdered as natural coagulant aid with alum on drinking water treatment *J Sci Technol*. 3:1222–1228.

Parmar KA, Prajapati S, Patel R, Dabhi Y. 2011. Effective use of ferrous sulfate and alum as a coagulant in treatment of dairy industry wastewater. ARPN *J Eng Appl Sci*. 6(9):42–45.

Patale V, Pandya J. 2012. Mucilage extract of *Coccinia indica* fruit as coagulant-flocculent for turbid water treatment. *Asian J Plant Sci Res*. 2(4):442–445.

Pidou M, Avery L, Stephenson T, Jeffrey P, Parsons SA, Liu S, Memon FA, Jefferson B. 2008. Chemical solutions for greywater recycling. *Chemosphere*. 71(1):147–155.

Pritchard M, Craven T, Mkandawire T, Edmondson AS, O'Neill JG. 2010. A study of the parameters affecting the effectiveness of *Moringa oleifera* in drinking water purification. *Phys Chem Earth, Parts A/B/C*. 35(13–14):791–797.

Rak AE, Ismail AAM. 2012. *Cassia alata* as a potential coagulant in water treatment. *Res J Recent Sci*. 1(2):28–33.

Ramamurthy C, Maheswari MU, Selvaganabathy N, Muthuvel SK, Sujatha V, Thirunavukkarasu C. 2012. Evaluation of eco-friendly coagulant from *Trigonella foenum-graecum* seed. *Adv Biol Chem*. 2(1):58.

Ramesh P, Padmanaban V, Sivacoumar R. 2015. Influence of homemade coagulants on the characteristics of surface water treatment: experimental study. *Int J Eng Res Technol*. 4(12):342–345.

Ramli SF, Aziz HA. 2015. Use of ferric chloride and chitosan as coagulant to remove turbidity and color from landfill leachate. *Appl Mech Mater* 773: 1163–1167.

Šćiban M, Klašnja M, Antov M, Škrbić B. 2009. Removal of water turbidity by natural coagulants obtained from chestnut and acorn. *Bioresour Technol*. 100(24):6639–6643.

Shilpaa B, Akankshaa K, Girish P. 2012. Evaluation of cactus and hyacinth bean peels as natural coagulants. *Int J Chem Environ Eng.* 3(3).

Tawakkoly B, Alizadehdakhel A, Dorosti F. 2019. Evaluation of COD and turbidity removal from compost leachate wastewater using *Salvia hispanica* as a natural coagulant. *Ind Crops Prod.* 137:323–331.

Thakur SS, Choubey S. 2014. Use of tannin based natural coagulants for water treatment: an alternative to inorganic chemicals. *Int J ChemTech Res.* 6(7):3628–3634.

Tinker SC, Moe CL, Klein M, Flanders WD, Uber J, Amirtharajah A, Singer P, Tolbert PE. 2010. Drinking water turbidity and emergency department visits for gastrointestinal illness in Atlanta, 1993–2004. *J Expo Sci Environ Epidemiol.* 20(1):19–28.

Unnisa SA, Deepthi P, Mukkanti K. 2010. Efficiency studies with *Dolichos lablab* and solar disinfection for treating turbid waters. *J Environ Prot Sci.* 4:8–12.

Vijayaraghavan G, Sivakumar T, Kumar AV. 2011. Application of plant based coagulants for waste water treatment. *Int J Adv Eng Res Stud.* 1(1):88–92.

Yongabi KA, Lewis DM, Harris PL. 2011. Indigenous plant based coagulants/disinfectants and sand filter media for surface water treatment in Bamenda, Cameroon. *African J Biotechnol.* 10(43):8625–8629.

Zurina AZ, Mohd Fadzli M, Ghani A, Abdullah L. 2014. Preliminary study of rambutan (*Nephelium lappaceum*) seed as potential biocoagulant for turbidity removal. *Adv Mater Res.* 917:96–105.

7 The Role of Nanomaterials in Wastewater Treatment

A. Thirunavukkarasu and R. Nithya

Government College of Technology, India

CONTENTS

7.1 INTRODUCTION

In spite of water being abundant and rich in our earth's crust, about 1% of the total reserves are available for human consumption [1, 2]. Because of growing potable water costs, various climatic and environmental conditions, as well as the increasing populations all over the world [2], an adequate amount of drinking water is not available for approximately 1.1 billion people and they suffer from water scarcity [3]. Day by day, fresh water resources are polluted by various organic and inorganic pollutants [4]. Hence the treatment of waste water to provide safe potable and drinking water needs to be prioritized at the top level [5]. However, the major challenge is that the traditional methods of treatment process are considered to be of low efficiency in removing the contaminants completely to meet water quality standards [6]. The available technologies which are currently used for water treatment also have certain drawbacks such as incomplete removal of pollutants and toxic sludge release, and high energy is required for its operation process [5]. The biological wastewater treatment process also shows some limitations such as low rate of action, toxic to

113

microorganisms, and the existence of non-biodegradable pollutants and so on which limits their application [7]. The physical treatment process, like filtration, is used to remove contaminants by altering the phases, but this process results in the formation of extremely concentrated sludge release which is toxic and quite challenging to dispose of [8, 9]. By considering the available methods, more reliable and advanced technologies are required for both waste water and municipal water treatment processes [10–12]. This can be accomplished by developing completely an efficient and new method through improvising the existing methodology. Nanotechnology is reported to be an emerging technology, and it has been proven as a potential technology to treat waste water and address other environmental issues [13–15].

Nanotechnology includes the materials which are smaller in their size and can be measured around few nanometers [16]. Otherwise, the nanoparticles are those components which measure one dimension less than 100 nm [17]. The different form of nanomaterials developed includes nanotubes, nanowires, quantumdots, particles, films, and colloids [18, 19]. Such developed nanomaterials are exclusively used for the treatment of wastewater since it possesses constructive attributes such as being eco-friendly, efficient, and cost-effective in removing contaminants in industrial effluents and municipal water and so on [20–22].Various literature studies highlighted the use of nanotechnology in the wastewater treatment process because of its wide applications. The developed nanomaterials are classified into three main groups such as nano-membranes, nano-catalysts, and nano-adsorbents. Many effective works proved using nano-adsorption technology to investigate the decontamination process from wastewater with different nano-adsorbents [23–27]. Nano-adsorbents are produced from atoms of those elements which have chemically active binding pockets with high adsorption capacity at the surface of the nanomaterial [27]. The materials used for the development of nano-adsorbents include silica, activated carbon, metal oxides, modified compounds as composite materials, and clay materials [28]. The development of metal oxides and semiconductors as nanomaterials gained the attention of researchers in developing wastewater technology. Besides, various nano-catalysts are developed for the degradation of contaminants from waste water such as electrocatalysts [29], Fenton-based catalysts [30] which improve chemical oxidation of organic pollutants [31], and catalysts which possess antimicrobial properties [32]. Nano-membranes were also used in the waste water treatment process. In this technology, pressure-driven treatment is considered as the best for wastewater treatment to improve the water quality [33]. Different types of membrane filtration are available, [34–36] amongst them, nanofiltration (NF) which has wide application for wastewater treatment in industries because of its tiny pore size, high efficiency with low cost, and ease of operation [37–40]. Nano metal particles, nano-carbon tubes, and non-metal particles are used for the development of nanomebranes.

The unique properties of nanomaterials such as large surface area, nano size, highly reactive [41], strong mechanical property, porosity, dispersibility, hydrophilicity, and hydrophobicity attract researchers to develop nanomaterials for the water treatment process[42–44]. Nanotechnology provides a treated water with low expense and high efficiency in removing pollutants and its ability to be reused [45]. Different heavy metals like Pb, Zn, Cr, and so on, and other organic and inorganic contaminants and certain harmful microbes were also removed using these nanomaterials

[46–50]. As per the reports of WHO, yearly 1.7 million people lose their lives due to water pollution and nearly 4 billion people are registered with various health issues relating to the waterborne infections [51]. Recently, more advancement can be seen in the development of nanomaterials in the fields of nanomotor science, nanophotocatalysis, nanofiltration, and nano-adsorption. Concisely, the use of the nano-sized materials in the field of wastewater processing is considered effective and ideal. Hence, the present chapter is intended to review the possible role of different nanomaterials in removing the organic contaminants in wastewater. Also, it summarizes the possible flaws of the materials or techniques and the challenges need to be studied systematically for the development of sustainable technologies in the near future.

7.2 ROLE OF NANOMATERIALS IN THE WASTEWATER TREATMENT

7.2.1 NANOPHOTOCATALYSTS

Photocatalysis is a term which comes from two Greek words: photo and catalysis, referring to the process of molecular degradation under the influence of light (ultraviolet, visible, or infrared). More precisely, photocatalysis can be defined as the process in which the radiant power of light stimulates or activates the process of decomposition of molecules. It is easier to recognize the role of key component of the process, photocatalysts, which are passively involved and are able to amend the rate of the molecular degradation of the several components [52]. The mechanistic aspects of this surface-oriented process can be viewed as the series of five transfer phases [53]:

a) Reactant diffusion to the photocatalyst surface;
b) Adsorption on the photocatalytic surface;
c) Possible reactive sites/hotspots on the photocatalytic surface;
d) Product desorption from the photocatalytic surface;
e) Product diffusion from the photocatalyst surface.

Nanophotocatalysts are the photocatalysts having their size in the range of 1 nm and 100 nm, which are able to accelerate reaction in the presence of light. The likely advantage of enhanced reactivity due to the improved surface area and shape-related attributes finds them as potential candidates for the wastewater treatment processes [54]. In general, the nano-sized particles will exhibit different characteristics than their bulk forms due to their discrete quantum effects and surface morphological features. These unique traits aid them in enhancing their chemical, mechanical, electric, optic and magnetic properties and thus these nanomaterials find their potential applications in varied fields [55, 56]. Further, the nano-sized photocatalysts are able to inflate oxidation capacity by producing oxidizing species on the nanomaterial surfaces and, hence, can be well exploited in the degradation of contaminants in wastewater [57].

Significant research has been undertaken on the use of nano-sized metals, zero-valent forms, mono metallic oxides, bimetallic oxides, semiconductors and so on in the degradation of contaminants like dyes or heavy metal ions in wastewater [12, 56, 70, 176, 179]. Of them, the nano-sized metal oxides, which include TiO_2, CeO_2, Fe_2O_4, SiO_2, ZnO, and Al_2O_3 provided excellent reports on the removal of pollutants like azo dyes, organochlorine pesticides, nitroaromatics, and other aromatic-based compounds and so on[58–70].TiO_2, the most effective and substantial photocatalyst is preferred for the treatment process due to its cost-effectiveness, low toxicity, being chemically stable, and highly accessible. It can exist in three different states, including anatase, rutile, and brookite, and significant reports were made on the anatase forms [71]. In spite of these, several strategies are continually adopted on the nano-sized metal oxide particles to increase their photocatalytic performances by doping with the other metals or metal ions [72], carbon-containing materials, dye sensitizers and so on [73].

In general, the nanophotocatalytic process can be homogeneous or heterogeneous in nature, depending on the physical states of the reactants and catalyst. Of these, the heterogeneous process finds wide scope in the area of wastewater treatment and other eco-related applications. Hence, there is the utmost need for comprehensive understanding of the interphase among wastewater (liquid) and photocatalysts (solid) is needed through which the masses of components are transferred through molecular diffusion [74, 75]. An ample number of reactor design research is focussing on the effective transfer of solutes across the interphase by reducing diffusional limitations and also on the proper exposure of catalysts to radiant energy. The use of light-emitting diodes, optical fibers, and solar-based photoreactors are yet to be efficiently explored [76–78].

One of the strongest advantages of nanophotocatalysis is that the potent nano-sized metal oxide particles can efficiently mineralize the highly toxic and most complex organic substances at room temperature (about 25°C) [79]. Also, the nano-sized catalyst is highly associated with the quantum effects, which in turn helps in improving the energy bandgap levels and might enhance the surface area [80]. Comprehensively, this photocatalytic process possesses excellent merits: a cheaper operational cost and effectiveness in the mineralization process. However, few challenges are yet to be tackled in the areas of the material recovery and toxicity-related issues and hence this limits their application in the vast areas of environmental clean-up [81]. Few innovations have been made in catalyst recovery by coupling them with magnetic elements so that the externally applied magnetic fields can do the recovery efficiently. Further, the most intricate task to the scientific community is in developing stable catalysts which can resist the change of their oxidation states during exposure to radiant energies. Also, the resistance in the transfer of solute across the phases and increased depletion of photons are also posing vital tasks to the researchers [82]. In addressing the issue of improved diffusional properties, the development of microfluidic reactors opened up a gateway for the efficient use of photocatalysis. It also focused on the operational aspects like providing a larger surface to volume ratio, enhanced mass transfer coefficient, stable hydrodynamics with low Reynold's number, and so on [83, 84]. These possible advantages of microfluidic reactors enabled them to be an alternate and viable option for the effective photocatalytic process over conventional reactors. However, its probable use is highly confined to pilot scale and further research still needs to be done in field applications. Hence, systematic and meticulous investigations are imperative to exploit their application on a largescale and in the search for novel or engineered materials to overcome the former issues and thus promote sustainable development in the area of photocatalysis. Further, greater emphasis can be given for the novel synthetic routes in developing noteworthy structures like nano-rod, nano-sphere, nano-flowers, nano-cones and nano-flakes, and so on, in improving their structural and functional properties, thereby resulting in effective photocatalysis. Eventually, the feasible use of solar-driven catalytic action with a minimal site of operation in the near future can serve this technology as the most reliable and effective technique in wastewater treatment.

Owing to the notable physico-chemical properties of nano-sized materials, a series of mineralization and degradation of organic contaminants can be carried out effectively [85, 86]. In the former stage, the complex organic contaminants are broken down or decomposed into simpler products whereas in the latter stage, the decomposed products are completely destructed to form the much simpler compounds like water, CO_2, and some other inorganic ionic forms [86]. Figure 7.1 illustrates the proposed photocatalytic mechanism of organic contaminants. Here, MX is a metal oxide absorbing incident radiation which is likely to be greater or equal to its bandgap width and thus forms electron-hole pairs. Further, these pairs might move to the exterior surface of the catalyst to mediate the redox reactions with the surface adsorbed ionic species. More precisely, the surface-bound h+ will react with the water molecules and thus generate hydroxyl radicals. Likewise, the e- will be able to form superoxide radicals with the combination of oxygen. Table 7.1 reports the use of nanophotocatalysts in the remediation of organic pollutants from wastewater.

7.2.2 NANO- AND MICROMOTORS

As discussed earlier, the field of nano-science and technology offers a wide variety of applications and hence the thrust in developing nano-sized materials has exponentially increased in recent decades. One such recent development is the nano- or micromotors which are able to convert energy into machine-driven force. Such novel materials are versatile and can be motorized with or without using fuel resources such as acoustics, electric, or magnetic field [102]. High speed of operation, specificity in their movements, self-mixing ability and so on are the significant features of

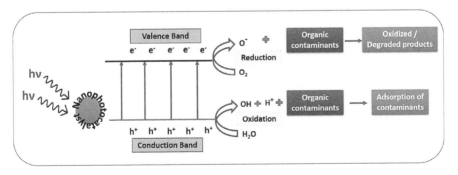

FIGURE 7.1   Photocatalytic mechanism of organic contaminants.

nano- or micromotors. Further, the use of this strategy primarily rules out the common challenge of diffusional resistance encountered in traditional wastewater treatment techniques. Energetic mixing of this method can easily overcome resistance across the boundary and also the enhanced surface area of these self-propelled nano- or micromotors would be the ideal choice in such treatment processes [103–106]. Also, the small-scaled dimensions might show their ability to reduce the clear-out duration of the treatment process and thus reduce the overall economy of the process [107]. More precisely, the constant motion of the nanomotors can be used to carry the functionally active or reactive species over longer distances and this can also be used for the study of their non-ideal mixing behavior in the treatment solution [108]. However, research on commercial exploitation is still required to completely replace conventional strategies, as the lifecycle of these nano- or micro-sized motors is very short due to their exhaustion during transit or possible oxidative processes. Also, their application is very limited for a highly viscous solution as it hinders the transit of motors. Few reports were made on the possible poisoning of the catalyst used in the process with their covalent bonding on its surfaces [109]. Hence, a large number of research avenues are still open for a more thorough and complete understanding of and for the commercial utilization in the field of wastewater treatment (Figure 7.2 and Table 7.2).

7.2.3 Nano-membranes

The process of nanofiltration is a pressure-driven operation with the use nano-membranes to selectively separate or reject the molecular or ionic species in the range of 0.5–1 nm. A typical nano-membrane with a pore size of 1 nm ideally allows the materials to pass through it with the molecular weight cut-off of 300 Da–500 Da. This method is highly efficient in the separation or removal of inorganic species and other smaller organic compounds. The greater rejection of divalent ionic species, improved diffusional flux, and smaller rejection of monovalent ionic species paved the way for exploiting this strategy in various sectors including biotechnology, food, pharmaceutical, and effluent treatment industries. This method has evolved drastically in separation technology as it removes or reduces the odor, hardness, and the color of the polluted water systems. Primarily, these nano-membranes are highly efficient in the desalination process and in the removal of certain heavy metal ions. In the treatment process, these nano-membranes will slightly develop charges on the surface due to physical contact with the solute species or due to the functional moieties' dissociation. Hence, in this context, these processes can be assumed as a charged ultrafiltration system [131, 132]. However, it showed a greater rejection rate of smaller organic molecules than the UF systems. This diffusion-driven and charge-effect separation mechanism can be elucidated as the series of five stages [133]:

    a. Wetted surface formation;
    b. Capillary rejection in heterogeneous and microporous membrane;
    c. Solute and solvent dissolves in homogeneous and nonporous membrane;
    d. Diffusion of solvent across the membrane;
    e. Preferential electrostatic attraction resulting in the rejection of solute.

TABLE 7.1

List of Nanophotocatalysts in the Removal of Organic Contaminants

S.No	Nanophotocatalyst	Synthetic Route	Organic Contaminant	Light Source	References
1	TiO_2 samples with different ratios of anatase/rutile phases	Microwave-assisted sol–gel method	Crystal Violet & Methylene Blue	UV	Karima Almashhori et al. [87]
2	Bi_2WO_6	Solid-state reaction route	Methylene Blue	Visible	Vishvendra Pratap Singh et al. [88]
3	$TiO_2/Fe_2O_3/PAC$	Chemical synthesis	Cyanide	UV	Parisa Eskandari et al. [89]
4	TiO_2 & ZnO	Chemical synthesis	Food Black 1	UV	Sajjad Khezrianjoo et al. [90]
5	$MgFe_2O_4$, $CaFe_2O_4$, $BaFe_{12}O_{19}$, $CuFe_2O_4$, and $ZnFe_2O_4$	Chemical Co-precipitation	Methylene Blue	Solar, UV-Vis	Nauman Ali et al. [91]
6	ZnS-ZnO/graphene	Solid-state mixing	Organic dyes &Phenol	Visible	Lonkar et al. [92]
7	ZnO/SnO_2	Chemical Co-precipitation	Methyl Orange	Visible	Wajid Ali et al. [93]
8	Co doped nano TiO_2	Chemical sol–gel method	Rose Bengal	Visible	Malini et al. [94]
9	$TiO_2/ZnTiO_3/\alpha Fe_2O_3$	Chemical sol–gel method	Methylene Blue & Methyl Orange	UV	Mehrabi et al. [95]
10	Magnetic activated carbon composite	Chemical precipitation	Methylene Blue, Methyl Orange, Rhodamine B,	Solar, UV-Vis	Mehdi Taghdiri et al. [96]
11	Bismuth oxychloride	Facile hydrolysis route	Methylene Blue & Methyl Orange	UV-Vis	Seddigi et al. [97]
12	TiO_2/Fe_3O_4	Chemical synthesis	Methyl Orange	UV	Amin Ahmadpour et al. [98]
13	Nano-TiO_2	Chemical synthesis	Reactive Red & Reactive Yellow	Solar	Jeni et al. [99]
14	CuO/ZnO	Chemical wet impregnation	Acid Red 88	Visible	Panneerselvam Sathishkumar et al. [100]
15	TiO_2&ZnO	Chemical synthesis	Reactive Black 5 & Reactive Orange 4	UV	Kansal et al. [101]

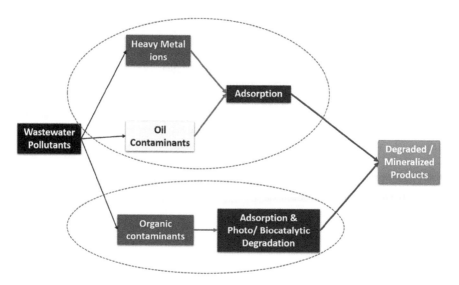

FIGURE 7.2 Possible contaminants removal mechanisms of nano/micro motors.

It is also observed that the separation of uncharged species occurs due to the size exclusion phenomenon. In the case of charged ionic species, preferential separation was assumed due to the combined effects of size exclusion and electrostatic interactions. More precisely, the anionic species rejection can be viewed as the formation of negative zeta potential on the membrane surface and hence the removal processes of anionic species were highly dependent of the pH and the concentration of the electrolytes [134]. Several reports were made on the fabrication of nano-membranes with the use of the surface engineering process, wherein functionally active moieties were employed resulting in the formation of nano-sized composites. By doing so, the resultant engineered nano-sized catalytic membranes would possess significant characteristics like better selectivity, rapid degradation rate, and enhanced resistance to the fouling process [135, 136].

Significant research was employed in the use of embedded nano-fillers in nanocomposites in order to improve surface area and thus the separation or rejection capacity of the treatment process [137, 138]. Also, the use of nano metal oxide particles like TiO_2 and Al_2O_3 can be doped with the membranes to improve the thermal, mechanical, and diffusional flux of the transport process. For instance, TiO_2 doped nano-membranes were found to be efficient in deactivating the microbial population and several organic contaminants [139]. Incorporating or the doping of active moieties or molecules show improved properties in the catalytic application. The iron-incorporated nano-membranes made feasible carrying over the oxidation reactions for the removal of organic contaminants in a toxic-free environment and thus the use of hazardous chemicals can be limited to greater extent. In field applications, the use of immobilized iron oxide facilitated the reaction with hydrogen peroxide to generate free radical ions and thus can be employed in the remediation of chlorine-based organic contaminants [140]. Adding to this, many other reports were made on the immobilization of cellulose acetate, chitosan, and so on, for enhancing the reactivity,

TABLE 7.2

List of Nano and Micromotors in the Removal of Organic Contaminants

Mechanism of Propulsion	Nano and Micromotors	Organic Contaminant	Removal Mechanism	References
	Biotin-functionalized Janus silica micromotor	Rhodamine 6G & Fluorescein Sodium	Charge adsorption	Xuan et al. [110]
	Au NPs/TiO$_2$/Pt nanomotor	Rhodamine B & Methyl Orange	Photocatalytic degradation	Zhang et al. [111]
	TiO$_2$-PtPd-Ni nanotube	Rhodamine B, Methyl Orange & Methylene Blue	Photocatalytic degradation	Mushtaq et al. [112]
	Zero-valent-iron/platinum (ZVI/Pt) Janus	Methylene Blue	Fenton reaction	Lee et al. [113]
	Polystyrene@ ZIF–Zn–Fe core-shell microparticles	Rhodamine B	Fenton reaction	Wang et al. [114]
	WO$_3$@CJanus micromotors	Sodium-2,6-dichloroindophenol & Rhodamine B	Photocatalytic degradation	Zhang et al. [115]
	CoNi@Pt nanorods	4-nitrophenol, Methylene Blue & Rhodamine B	Catalytic degradation	García-Torres et al.[116]
	rGO-SiO$_2$–Pt Janus magnetic micromotors	Polybrominated diphenylethers & triclosan	Adsorption	Orozco et al. [117]
Bubble propulsion (Pt-H$_2$O$_2$)	Pt coating activated carbon-based Janus particle micromotors	Pb^{2+}	Adsorption	Jurado-Sánchez et al. [118]
	Ag-incorporated zeolite-based micromotors	Chemical & biological warfare agents	Adsorptive detoxification	Singh et al. [119]
	TiO$_2$/Au/Mg microspheres	Organophosphate nerve agents, bis(4-nitrophenyl) phosphate & methyl paraoxon	Photocatalyticdegradation	Pourrahimi et al. [120]
	3D printed motors (TSM)	Oil droplets	Adsorption by Induced hydrophobicity	Yu et al. [121]
	Alkanethiols-coated Au/Ni/PEDOT/Pt microsubmarine	Oil droplets	Adsorption by Surface hydrophobicity	Guix et al. [122]
	Pot-like MnFe$_2$O$_4$ micromotorsoleic acid microparticles	Oil droplets	Adsorption	Mou et al. [123]
	DNA-functionalized Au/Pt microtubes	Hg^{2+}	Adsorption	Wang et al. [124]
	GOx-Ni/Pt	Pb^{2+}	Adsorption	Vilela et al. [125]
	Fe/Pt tubular micromotors	Xanthene dyes	Fenton reaction	Soler et al. [126]

(Continued)

TABLE 7.2
Continued

Mechanism of Propulsion	Nano and Micromotors	Organic Contaminant	Removal Mechanism	References
	PSF millimeter-sized motor	Oil droplets	Surface tension	Seah et al. [127]
	SDS/PSF capsule	Oil droplets	Surface tension	Zhao et al. [128]
Marangoni effect	SA-modified polyvinyl alcohol (PVA) Janus foam motors	Oil droplets	Adsorption	Li et al. [129]
	Commercial pipette tips with laccase enzyme or EDTA	Phenol oxidases, Eriochorome black-T	Enzymatic Biocatalytic decontamination	Orozco et al. [130]

to prevent agglomeration, and to overcome the hindrance effects in the wastewater treatment process [141, 142].

Several other vital findings were also reported with the use of silver nanoparticles, titanium oxide nanoparticles, and carbon nanotubes (CNT) to increase the anti-bacterial, photocatalytic performance and to overcome the fouling resistance. For instance, CNTs are able to exhibit anti-bacterial properties which can greatly reduce the formation of biofilms, fouling effects, and accidental mechanical failures [143]. Further, the selectivity and permeability of nano-membranes are highly associated with the quantity and dimension of the nanoparticles, and hence the ample volume of research has triggered the control of the mentioned limitations. In some cases, the biological membranes can be used to exhibit high permeability and selective rejection [144, 145]. The assembly of nano-films onto active thin film composite through the doping process is often limited by the clogging and fouling effects. In such cases, the incorporation of super hydrophilic nanomaterials in the fabrication is highly imperative. For example, the use of zeolites impregnated into the membranes showed increased hydrophilicity and permeability of the separation process [146–148]. Also, an interesting interaction occurred between the hydrogen and palladium nanomaterial and thus enhanced removal efficiency was observed when the combination of palladium acetate and poly-etherimide was used in the preparation of nano-films [149]. The possible use of in-situ and ex-situ methods showed better results in the fabrication of nano-films under the varied operating conditions [150–152]. Recent research is often hybridized with other strategies and employs bottom-up approaches to exploit their versatile application in the environmental sector [153]. These constructive properties like uniformity, homogeneity, reactivity, rapidity, and the ability to control and optimize the design variables, making it accessible to the hybridized process, recognized the use of nano-membranes as effective in the wastewater treatment [154]. In spite of these productive results, the primary challenges that needxto be focused on very carefully are membrane selectivity, fouling, and stability in order to attain sustainable development in the process of wastewater treatment.

Since membrane stability or the life-period is short, it requires replacement procedures and hence it increases the overall cost of the process and there is possible contamination during the replacement [155]. Also, it suffers with other drawbacks like poor reproducibility, slow run-time, high maintenance cost, and decreased efficacy over the run-time and scale-up issues and so on. Hence, there is a wide area of scope available for scientific researchers in the development of new generation multifunctional nano-membranes targeted for industrial use (Table 7.3).

7.2.4 Nano-adsorbents

The most powerful tool in wastewater treatment is adsorbents and modern adsorption science has heavily relied on nano-sized adsorbents for their versatile properties. The most common and heavily exploited nano-adsorbents are either carbon-based or metal/metal oxide nanomaterials [172]. Also, significant efforts have been made in the toxicity reduction process by making nanocomposites like silver/carbon, carbon/titanium oxide, and so on; single-walled and multi-walled CNTs also find a prominent role in the clean-up strategy as they possess a higher surface area and multiple adsorption sites and so on. However, prior measures need to be taken for the stabilization of CNTs as they have hydrophobic surface properties which might lead them to aggregate themselves and reduce surface-active moieties. Several promising results were also reported on the use of polymer-based nano-adsorbents to effectively eliminate organic contaminants and heavy metal ions from wastewater [173]. Implantable zeolites also showed promising results in the removal of organic dyes and heavy metal ions in which nano-sized metallic ions (Cu^{2+}, Ag^{2+}) played a significant role as implants [174]. These implanted zeolites also exhibited strong antimicrobial properties against the wide range of microbial population [175]. For the ease of downstream recovery, magnetic nano-adsorbents were used in the removal of a wide range of organic contaminants and these materials can be effectively regenerated for the sustainable process of wastewater treatment.

Few innovative developments were reported in the literature in the synthesis of novel nano-adsorbents from the electronic wastes. For instance, copper oxide nano-adsorbent was prepared from the hydro-metallurgical leachate of electronic waste which showed effective results on the removal of Methylene Blue [176]. The use of nano-adsorbents is the most encouraging tool in wastewater treatment as it is cost-effective, has biocompatibility and ease of commercialization, is a toxic-free process and is biodegradable. It also requires less trained manpower, has selective separation and ease of recovery, and, above all, is highly efficient in removing the pollutants.

The surface binding of adsorbate species with the adsorbent depends on the nature of the active sites on both molecules. As the nano-adsorbents have a larger surface area and greater adsorption ability, application is highly inevitable in solute removal. In general, organic pollutants, either of polar or non-polar in nature, can be attracted to the surface of the adsorbent either through physical interactions (physisorption) or through chemical bonding (chemisorption). The possible physical forces that exist among the adsorbates and adsorbents are weak van der Waal's forces, hydrophobic interactions, hydrogen bonding, π-π stacking interactions, and so on. These primary forces, either alone or in combination, were solely responsible for the surface

TABLE 7.3

List of Nano-membranes in the Removal of Organic Contaminants

S. No	Nano-membranes	Organic Contaminant	References
1	Chitosan Nanofiltration Membranes	Methyl viologen, Methylene Blue, Methyl Orange, Orange G, Rose Bengal, Brilliant Blue, Neutral Methyl Red, $CaCl_2$ & $MgSO_4$	Qingwu Long et al. [156]
2	Fe-modified MMT nano-membranes	Hg^{2+}	Lodo and Diaz [157]
3	asymmetric nanofiltration-surfactant (NFS) membrane	Methyl Violet, Methyl Blue and Acid Orange 74	Abdul Rahman Hassan et al. [158]
4	polyetherimide (PEI)-based nanofiltration (NF) membrane	Reactive Red	Felicia Marcella Gunawan et al. [159]
5	Hydracore10 and Hydracore50	Cibacron Yellow S-3R	Valentina Buscio et al. [160]
6	Nano-membrane prepared from coating c-alumina and titania nanocrystallites	Microorganisms and ions rejection from wastewater	Shayesteh et al. [161]
7	Sodium titanate nanobelt membrane (Na-TNB)	Removal of oil and radioactive Cs^+ ions and Sr^{2+}	Wen et al. [162]
8	polyamide nanofilter membrane	Anthraquinone dyes	Najmeh Askari et al. [163]
9	Carbon nanofiber membrane	Metal and metal oxide nanoparticles	Faccini et al. [164]
10	TFC commercial polyamide nanofilter	Reactive Blue 19 and Acid Black 172	Najmeh Askari et al. [165]
11	ZrO2 micro-filtration membrane	Dimethylformamide	Zhang et al. [166]
12	polyamide-6/chitosan	Solophenyl Red 3BL & Polar Yellow GN	Mozhdeh Ghani et al. [167]
13	Nanofiltration membrane bioreactor	NH3-N, NO3-N, & PO4-P	Kootenaei and Rad [168]
14	Integrated carbon nanotube (CNT) polymer composite membrane with polyvinyl alcohol layer	Oil contaminants	Maphutha et al. [169]
15	Hydrophilic electrospun nanofiber membrane	Suspended particles	Asmatulu et al. [170]
16	Nano-membrane combined with biodegradable poly-gamma-glutamic acid (c-PGA)	Pb^{2+} ion	Hajdu et al. [171]

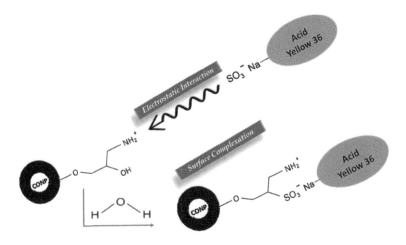

FIGURE 7.3 Proposed adsorption mechanism of anionic azo dye using nanoadsorbent.

adsorption in the case of physisorption. On the contrary, stable covalent and irreversible bonds under the applied conditions will be formed on the surface in case of chemisorptive nature of adsorption process. Previously, the authors had elucidated the mechanistic aspects of anionic azo dye removal using functionalized cerium oxide nanoparticles [56]. The adsorption reported was mediated by electrostatic interactions followed by surface complexation mechanisms and is illustrated in Figure 7.3.

In this chapter, the numerous scholarly reports were presented on four major classes of nano-adsorbents: carbon-based, silica-based, metal oxide-based, and chitosan-based nano-adsorbents.

7.2.4.1 Carbon-based Nano-adsorbents

Carbon-based nano-adsorbents are the allotropes of carbon which may be in 3-D nanotubes or in 2-D graphene sheets. Based on the number of sheets or layers folded, it may be classified as single-walled CNTs (SWCNTs), double-walled CNTs (DWCNTs) or multi-walled CNTs (MWCNTs). These nano-adsorbents possess greater surface area and improved adsorption capacity than the graphene. With simple hydrophobic interactions, adsorption of solute can be mediated on the surface of the adsorbent. Significant research has been made on functionalization with the specified functional moieties like carbonyl or carboxyl groups to mediate the electrostatic attraction of organic pollutants. Table 7.4 lists the reported works pertaining to the role of CNTs in the removal of contaminants in wastewater.

7.2.4.2 Silica-based Nano-adsorbents

The most abundant element on our earth's crust is silica and is the major constituent of sand. As silica is cheap and weightless, its application is in various fields of electronics and communication. Conventionally, silica has been in use for the separation of complex mixtures in chromatographic techniques. The biocompatible nature, ability to degrade, increased surface area, better adsorptive capacity, and so on, of silica-based nanomaterials assures their role in wastewater treatment. Several scholarly works were reported on the use of silica-based nano-adsorbents (see Table 7.4).

TABLE 7.4
List of Nano-adsorbents Used in the Removal of Organic Contaminants

S.No	Adsorbent	Organic Contaminant	Adsorption Capacity/mg g	References
Carbon-based Nano-adsorbents				
1	MWCNT-NH$_2$	Methyl Orange	96.0	Sadegh et al. [180]
2	Amino-functionalized MWCNTs@Fe$_3$O$_4$	Methylene Blue	178.5	Ahamad et al. [181]
3	ZnO-MWCNTs	Congo Red	249.5	Arabi et al. [182]
4	pTSA doped polyaniline@graphene oxide-MWCNTs	Congo Red	60.0	Ansari et al. [183]
5	Polypyrrole/CNTs-CoFe$_2$O$_4$magnetic nanocomposites	Methyl Blue	137.0	Li et al. [184]
6	Al-doped carbon nanotubes	Methyl Orange	69.7	Kang et al. [185]
7	MWCNTs	Congo Red	165.2	Kamil et al. [186]
8	Peroxymonosulfate activated by CNT	Acid Orange 7		Chen et al. [187]
9	MWCNTs/LDHs nanohybrids	Congo Red	595.8	Long et al. [188]
10	Halloysite/carbon nanocomposites	Phenol	5.23	Wu et al. [189]
11	MWCNTs decorated on calcined eggshell	Congo Red	136.9	Seyahmazegi et al. [190]
12	MWCNT-k-carrageenan-Fe$_3$O$_4$	Methylene Blue	0.124	Duman et al. [191]
13	O-MWCNTs	Congo Red	357.1	Sheibani et al. [192]
14	SWCNTs	Bisphenol A	71.0	Dehghani et al. [193]
15	Magnetic graphene-carbon nanotube	Methylene Blue	65.8	Wang et al. [194]
16	MWCNTs O-SWCNTs	p-Nitrophenol	206.0	Yao et al. [195]
17	MWNTs- coated biochars	Methylene Blue	6.2	Inyang et al. 2014 [196]
18	Magnetic polymer MWCNTs	Sunset Yellow FCF	85.5	Gao et al. [197]
19	SWCNT	Reactive Red 120	426.0	Bazrafshan et al. [198]
20	O-SWCNTs	Basic Red 46	49.4	Moradi et al. [199]
21	MWCNTs	Methylene Blue	400.0	Szlachta et al. [200]
22	MWCNTs	Methyl Orange	52.9	Zhao et al. [201]
23	MWCNTs	Direct Blue53	409.4	Prola et al. [202]
Silica-based Nano-adsorbents				
24	Multimorphological mesoporous silica NPs	RhodamineB	234.6	Chen et al. [203]
25	Silica nanoparticle	Acid Orange8	230.0	Suzimara et al. [204]
26	Mesoporous silica nanoparticles	MethyleneBlue	34.2	Qin et al. [205]
27	Amphi-functional mesoporous silica NPs	Rhodamine6G	9.9*	Shinde et al. [206]

TABLE 7.4
Continued

S.No	Adsorbent	Organic Contaminant	Adsorption Capacity/mg g	References
28	Iron oxide/Silica Gel Nanocomposite	Sulfur dyes	11.1	Tavassolia et al. [207]
29	Modified nano-SiO_2	Methylene Blue	9.5	Salimi et al. [208]
30	Cation functionalized silica nanoparticles	Xylenol Orange	9.0	Zhang et al. [209]
31	Silica nanoparticle	Basic Violet	416.0	Mahmoodi et al. [210]
32	Chitosan coated magnetic mesoporoussilica NPs	Methylene Blue	43.0	Li et al. [211]
33	Amino-functionalized silica magnetite NPs	Reactive Black 5	83.3	Araghi et al. [212]
34	Primary–secondary amino silica nanoparticle	Direct Red31	114.9	Mahmoodi et al. [213]
35	Nanoporous silica nanoparticles	Basic Blue	345.0	Zarezadeh-Mehrizi et al. [214]
Metal oxide Nano-adsorbents				
36	Amine functionalized CeO_2 nanoparticles	Fenalan Yellow G	25.58	Nithya et al. [179]
37	Amine functionalized CeO_2 nanoparticles	Acid Yellow 36	26.95	Thirunavukkarasu et al. [56]
38	Copper oxide NPs	Methylene Blue	163.93	Nithya et al. [176]
39	ZnO_2 nanoparticles	Congo Red	208.0	Chawla et al. [215]
40	MgO/GO nanocomposites	Methylene Blue	833.0	Heidarizad et al. [216]
41	MnO_2 and Fe_3O_4 and ionic polymercomposite	Crystal Violet	1200.0	Atta et al. [217]
42	MgO/Fe_3O_4 nanoparticles	Amaranth	37.9	Salem et al. [218]
43	Magnetic iron oxide/ carbon nanocomposites	Methyl Orange	72.7	Istratie et al. [219]
44	Fe_3O_4/ montmorillonite nanocomposite	Methylene Blue	106.4	Chang et al. [220]
45	Silica-Fe_3O_4 nanospheres	Congo Red	54.6	Wang et al. [221]
46	Quercetin α-Fe_2O_3 nanoparticles	Congo Red	427.4	Satheesh et al. [222]
47	Trisodium citrate (TSC) Fe_3O_4 nanocomposite	Malachite Green	435	Alqadami et al. [223]
48	Co_3O_4/SiO_2 nanocomposite	Methylene Blue	53.9	Abdel Ghafar et al. [224]
49	CuO-NP-AC	Acid Blue 129	65.4	Nekouei et al. [225]
50	α-Fe_2O_3	Congo Red	253.8	Hao et al. [226]
51	Fe_3O_4@C nanoparticles	Methylene Blue	117.0	Wu et al. [227]
52	3-Glycidoxypropyl-trimethoxysilane-lysine-Fe_3O_4 nanoparticles	Methyl Blue	185.0	Zhang et al. [228]

(Continued)

TABLE 7.4

List of Nano-adsorbents Used in the Removal of Organic Contaminants (Continued)

S.No	Adsorbent	Organic Contaminant	Adsorption Capacity/mg g	References
53	SiO_2-coated Fe_3O_4 nanoparticles	Methyl Orange	53.2	Shariati-Rad et al. [229]
54	Sodium alginate-Fe_3O_4 nanoparticles	Malachite Green	47.8	Mohammadi et al. [230]
55	SiO_2-decorated Fe_3O_4 nanoparticles	Methylene Blue	4.9	Che et al. [231]
56	α-Fe_2O_3 nanocrystals	Congo Red	93.5	Wu et al. [232]
57	ZnO	Malachite Green	310.5	Kumar et al. [233]
58	Fe_2O_3–Al_2O_3	Congo Red	498.0	Mahapatra et al. [234]
59	Aminoguanidine-Fe_3O_4 nanoparticles	Acid Green 25	121.1	Li et al. [235]
60	Humic acid- Fe_3O_4 nanoparticles	Methylene Blue	0.29*	Zhang et al. [236]
Chitosan-based Nano-adsorbents				
61	H_2SO_4 cross-linked magnetic chitosan	Methylene Blue	20.4	Mustafa et al. [237]
62	Magnetic chitosan nanocomposite	Crystal Violet	333.3	Massoudinejad et al. [238]
63	Nano-chitosan	Congo Red	5997.7	Rezaei et al. [239]
64	Magnetic polydopamine–chitosan nanoparticles	Methylene Blue	204.0	Wang et al. [240]
65	Chitosan/Al_2O_3/ magnetite nanoparticles	Methyl Orange	417.0	Tanhaei et al. [241]
66	Glutaraldehyde cross-linked magnetic chitosan	Acid Red 2	90.1	Kadam et al. [242]
67	Poly HEMA-chitosan-MWCNT nanocomposite	Methyl Orange	416.7	Mahmoodian et al. [243]
68	Glutaraldehyde cross-linked magnetic chitosan	FD&C Blue 1	475.6	Zhou et al. [244]
69	magnetic Fe_3O_4/chitosan nanoparticles	Reactive Brilliant Red X-3B	76.8	Cao et al. [245]
70	Nano-chitosan	Reactive Blue T	357.1	Sarkheil et al. [246]
71	Chitosan modified magnetic graphitized multi-walled carbon nanotubes	Congo Red	262.9	Zhu et al. [247]
72	Magnetic β-cyclodextrin–chitosan/graphene oxide	Methylene Blue	84.3	Fan et al. [248]
73	Bio-silica/chitosan nanocomposite	Acid Red 88	25.8	Soltani et al. [249]

7.2.4.3 Metal Oxide-based Nano-adsorbents

Owing to charges on the surface, metal oxide-based nano-adsorbents are inevitable in the wastewater remediation process. Improved selectivity, larger surface area, high adsorptive capacity, ease of synthesis and low cost makes the use of these materials feasible in the removal of organic pollutants. However, the enhanced energy levels on the surface might lead to the aggregation of nanomaterials, which in turn reduces the active sites on the surface. Hence, the stability of the nano-adsorbents needs to focused without the loss of active adsorption sites or binding pockets. Also, these materials suffer low mechanical strength and are highly unstable during continuous flow studies due to excessive pressure drops. So, numerous research has reported on the use of modified or engineered metal oxide nanoparticles in order to enhance their mechanical properties and stability in varied conditions. For instance, the authors functionalized cerium oxide with amine moieties on the surface which showed improved adsorption capacity and found high stability in varied pH [56, 179].

7.2.4.4 Chitosan-based Nano-adsorbents

Chitosan, the deacetylated product of chitin and a biocompatible polymer, has three functional moieties on the surface: amine, primary hydroxyl, and secondary hydroxyl. These surface groups aid the attraction of ionic or polar species towards their surface. It has many advantages like low cost, ease of synthesis, biodegradable, no toxicity, and richly available. As it is highly fragile and acid sensitive but with poor selectivity, application is limited when compared with the other class of adsorbents. Hence, several approaches were made on the physical or chemical modification on its surface to overcome the former challenges.

Despite these promising features, some nano-adsorbents showed toxicity, depending on surface morphology or with the use of chemical stabilizers or with the process of surface modifications [177, 178]. Hence, it is highly imperative to focus on the synthetic strategy to overcome these toxicity-related issues which will highly limit their application. To address this, several biological routes of synthesis were developed with the use of plants or microbes or biomass, and so on, to produce non-toxic, biodegradable, and highly stable nano-adsorbents. However, research avenues are still open for the generation of nano-adsorbents targeting the removal of more than one contaminant from wastewater and attention also needs to be paid to the comprehensive elucidation on the equilibrium mechanism to improve efficiency significantly.

7.3 THE CHALLENGES OF NANOMATERIALS IN THE WASTEWATER TREATMENT

It is obvious to understand that the currently developed nanomaterials pose several challenges in wastewater treatment [250, 251]. Yet, the field of nanotechnology has opened various options into the ways that distinct nanomaterials have differed in their particle size, shape, and physico-chemical parameters. The elevation from the laboratory to the industrial scale seems to have grown rapidly across the globe [252], as well as the primary concern with nanomaterials which facilitate the processes of photocatalysis, nanofiltration, and nano-adsorption of wastewater treatment [253, 254].

However, systematic and continuous upgradation are required in such techniques to improve process efficiency and to achieve the goal of sustainability. Further, the possible risk of using nano-sized materials in photocatalysis is not yet fully understood by any scientific researchers to date. The fate of these nanomaterials in the atmosphere or in the water system, their transit behavior, and their persistence are the greatest challenges that need to be addressed comprehensively [255]. The United State Environmental Protection Agency (USEPA) has raised significant arguments pertaining to the use of the nanomaterials in the wastewater treatment. Their primary points of consideration were how the removal of contaminants can be viewed and what are the likely intermediates it can produce during such solute transfer process. It is also strongly suggested to study systematically the influence of the nanomaterials on the other contaminants of complex wastewater. In spite of the research which has reported on the minimization of toxicity associated with the nanomaterials, it is still considered a potent threat to society. Also, in case of the nanofiltration, membrane fouling and stability issues strongly inhibit their commercial exploitation. Further, the ability to reuse nano-adsorbents and the search for versatile materials are still the hotspots for the scientific community.

7.4   CONCLUSION AND FUTURE PERCEPTIONS

In the present scenario, it is imperative to search for sustainable water treatment technologies to effectively remove contaminants and exploit the commercial scale to provide ahigh quality of water to every individual in society. Concerning this, the field of nano-science and technology can be viewed as one of the promising solutions in the area of wastewater treatment. Having a wide range of nano-sized materials with their distinct physico-chemical properties, this technology showed effective results in the removal of all possible contaminants in wastewater. These range of materials include: nanophotocatalysts (TiO_2, ZnO, CeO_2, $ZnSCu$, CdS/Eu, CdS/Mn), nano/micromotors, nano-membranes (single- or multi-walled CNTs, electrospun PVDF, PVC, Na-TNB), and nano-adsorbents (carbon-based, metal/metal oxides, silica-based, and chitosan-based). In some cases, these materials were hybridized with biological processes to enhance removal efficiency. All these methods are highly efficient, not energy intensive, ecologically safe, and require least manpower. However, there are no real-time monitoring strategies available as yet to continually track the concentration of these nanomaterials in the treated water system. Also, proposed generic guidelines are required globally in the use of these ultra-fine materials in the treatment of wastewater systems. Furthermore, research avenues challenging material toxicity, membrane stability, fouling, reusability, and so on, are yet to be successfully tackled either with the engineering of existing nanomaterials or with the development of a new generation of nanomaterials.

REFERENCES

1. Grey, D., Garrick, D., Blackmore, D., Kelman, J., Muller, M., Sadoff, C., 2013. Water security in one blue planet: twenty-first century policy challenges for science. *Philos. Trans. Roy. Soc. London A: Math., Phys. Eng. Sci.* 371, 20120406.

2. Adeleye, A.S., Conway, J.R., Garner, K., Huang, Y., Su, Y., Keller, A.A., 2016. Engineered nanomaterials for water treatment and remediation: costs, benefits, and applicability. *Chem. Eng. J.* 286, 640–662.

3. WHO (World Health Organization),2015. Drinking-water: Fact sheet No.391<http://www.who.int/mediacentre/factsheets/fs391/en/>.

4. Schwarzenbach, R.P., Escher, B.I., Fenner, K., Hofstetter, T.B., Johnson, C.A., Von Gunten, U., Wehrli, B., 2006. The challenge of micropollutants in aquatic systems. *Science* 313, 1072–1077

5. Ferroudj, N., Nzimoto, J., Davidson, A., Talbot, D., Briot, E., Dupuis, V., Abramson, S., 2013. Maghemite nanoparticles and maghemite/silica nanocomposite microspheres as magnetic Fenton catalysts for the removal of water pollutants. *Appl. Catal. B: Environ.* 136, 9–18.

6. Qu, X., Brame, J., Li, Q., Alvarez, P.J., 2012. Nanotechnology for a safe and sustainable water supply: enabling integrated watertreatment and reuse. *Acc. Chem. Res.* 46 (3), 834–843.

7. Zelmanov, G., Semiat, R., 2008. Phenol oxidation kinetics in water solution using iron (3)-oxide-based nano-catalysts. *Water Res.* 42, 3848–3856.

8. Catalkaya, E.C., Bali, U., Sengul, F., 2003. Photochemical degradation and mineralization of 4-chlorophenol. *Environ. Sci. Poll. Res. Int.* 10, 113–120.

9. Bali, U., Catalkaya, E.C., Sengul, F., 2003. Photochemical degradation and mineralization of phenol: a comparative study. *J. Environ. Sci. Health* 38, 2259–2275

10. Burkhard, R., Deletic, A., Craig, A., 2000. Techniques for water and wastewater management: a review of techniques and their integration in planning. *Urban Water* 2 (3), 197–221.

11. Karthika, T., Thirunavukkarasu, A., Ramesh, S., 2010, Biosorption of copper from aqueous solutions using Tridax Procumbens. *Recent Res. Sci. Technol.* 2(3): 86–91

12. Thirunavukkarasu, A., Nithya, R., 2019, Response surface optimization of Cu (II) biosorption onto Candida tropicalis immobilized strontium alginate beads by Box-Behnken experimental design. *J. Environ. Biotechnol. Res.* 8 (2), 14–21

13. Zare, K., Najafi, F., Sadegh, H., 2013. Studies of ab initio and Monte Carlo simulation on interaction of fluorouracil anticancer drug with carbon nanotube. *J. Nanostruct. Chem.* 3, 1–8.

14. Sadegh, H., Shahryari-Ghoshekandi, R., Kazemi, M., 2014. Study in synthesis and characterization of carbon nanotubes decorated by magnetic iron oxide nanoparticles. *Int. Nano Lett.* 4, 129–135

15. Gupta, V.K., Tyagi, I., Sadegh, H., Shahryari-Ghoshekand, R., Makhlouf, A.S.H., Maazinejad, B., 2015. Nanoparticles as adsorbent; a positive approach for removal of noxious metal ions: a review. *Sci. Technol. Dev.* 34, 195.

16. Chaturvedi, S., Dave, P.N., Shah, N.K., 2012. Applications of nanocatalyst in new era. *J. Saudi Chem. Soc.* 16, 307–325.

17. Amin, M.T., Alazba, A.A., Manzoor, U., 2014. A review of removal of pollutants from water/wastewater using different types of nanomaterials. *Adv. Mater. Sci. Eng.* 825910

18. Edelstein, A.S., Cammaratra, R.C., 1998. *Nanomaterials: Synthesis, Properties and Applications.* London: CRC Press.

19. Lubick, N., Betts, K., 2008. Silver socks have cloudy lining Court bans widely used flame retardant. *Environ. Sci. Technol.* 42(11), 3910.

20. Brumfiel, G., 2003. Nanotechnology: a little knowledge. *Nature* 424, 246–248.

21. Theron, J., Walker, J.A., Cloete, T.E., 2008. Nanotechnology and water treatment: applications and emerging opportunities. *Crit. Rev. Microbiol.* 34, 43–69.

22. Gupta, V.K., Tyagi, I., Sadegh, H., Shahryari-Ghoshekand, R.,Makhlouf, A.S.H., Maazinejad, B., 2015. Nanoparticles as adsorbent; a positive approach for removal of noxious metal ions: a review. *Sci. Technol. Dev.* 34, 195.

23. Zhang, Q., Xu, R., Xu, P., Chen, R., He, Q., Zhong, J., Gu, X., 2014a. Performance study of ZrO2 ceramic micro-filtration membranes used in pretreatment of DMF wastewater. *Desalination* 346,1–8.

24. Zhang, Y., Yan, L., Xu, W., Guo, X., Cui, L., Gao, L., Wei, Q., Du, B., 2014b. Adsorption of Pb (II) and Hg (II) from aqueous solution using magnetic CoFe2O4 reduced graphene oxide. *J. Mol. Liq.* 191, 177–182.

25. Tang, X., Zhang, Q., Liu, Z., Pan, K., Dong, Y., Li, Y., 2014. Removal of Cu (II) by loofah fibers as a natural and low-cost adsorbent from aqueous solutions. *J. Mol. Liq.*199, 401–407.

26. Shamsizadeh, A.A., Ghaedi, M., Ansari, A., Azizian, S., Purkait, M. K., 2014. Tin oxide nanoparticle loaded on activated carbon as new adsorbent for efficient removal of malachite green-oxalate: nonlinear kinetics and isotherm study. *J. Mol. Liq.* 195, 212–218.

27. Kyzas, G.Z., Matis, K.A., 2015. Nanoadsorbents for pollutants removal: a review. *J. Mol. Liq.* 203, 159–168

28. El Saliby, I.J., Shon, H., Kandasamy, J., Vigneswaran, S., 2008. Nanotechnology for wastewater treatment: in brief. *EOLSS* 7.

29. Dutta, A.K., Maji, S.K., Adhikary, B., 2014. C-Fe2O3 nanoparticles: an easily recoverable effective photo-catalyst for the degradation of rose bengal and methylene blue dyes in the waste-water treatment plant. *Mater. Res. Bull.* 49, 28–34.

30. Kurian, M., Nair, D.S., 2015. Heterogeneous Fenton behavior of nano nickel zinc ferrite catalysts in the degradation of 4-chlorophenol from water under neutral conditions. *J. Water Process Eng.* 8, 37–49.

31. Ma, H., Wang, H., Na, C., 2015. Microwave-assisted optimization of platinum-nickel nanoalloys for catalytic water treatment. *Appl. Catal. B: Environ.* 163, 198–204.

32. Arunachalam, T., Karpagasundaram, M., Rajarathinam, N.,2017. Ultrasound assisted green synthesis of cerium oxide nanoparticles using*Prosopis juliflora* leaf extract and their structural, optical and antibacterial properties. *Materials Sci. Poland* 35 (4), 791–798

33. Rao, L.N., 2014. Nanotechnological methodology for treatment of wastewater. *Int. J. Chem. Tech. Res.* 6 (4), 2529.

34. Lau, W.J., Ismail, A., 2009. Polymeric nanofiltration membranes for textile dye wastewater treatment: preparation, performance evaluation, transport modelling, and fouling control – a review. *Desalination* 245, 321–348.

35. Ouyang, X., Li, W., Xie, S., Zhai, T., Yu, M., Gan, J., Lu, X., 2013. Hierarchical CeO2 nanospheres as highly-efficient adsorbents for dye removal. *New J. Chem.* 37 (3), 585–588.

36. Blanco, J., Torrades, F., De la Varga, M., Garcı´a-Montan˜ o, J., 2012. Fenton and biological-Fenton coupled processes for textile wastewater treatment and reuse. *Desalination* 286, 394–399.

37. Petrinic, I., Andersen, N.P.R., Sostar-Turk, S., Le Marechal, A.M., 2007. The removal of reactive dye printing compounds using nanofiltration. *Dyes Pigment.* 74, 512–518.

38. Hilal, N., Al-Zoubi, H., Darwish, N.A., Mohamma, A.W., Arabi, M. A., 2004. A comprehensive review of nanofiltration membranes: treatment, pretreatment, modelling, and atomic force microscopy. *Desalination* 170 (3), 281–308.

39. Babursah, S., Cakmakci, M., Kinaci, C., 2006. Analysis and monitoring: costing textile effluent recovery and reuse. *Filtr. Separat.* 43 (5), 26–30.

40. Rashidi, H.R., Sulaiman, N.M.N., Hashim, N.A., Hassan, C.R.C., Ramli, M.R., 2015. Synthetic reactive dye wastewater treatment by using nanomembrane filtration. *Desalin. Water Treat.* 55 (1), 86–95.

41. Wu, Y., Pang, H., Liu, Y., Wang, X., Yu, S., Fu, D., Chen, J., Wang, X., 2019. Environmental remediation of heavy metal ions by novel-Nanomaterials: A review. *Environ. Pollut.* 246, 608–620

42. Daer, S., Kharraz, J., Giwa, A., Hasan, S.W., 2015. Recent applications of nanomaterials in water desalination: A critical review and future opportunities. *Desalination* 367, 37–48

43. Yaqoob, A.A., Ibrahim, M.N.M., 2019. A review article of nanoparticles: Synthetic approaches and wastewater treatment methods. *Int. Res. J. Eng. Technol.* 6, 1–7.

44. Tang, W.W., Zeng, G.M., Gong, J.L., 2014. Impact of humic/fulvic acid on the removal of heavy metals from aqueous solutions using nanomaterials: A review. *Sci. Total Environ.* 468, 1014–1027

45. Baruah, S., Khan, M.N., Dutta, J., 2016. Perspectives and applications of nanotechnology in water treatment. *Environ. Chem. Lett.* 14, 1–14.

46. Mir, N.A., Haque, M.M., Khan, A., Umar, K., Muneer, M., Vijayalakshmi, S., 2012. Semiconductor mediated photocatalysed reaction of two selected organic compounds in aqueous suspensions of Titanium dioxide. *J. Adv. Oxid. Technol.* 15, 380–391

47. Umar, K., Water Contamination by Organic-Pollutants: TiO2 Photocatalysis. In *Modern Age Environmental Problem and Remediation*, Oves, M., Khan, M.Z., Ismail, I.M.I., Eds. Basel, Switzerland: Springer Nature, 2018, 95–109.

48. Kalhapure, R.S., Sonawane, S.J., Sikwal, D.R., 2015. Solid lipid nanoparticles of clotrimazole silver complex: An e_cient nano antibacterial against *Staphylococcus aureus* and MRSA. *Colloid Surf. B* 136, 651–658.

49. Fang, X., Li, J., Li, X., Pan, S., Zhang, X., Sun, X., Han, J.S.W., Wang, L., 2017. Internal pore decoration with polydopamine nanoparticle on polymeric ultrafiltration membrane for enhanced heavy metal removal. *Chem. Eng.* 314, 38–49.

50. Sekoai, P.T., Ouma, C.N.M., Du Preez, S.P., Modisha, P., Engelbrecht, N., Bessarabov, D.G., Ghimire, A. Application of nanoparticles in biofuels: An overview, 2019. *Fuel* 237, 380–397.

51. Briggs, A.M., Cross, M.J., Hoy, D.G., Blyth, F.H., Woolf, A.D., March, L., 2016. Musculoskeletal Health Conditions Represent a Global Threat to Healthy Aging: A Report for the 2015World Health OrganizationWorld Report on Ageing and Health. *Gerontologist* 56, 243–255.

52. McNaught, A.D., Wilkinson, A., *IUPAC Gold Book*. Oxford: Blackwell Scientific Publications, 1997.

53. Pirkanniemi, K., Sillanpaa, M., 2002. Heterogeneous water phase catalysis as an environmental application: a review. *Chemosphere* 48:1047–1060

54. Chen, W., Liu, Q., Tian, S., Zhao, X., 2019. Exposed facet dependent stability of ZnO micro/nano crystals as a photocatalyst. *App. Surf. Sci* 470, 807–816

55. Ong, C.B., Ng, L.Y., Mohammad, A.W. A review of ZnO nanoparticles as solar photocatalysts: Synthesis,mechanisms and applications. *Renew. Sustain. Energy Rev.*2018, 81, 536–551.

56. Thirunavukkarasu, K., Muthukumaran, K., Nithya, R., 2018. Adsorption of acid yellow 36 onto green nanoceria and amine functionalized green nanoceria: Comparative studies on kinetics, isotherm, thermodynamics, and diffusion analysis. *J. Taiwan Inst. Chem. Eng.* 93, 211–225

57. Gómez-Pastora, J., Dominguez, S., Bringas, E., Rivero, M.J., Ortiz, I., Dionysiou, D.D., 2017. Review and perspectives on the use of magnetic nanophotocatalysts (MNPCs) in water treatment. *Chem. Eng. J.* 310, 407–427

58. Umar, K., Aris, A., Parveen, T., Jaafar, J., Majid, Z.A., Reddy, A.V.B., Talib, J., 2015. Synthesis, Characterization of Mo and Mn doped Zno and their photocatalytic activity for the decolorization of two di_erent chromophoric dyes. *Appl. Catal A* 505, 507–514.

59. Loeb, S.K., Alvarez, P.J., Brame, J.A., Cates, E.L., Choi, W., Crittenden, J., Dionysiou, D.D., Li, Q., Li-Puma, G., Quan, X. et al., 2019. The technology horizon for photocatalytic water treatment: Sunrise or sunset? *Environ. Sci. Technol.*, 53, 2937–2947.

60. Reddy, A.V.B., Jaafar, J., Majid, Z.A., Aris, A., Umar, K., Talib, J., Madhavi, G., 2015. Relative efficiency comparison of carboxymethyl cellulose (cmc) stabilized fe0 and fe0/ag nanoparticles for rapid degradation of chlorpyrifos in aqueous solutions. *Dig. J. Nanomater. Bios.* 10, 331–340.

61. Samanta, H.S., Das, R., Bhattachajee, C., 2016. Influence of nanoparticles for wastewater treatment-A short review. *Austin Chem. Eng.* 3, 1036–1045.

62. Qu, X., Alvarez, P.J., Li, Q., 2013. Applications of nanotechnology in water and wastewater treatment. *Water Res.* 47, 3931–3946.

63. Sadegh, H., Ali, G.A.M., Gupta, V.K., Makhlouf, A.S.H., Nadagouda, M.N., Sillanpaa, M., Megiel, E., 2017. The role of nanomaterials as effective adsorbents and their applications in wastewater treatment. *J. Nanostructure Chem.* 7, 1–14.

64. Raliya, S.R., Avery, C., Chakrabarti, S., Biswas, P., 2017. Photocatalytic degradation of methyl orange dye by pristine TiO2, ZnO, and graphene oxide nanostructures and their composites under visible light irradiation. *Appl. Nano Sci.* 7, 253–259.

65. Liang, X., Cui, S., Li, H., Abdelhady, A., Wang, H., Zhou, H., 2019. Removal e_ect on stormwater runo_ pollution of porous concrete treated with nanometre titanium dioxide. *Transp. Res. D* 73, 34–45.

66. Bhatia, D., Sharma, N.R., Singh, J., Kanwar, R.S., 2017. Biological methods for textile dye removal from wastewater: A review. *Critcal Rev. Environ. Sci. Technol.*, 47, 1836–1876.

67. Sherman, J., Nanoparticulate Titanium Dioxide Coatings, and Processes for the Production and Use Thereof, U.S. Patent No, 6653356B2, 25 November 2003.

68. Ali, I., Ghamdi, K.A., Wadaani, F.T.A., 2019. Advances in iridium nano catalyst preparation, characterization and applications. *J. Mol. Liq.*, 280, 274–284.

69. Bhanvase, B.A., Shende, T.P., Sonawane, S.H., 2017. A review on grapheme-TiO2 and doped grapheme-TiO2 nanocomposite photocatalyst for water and wastewater treatment. *Environ. Technol. Rev.* 6, 1–14.

70. Sivashankar, R., Thirunavukkarasu, A., Nithya, R., Kanimozhi, J., Sathya, A.B., 2020. Sequestration of methylene blue dye from aqueous solution by magnetic biocomposite: Three level Box–Behnken experimental design optimization and kinetic studies. *Separation Sci. Technol.* 55(10), 1752–1765.

71. Yamakata, A., Junie Jhon, M.V., 2019. Curious behaviors of photogenerated electrons and holes at the defects on anatase, rutile, and brookite TiO2 powders: A review. *J. Photochem. Photobiol C Phtotochem. Rev.* 40, 234–243.

72. Umar, K., Ibrahim, M.N.M., Ahmad, A., Rafatullah, M., 2019. Synthesis of Mn-Doped TiO2 by novel route and photocatalytic mineralization/intermediate studies of organic pollutants. *Res. Chem. Intermediat.*, 45, 2927–2945.

73. Malik, A., Hameed, S., Siddiqui, M.J., Haque, M.M., Umar, K., Khan, A., Muneer, M., 2014. Electrical and optical properties of nickel-and molybdenum-doped titanium dioxide nanoparticle: Improved performance in dye-sensitized solar cells. *J. Mater. Eng. Perform.* 23, 3184–3192.

74. Serrà, A., Grau, S., Gimbert-Suriñach, C., Sort, J., Nogués, J., Vallés, E., 2017. Magnetically-Actuated mesoporous nanowires for enhanced heterogeneous catalysis. *App. Catal B Environ.* 217, 81–91

75. Ahmed, S.N., Haider, W., 2018. Heterogeneous photocatalysis and its potential applications in water and wastewater treatment: A review. *Nanotechnology* 29, 342001

76. Kanmani, S., Sundar, K.P., 2020. Progression of photocatalytic reactors and it's comparison: A Review. *Chem. Eng. Res. Des.*, 154, 135–150.

77. Parrino, F., Loddo, V., Augugliaro, V., Camera-Roda, G., Palmisano, G., Palmisano, L., Yurdakal, S., 2019. Heterogeneous photocatalysis: Guidelines on experimental setup, catalyst characterization, interpretation, and assessment of reactivity. *Catal Rev.* 61, 163–213

78. Chong, M.N., Jin, B., Chow, C.W., Saint, C., 2010. Recent developments in photocatalytic water treatment technology: A review. *Water Res.* 44, 2997–3027

79. Tahir, M.B., Kiran, H., Iqbal, T., 2019. The detoxification of heavy metals from aqueous environment using nano-photocatalysis approach: A review. *Environ. Sci. Pollut. Res.* 26, 10515–10528.

80. Rajabi, H.R., Shahrezaei, F., Farsi, M., 2016. Zinc sulfide quantum dots as powerful and efficient Nanophotocatalysts for the removal of industrial pollutant. *J. Mater. Sci. Mater. Electron.* 27, 9297–9305.

81. Mahmoodi, N.M., Arami, M., 2009. Degradation and toxicity reduction of textile wastewater using immobilized titania nanophotocatalysis. *J. Photoch. Photobio. B* 94, 20–24

82. Van Gerven, T., Mul, G., Moulijn, J., Stankiewicz, A., 2007. A review of intensification of photocatalytic processes. *Chem. Eng. Process. Process Intensif.* 46, 781–789.

83. Lin, W.Y., Wang, Y., Wang, S., Tseng, H.R., 2009. Integrated microfluidic reactors. *Nano Today* 4, 470–481.

84. Wang, N., Zhang, X., Wang, Y., Yu, W., Chan, H.L., 2014. Microfluidic reactors for photocatalytic water purification. *Lab Chip* 14, 1074–1082.

85. Umar, K., Haque, M.M., Mir, N.A., Muneer, M., 2013. Titanium dioxide-Mediated photocatalyzed mineralization of Two Selected organic pollutants in aqueous suspensions. *J. Adv. Oxid. Technol.* 16, 252–260.

86. Umar, K., Dar, A.A., Haque, M.M., Mir, N.A., Muneer, M., 2012. Photocatalysed decolourization of two textile dye derivatives, Martius Yellow and Acid Blue 129 in UV-irradiated aqueous suspensions of Titania. *Desal. Water Treat.* 46, 205–214

87. Karima, A., Ali, T.T., Saeed, A., Alwafi, R., Aly, M., Al-Hazmi, F.E., 2020. Antibacterial and photocatalytic activities of controllable (anatase/rutile) mixed phase TiO_2 nanophotocatalysts synthesized *via* a microwave-assisted sol–gel method. *New J. Chem.*, 44, 562–570

88. Pratap Singh, V., Sharma, M., Vaish, R., 2020. Enhanced dye adsorption and rapid photo catalysis in candle soot coated Bi_2WO_6 ceramics. *Eng. Res. Express* 1, 025056

89. Eskandari, P., Farhadian, M., SolaimanyNazar, A.S., Jeon, B.H., 2019. Adsorption and Photodegradation Efficiency of $TiO_2/Fe_2O_3/PAC$ and $TiO_2/Fe_2O_3/Zeolite$ Nanophotocatalysts for the Removal of Cyanide. *Ind. Eng. Chem. Res.*, 58(5), 2099–2112

90. Khezrianjoo, S., Lee, J., Kim, K.H., Kumar, V., 2019. Eco-Toxicological and Kinetic Evaluation of TiO2 and ZnO Nanophotocatalysts in Degradation of Organic Dye, *Catalysts* 9, 871, doi:10.3390/catal9100871

91. Ali, N., Zada, A., Zahid, M., Ismail, A., Rafiq, M., Riaz, A., Khan, A., 2019. Enhanced photodegradation of methylene blue with alkaline and transition-metal ferrite nanophotocatalysts under direct sun light irradiation, *J. Chinese Chem.Soc.* 66(4), 402–408

92. Lonkar, S.P., Pillai, V.V., Alhassan, S.M., 2018. Facile and scalable production of heterostructured ZnS-ZnO/Graphene nano-photocatalysts for environmental remediation. *Sci. Rep.* 8, 13401

93. Ali, W., Ullah, H., Zada, A., Alamgir, M.K., Ahmad, W.M.M.J., Nadhman, A., 1 July, 2018. Effect of calcination temperature on the photoactivities of ZnO/SnO_2 nanocomposites for the degradation of methyl orange, *Mater. Chem. Phys.* 213, 259–266

94. Malini, B., Gnana Raj, G.A., 2018. Synthesis, characterization and photocatalytic activity of cobalt doped TiO2 nanophotocatalysts for Rose Bengal Dye Degradation under Day Light Illumination, *Chem. Sci. Trans.* 7(4), 687–695

95. Mehrabi, M., Javanbakht, V., 2018. Photocatalytic degradation of cationic and anionic dyes by a novel nanophotocatalyst of $TiO_2/ZnTiO_3/\alpha Fe_2O_3$ by ultraviolet light irradiation. *J. Mater. Sci.* 29, 9908–9919

96. Taghdiri, M., Selective Adsorption and Photocatalytic Degradation of Dyes Using Polyoxometalate Hybrid Supported on Magnetic Activated Carbon Nanoparticles under Sunlight, Visible, and UV Irradiation ID8575096, 2017, https://doi.org/10.1155/2017/8575096

97. Seddigi, Z.S., Gondal, M.A., Baig, U., Ahmed, S.A., Abdulaziz, M.A., Danish, E.Y. et al., 2017. Facile synthesis of light harvesting semiconductor bismuth oxychloride nano photo-catalysts for efficient removal of hazardous organic pollutants. *PLoS ONE* 12(2), e0172218. https://doi.org/10.1371/journal.pone.0172218

98. Ahmadpour, A., Zare, M., Behjoomanesh, M., Avazpour, M., 2015. Photocatalytic decolorization of methyl orange dye using nano- photocatalysts. *Adv. Environ. Technol.* 3, 121–127

99. Jeni, J., Kanmani, S., 2011. Solar nanophotocatalytic decolorisation of reactive dyes using titanium dioxide iran. *J. Environ. Health. Sci. Eng.* 8(1),15–24

100. Sathishkumar, P., Sweena, R., Wu, J.J., Anandan, S., 15 June 2011. Synthesis of CuO-ZnO nanophotocatalyst for visible light assisted degradation of a textile dye in aqueous solution. *Chem. Eng. J.* 171(1),136–140

101. Kansal, S.K., Kaur, N., Singh, S., 2009. Photocatalytic degradation of two commercial reactive dyes in aqueous phase using nanophotocatalysts. *Nanoscale Res. Lett.* 4, 709. https://doi.org/10.1007/s11671-009-9300-3

102. Jurado-Sánchez, B., Wang, J., 2018. Micromotors for environmental applications: A review. *Environ. Sci. Nano* 5, 1530–1544

103. Pacheco, M., López, M.Á., Jurado-Sánchez, B., Escarpa, A., 2019. Self-Propelled micromachines for analytical sensing: A critical review. *Anal. Bioanal. Chem.* 411, 6561–6573.

104. Chi, Q., Wang, Z., Tian, F., You, J.A., Xu, S., 2018. A review of fast bubble-Driven micromotors powered by biocompatible fuel: Low-Concentration fuel, bioactive fluid and enzyme. *Micromachine* 9, 537.

105. García-Torres, J., Serrà, A., Tierno, P., Alcobé, X., Vallés, E., 2017. Magnetic propulsion of recyclable catalytic nanocleaners for pollutant degradation. *ACS Appl. Mater. Interfac.* 9, 23859–23868.

106. Pourrahimi, A.M., Pumera, M., 2018. Multifunctional and self-Propelled spherical Janus nano/micromotors: Recent advances. *Nanoscale* 10, 16398–16415

107. Safdar, M., Simmchen, J., Jänis, J., 2017. Correction: Light-Driven micro- and nanomotors for environmental remediation. *Environ. Sci. Nano* 4, 2235.

108. Fu, P.P., Xia, Q., Hwang, H.M., Ray, P.C., Yu, H., 2014. Mechanisms of nanotoxicity: Generation of reactive oxygen species. *J. Food Drug Anal.* 22, 64–75.

109. Ying, Y.; Pumera, M., 2019. Micro/nanomotors for water purification. *Chem–Eur. J.* 25, 106–121.

110. Xuan, M. J., Lin, X. K., Shao, J. X., Dai, L. R., He, Q., 2015. Self-propelled Janus mesoporous silica nanomotors with sub-100 nm diameters for drug encapsulation and delivery. *ChemPhysChem* 8(16), 147–151.

111. Zhang, Z. J., Zhao, A. D., Wang, F. M., Ren, J. S., Qu, X. G., 2016. Design of a plasmonic micromotor for enhanced photo-remediation of polluted anaerobic stagnant waters. *Chem. Commun.* 52, 5550–5553

112. Mushtaq, A., Asani, M., Hoop, X., Chen, Z., Ahmed, D., Nelson, B. J., Pane, S., 2016. Highly Efficient Coaxial TiO_2-PtPd Tubular Nanomachines for Photocatalytic Water Purification with Multiple Locomotion Strategies. *Adv. Funct. Mater.* 26, 6995–7002.

113. Chung-Seop Lee, J. G., Oh, D. S., Jeon, J. R., Chang, Y. S., 2018. Zerovalent-Iron/ Platinum Janus Micromotors with Spatially Separated Functionalities for Efficient Water Decontamination. *ACS Appl. Nano Mater.*, DOI: 10.1021/acsanm.1027b00223

114. Wang, R. Q., Guo, W. L., Li, X. H., Liu, Z. H., Liu, H., Ding, S. Y., 2017. Highly efficient MOF-based self-propelled micromotors for water purification. *Rsc Adv.* 4(7), 42462–42467.

115. Zhang, Q. L., Dong, R. F., Wu, Y. F., Gao, W., He, Z. H., Ren, B. Y., 2017. Light-Driven Au-WO_3@C Janus Micromotors for Rapid Photodegradation of Dye Pollutants. *ACS Appl. Mater. Interfaces*, 9, 4674–4683

116. Garcia-Torres, J., Serra, A., Tierno, P., Alcobe, X., Valles, E., 2017. Magnetic Propulsion of Recyclable Catalytic Nanocleaners for Pollutant Degradation. *ACS Appl.Mater. Interfaces*, 9, 23859–23868.

117. Orozco, J., Mercante, L. A., Pol, R., Merkoci, A., 2016. Graphene-based Janus micromotors for the dynamic removal of pollutants. *J. Mater.Chem. A*, 4, 3371–3378.

118. Jurado-Sanchez, B., Sattayasamitsathit, S., Gao, W., Santos, L., Fedorak, Y., Singh, V. V., Orozco, J., Galarnyk, M., Wang, J., 2015. Self-propelled activated carbon Janus micromotors for efficient water purification. *Small* 11, 499–506.

119. Singh, V. V., Jurado-Sanchez, B., Sattayasamitsathit, S., Orozco, J., Li, J.X., Galarnyk, M., Fedorak, Y., Wang, J., 2015. Multifunctional Silver-Exchanged Zeolite Micromotors for Catalytic Detoxification of Chemical and Biological Threats. *Adv. Funct. Mater.* 25, 2147–2155

120. Pourrahimi, A.M., Liu, D., Pallon, L. K. H., Andersson, R. L., Abad, A., Lagaron, J. M., Hedenqvist, M. S., Strom, V., Gedde, U. W., Olsson, R. T., 2014. Water-based synthesis and cleaning methods for high purity ZnO nanoparticles – comparing acetate, chloride, sulphate and nitrate zinc salt precursors. *Rsc Adv* 4, 35568–35577.

121. Yu, F., Hu, Q. P., Dong, L., Cui, X., Chen, T. T., Xin, H. B., Liu, M. X., Xue, C. W., Song, X. W., Ai, F. R., Li, T., Wang, X. L., 2017. 3D printed self-driven thumb-sized motors for in-situ underwater pollutant remediation, *Sci. Rep.*, 7, 41169

122. Guix, M., Orozco, J., Garcia, M., Gao, W., Sattayasamitsathit, S., Merkoci, A., Escarpa, A., Wang, J., 2012. Superhydrophobic Alkanethiol-Coated Microsubmarines for Effective Removal of Oil. *ACS Nano* 6, 4445–4451.

123. Mou, F. Z., Pan, D., Chen, C. R., Gao, Y. R., Xu, L. L., Guan, J. G., 2015. Magnetically Modulated Pot-Like $MnFe_2O_4$ Micromotors: Nanoparticle Assembly Fabrication and their Capability for Direct Oil Removal. *Adv. Funct. Mater.* 25, 6173–6181

124. Wang, H., Khezri, B., Pumera, M., 2016. Catalytic DNA-Functionalized Self-Propelled Micromachines for Environmental Remediation. *Chem* 1, 473–481

125. Vilela, D., Parmar, J., Zeng, Y. F., Zhao, Y. L., Sanchez, S., Graphene-Based Microbots for Toxic Heavy Metal Removal and Recovery from Water. 2016. *Nano Lett.* 16, 2860–2866

126. Soler L, Magdanz V, Fomin VM, Sanchez S, Schmidt OG, 2013 Nov 26. Self-propelled micromotors for cleaning polluted water. *ACS Nano* 7(11):9611–9620.

127. Seah, T. H., Zhao, G. J., Pumera, M., 2013. Surfactant capsules propel interfacial oil droplets: An environmental cleanup strategy. *Chempluschem* 78, 395–397.

128. Zhao, G. J., Seah, T. H., Pumera, M., 2011. External-Energy-Independent Polymer Capsule Motors and Their Cooperative Behaviors. *Chemistry: A Eur. J.* 17, 12020–12026.

129. Li, X., Mou, F., Guo, J., Deng, Z., Chen, C., Xu, L., Luo, M., Guan, J., 2018. Hydrophobic Janus Foam Motors: Self-Propulsion and On-The-Fly Oil Absorption. *Micromachines* 9, 23.

130. Orozco, J., Vilela, D., Valdes-Ramirez, G., Fedorak, Y., Escarpa, A., Vazquez-Duhalt, R., Wang, J., 2014. Efficient Biocatalytic Degradation of Pollutants by Enzyme-Releasing Self-Propelled Motors. *Chemistry: A Eur. J.* 20, 2866–2871

131. Simpson, A.E., Kerr, C.A., Buckley, C.A., 1987. The effect of pH on the nanofiltration of the carbonate system in solution. *Desalination* 64, 305–319.

132. Rohe, D.L., Blanton, T.C., Marinas, B.J., 1990. *Drinking water treatment by nanofiltration, National Conference on Environmental Engineering*, Arlington, VA, USA.

133. Macoun, R.G., 1998. The mechanisms of ionic rejection in nanofiltration, Chemical Engineering. Ph.D. Thesis. University of New South Wales, Sydney, Australia.

134. Choi, J.H., Cockko, S., Fukushi, K., Yamamoto, K., 2002. Anovelapplicationofasubmergednanofiltrationmembranebioreactor(NFMBR)forwastewatertreatment. *Desalination* 146(5), 413–420.

135. Liu, S., Wang, Y., Zhou, Z., Hana, W., Li, J., Shen, J., Wang, I., 2017. Improved degradation of the aqueous flutriafol using a nanostructure microporous PbO2 as reactive electrochemical membrane. *Electrochim. Acta* 253, 357–367.

136. Shetti, N.P., Bukkitgar, S.D., Reddy, K.R., Aminabhavi, T.M., 2019. Nanostructured titanium oxide hybrids-Based electrochemical biosensors for healthcare applications. *Colloids Surf. B Biointerfaces* 178, 385–394.

137. Majd, S., 2010. Applications of biological pores in nanomedicine, sensing, and nano-electronics. *Curr.Opin.Biotechnol.* 21, 439–476

138. Dekker, C., 2007. Solid-statenanopores. *Nat. Nanotechnol.* 2, 209–215.

139. Gopalakrishnan, I., Samuel, S.R., Sridharan, K., 2018. Nanomaterials-Based adsorbents for water and waste water treatment. *Emerg. Nanotechnol. Environ. Sustain.* 6, 89–98

140. Ibrahim, R.K., Hayyan, M., Al-Saadi, M.A., Hayyan, A., Ibrahim, S., 2016. Environmental application of nanotechnology; air, soil, and water. *Environ. Sci. Pollut. R* 23, 13754–13788.

141. Mekaru, H., Lu, J., Tamanoi, F., 2015. Development of mesoporous silica-Based nanoparticles with controlled release capability for cancer therapy. *Adv. Drug Deliv. Rev.*, 95, 40–49.

142. Jawed, A., Saxena, V., Pandey, L.M., 2020. Engineered nanomaterials and their surface functionalization for the removal of heavy metals: A review. *J. Wat. Process. Eng.* 33, 101009.

143. Hogen-Esch, T., Pirbazari, M., Ravindran, V., Yurdacan, H.M., Kim,W., High Performance Membranes for Water Reclamation Using Polymeric and Nanomaterials. U.S. Patent No.20160038885A, 29 October 2019.

144. Yin, J., Yang, Y., Hu, Z., Deng, B., 2013. Attachment of silver nanoparticles (AgNPs) onto thin-Film composite (TFC) membranes through covalent bonding to reduce membrane biofouling. *J. Membr. Sci.* 441, 73–82.

145. Zhang, M., Field, R.W., Zhang, K., 2014. Biogenic silver nanocomposite polyethersulfone UF membranes with antifouling properties. *J. Membr. Sci.* 471, 274–284.

146. Waduge, P., Larkin, J., Upmanyu, M., Kar, S., Wanunu, M., 2015. Programmed Synthesis of Freestanding Graphene Nanomembrane Arrays. *Nano Microphone* 11, 597–603.

147. Bassyouni, M., Abdel-Aziz, M.H., Zoromba, M.S., Abdel Hamid, S.M.S., Drioli, E., 2019. A review of polymeric nanocomposite membranes for water purification. *J. Ind. Eng. Chem.* 73, 19–46.

148. Zahid, M., Rashid, A., Akram, S., Rehan, Z.A., Razzaq, W., 2018. A Comprehensive Review on Polymeric Nano-Composite Membranes for Water Treatment. *J. Membr. Sci. Technol.* 8, 179–190

149. Umar, K., Aris, A., Ahmad, H., Parveen, T., Jaafar, J., Majid, Z.A., Reddy, A.V.B., Talib, J., 2016. Synthesis of visible light active doped TiO2 for the degradation of organic pollutants-Methylene blue and glyphosate. *J. Anal. Sci. Technol.* 7, 29–36.

150. Kumar, M., Patil, P., Kim, G.D., 2018. Marine microorganisms for synthesis of metallic nanoparticles and their biomedical applications. *Coll. Surface B* 172, 487–495.

151. Al-Ghouti, M.A., Kaabi, M.A.A., Ashfaq, M.Y., Dana, D.A., 2019. Produced water characteristics, treatment and reuse: A review. *J. Water Proc. Eng.* 28, 222–239.

152. Ahn, Y.Y., Yun, E.T., Seo, J.W., Lee, C., Kim, S.H., Lee, J., 2016. Activation of peroxymonosulfate by surface loaded Nobel metal nanoparticles for oxidative degradation of organic compounds. *Environ. Sci. Technol.* 50,10187–10197.

153. Abdullah, N., Yusof, N., Lau, W.J., Jaafar, J., Ismail, A.F., 2019. Review Recent trends of heavy metal removal from water/wastewater by membrane technologies. *J. Ind. Eng. Chem.* 76, 17–38.

154. Hirata, K., Watanabe, H., Kubo, W., 2019. Nanomembranes as a substrate for ultra-thin lightweight devices. *Thin Solid Film* 676, 8–11

155. Bello, N. T., Polezhaev, P., Vobecká, L., Slouka, Z., 2019. Fouling of a heterogeneous anion-Exchange membrane and single anion-Exchange resin particle by ssdna manifests differently. *J. Membr. Sci.* 572, 619–631

156. Long, Q., Zhang, Z., Qi, G., Zhu, W., Chen, Y., Liu, Z.Q., 2020. Fabrication of Chitosan Nanofiltration Membranes by the Film Casting Strategy for Effective Removal of Dyes/ Salts in Textile Wastewater. *ACS Sustainable Chem. Eng.* 8(6), 2512–2522

157. Lodo, M. J. and Diaz, L. J., 2019. Reusability of Fe-modified MMT nanomembranes and the retrieval of the adsorbed mercury metal. *Earth Environ. Sci.* 345, 012011

158. Hassan, A.R., Rozali, S., Safari, N., Besar, B., 2018, The roles of polyethersulfone and polyethylene glycol additive on nanofiltration of dyes and membrane morphologies. *Environ. Eng. Res.* 23(3): 316–322.

159. Gunawan, F.M., Mangindaan, D., Khoiruddin, K., Wenten, I.G., 2019. Nanofiltration membrane cross-linked by m-phenylenediamine for dye removal from textile wastewater, *Polymers Adv. Technol.* 30(2), 360–367.

160. Buscio, V., García-Jiménez, M., Vilaseca, M., López-Grimau, V., Crespi, M., Gutiérrez-Bouzán, C., 2016. Reuse of Textile Dyeing Effluents Treated with Coupled Nanofiltration and Electrochemical Processes 1, *Materials*, 9, 490; doi:10.3390/ma9060490

161. Shayesteh, M., Samimi, A., Shafiee Afarani, M., Khorram, M., 2016. Synthesis of titania–c-alumina multilayer nanomembranes on performance-improved alumina supports for wastewater treatment. *Desalin Water Treat.* 57(20), 9115–9122.

162. Wen, T., Zhao, Z., Shen, C., Li, J., Tan, X., Zeb, A., Xu, A.W., 2016. Multifunctional flexible free-standing titanate nanobelt membranes as efficient sorbents for the removal of radioactive 90Sr2+ and 137Cs+ ions and oils. *Sci. Rep.*6.

163. Askari, N., Farhadian, M., Razmjou, A., Hashtroodi, H., 2016. Nanofiltration performance in the removal of dye from binary mixtures containing anthraquinone dyes. *Desalin. Water Treatment* 57(39), 18194–18201, 10.1080/19443994.2015.1090917

164. Faccini, M., Borja, G., Boerrigter, M., Martin, D.M., Crespiera, S. M., Vazquez-Campos, S., Amantia, D., 2015. Electrospun carbon nanofiber membranes for filtration of nanoparticles from water. *J. Nanomater.* 2, 1–9.

165. Askari, N., Farhadian, M., Razmjou, A., 2015. Decolorization of ionic dyes from synthesized textile wastewater by nanofiltration using response surface methodology. *Adv. Environ. Technol.* 2, 85–92

166. Zhang, Q., Xu, R., Xu, P., Chen, R., He, Q., Zhong, J., Gu, X., 2014. Performance study of ZrO2 ceramic micro-filtration membranes used in pretreatment of DMF wastewater. *Desalination* 346, 1–8.

167. Ghani, M., Gharehaghaji, A.A., Arami, M., Takhtkuse, N., Rezaei, B., 2014. Fabrication of Electrospun Polyamide-6/Chitosan Nanofibrous Membrane toward Anionic Dyes Removal. *J. Nanotechnol.* 2014, Article ID 278418

168. Kootenaei, F.G., Rad, H.A., 2013. Treatment of hospital wastewater by novel nano-filtration membrane bioreactor (NF-MBR). *Iranica J. Energy Environ.* 4(1), 60–67.

169. Maphutha, S., Moothi, K., Meyyappan, M., Iyuke, S.E., 2013. A carbon nanotube-infused polysulfone membrane with polyvinyl alcohol layer for treating oil-containing waste water. *Sci. Rep.* 3, 1509.

170. Asmatulu, R., Muppalla, H., Veisi, Z., Khan, W.S., Asaduzzaman, A., Nuraje, N., 2013. Study of hydrophilic electrospun nanofiber membranes for filtration of micro and nano-size suspended particles. *Membranes* 3(4), 375–388.

171. Hajdu, I., Bodnar, M., Csikos, Z., Wei, S., Daroczi, L., Kovacs, B., Borbely, J., 2012. Combined nano-membrane technology for removal of lead ions. *J. Membr. Sci.* 409, 44–53.

172. Yu, L., Ruan, S., Xu, X., Zou, R., Hu, J., 2017. Review One-Dimensional nanomaterial-Assembled macroscopic membranes for water treatment. *Nano Today* 17, 79–95.

173. Fuwad, A., Ryu, H., Malmstadt, N., Kim, S.M., Jeon, T.J., 2019. Biomimetic membranes as potential tools for water purification: Preceding and future avenues. *Desalination* 458, 97–115

174. Giwa, A., Hasan, S.W., Yousaf, A., Chakraborty, S., Johnson, D.J., Hilal, N., 2017. Biomimetic membranes: A critical review of recent progress. *Desalination* 420, 403–424.

175. Diallo, M.S., Water Treatment by Dendrimer-Enhanced Filtration. U.S. Patent No. 2009/0223896, 10 September 2009.

176. Nithya, R., Sivasankari, C., Thirunavukkarasu, A, Selvasembian, R., Novel adsorbent prepared from bio-hydrometallurgical leachate from waste printed circuit board used for the removal of methylene blue from aqueous solution, *Microchem. J.* 142, 321–328

177. Manikam, M.K., Halim, A.A., Hanafiah, M.M., Krishnamoorthy, R.R., 2019. Removal of ammonia nitrogen, nitrate, phosphorus and COD from sewage wastewater using palm oil boiler ash composite adsorbent, *Desal. Water Treat.* 149, 23–30.

178. Charee, S.W., Aravinthan, V., Erdei, L., Raj, W.S., 2017. Use of macadamia nut shell residues as magnetic nanosorbents. *Int. Biodeter. Biodegr.* 124, 276–287.

179. Rajarathinam, N., Arunachalam, T., Raja, S., Selvasembian, R., Fenalan Yellow G adsorption using surface-functionalized green nanoceria: An insight into mechanism and statistical modelling. *Environ. Res.* 181, 108920

180. Sadegh, H., Ali, G. A. M., Agarwal, S., Gupta, V. K., 2019. Surface Modification of MWCNTs with Carboxylic-to-Amine and Their Superb Adsorption Performance, *Int. J. Environ. Res.* 13(3), 523–531

181. Ahamad, T., Naushad, M., Eldesoky, G. E., Al-Saeedi, S. I., Nafady, A., Al-Kadhi, N. S., Khan, A., 2019. Effective and fast adsorptive removal of toxic cationic dye (MB) from aqueous medium using amino-functionalized magnetic multiwall carbon nanotubes. *J. Mol. Liq.* 282, 154–161.

182. Arabi, S. M. S., Lalehloo, R. S., Olyai, M. R. T. B., Ali, G. A., Sadegh, H., 2019. Removal of congo red azo dye from aqueous solution by ZnO nanoparticles loaded on multiwall carbon nanotubes, *Physica E Low Dimens. Syst. Nanostruct.* 106 150–155.

183. Ansari, M. O., Kumar, R., Ansari, S. A., Ansari, S. P., Barakat, M. A., Alshahrie, A., Cho, M. H., 2017. Anion selective *p*TSA doped polyaniline@graphene oxide-multi-walled carbon nanotube composite for Cr(VI) and Congo red adsorption. *J. Colloid Interface Sci.* 496, 407

184. Li, X., Lu, H., Zhang, Y., He, F., 2017. Efficient removal of organic pollutants from aqueous media using newly synthesized polypyrrole/CNTs-CoFe$_2$O$_4$ magnetic nanocomposites, *Chem. Eng. J.* 316, 893–902.

185. Kang, D., Yu, X., Ge, M., Xiao, F., Xu, H., 2017. Novel Al-doped carbon nanotubes with adsorption and coagulation promotion for organic pollutant removal. *J. Environ. Sci.* 54, 1.

186. Kamil, A.M., Mohammed, H. T., Alkaim, A. F., Hussein, F. H., 2016. Adsorption of Congo red on multiwall carbon nanotubes: Effect of operational parameters, *J. Chem. Pharm. Sci.* 9, 1128–1133

187. Chen, J., Zhang, L., Huang, T., Li, W., Wang, Y., Wang, Z., 2016. Decolorization of azo dye by peroxymonosulfate activated by carbon nanotube: Radical versus non-radical mechanism. *J. Hazard. Mater.* 320, 571

188. Long, Y. L., Yu, J. G., Jiao, F. P., Yang, W., 2016. Preparation and characterization of MWCNTs/LDHs nanohybrids for removal of Congo red from aqueous solution. *J. Trans. Nonferrous Met. Soc.China* 26, 2701–2710.

189. Wu, X., Liu, C., Qi, H., Zhang, X., Dai, J., Zhang, Q., Peng, X., 2016. Synthesis and adsorption properties of halloysite/carbon nanocomposites and halloysite-derived carbon nanotubes. *Appl. Clay Sci.* 119, 284

190. Seyahmazegi, E. N., Mohammad-Rezaei, R., Razmi, H., 2016. Multiwall carbon nanotubes decorated on calcined eggshell waste as a novel nano-sorbent: Application for anionic dye Congo red removal *Chem. Eng. Res. Des.* 109, 824–834.

191. Duman, O., Tunc, S., Polat, T. G., Bozoğlan, B. K., 2016. Synthesis of magnetic oxidized multiwalled carbon nanotube-κ-carrageenan-Fe3O4 nanocomposite adsorbent and its application in cationic Methylene Blue dye adsorption. *Carbohydr. Polym.* 147, 79–88.

192. Sheibani, M., Ghaedi, M., Marahel, F., Ansari, A., 2015. Congo red removal using oxidized multiwalled carbon nanotubes: kinetic and isotherm study. *Desalin. Water Treat.* 53, 844–852.

193. Dehghani, M. H., Niasar, Z. S., Mehrnia, M. R., Shayeghi, M., Al-Ghouti, M. A., Heibati, B., Yetilmezsoy, K., 2017. Optimizing the removal of organophosphorus pesticide malathion from water using multi-walled carbon nanotubes. *Chem. Eng. J.* 310, 22.

194. Wang, P., Cao, M., Wang, C., Ao, Y., Hou, J., Qian, J., 2014. Kinetics and thermodynamics of adsorption of methylene blue by a magnetic graphene-carbon nanotube composite *Appl. Surf. Sci.* 290, 116–124.

195. Yao, Y. X., Li, H. B., Liu, J. Y., Tan, X. L., Yu, J. G., Peng, Z. G., 2014. Removal and Adsorption of *p*-Nitrophenol from Aqueous Solutions Using Carbon Nanotubes and Their Composites. *J. Nanomater.* 2014, 84

196. Inyang, M., Gao, B., Zimmerman, A., Zhang, M., Chen, H., 2014. Synthesis, characterization, and dye sorption ability of carbon nanotube–biochar nanocomposites. *Chem. Eng. J.* 236, 39–46.

197. Gao, H., Zhao, S., Cheng, X., Wang, X., Zheng, L., 2013. Removal of anionic azo dyes from aqueous solution using magnetic polymer multi-wall carbon nanotube nanocomposite as adsorbent. *Chem. Eng. J.* 223, 84–90.

198. Bazrafshan, E., Mostafapour, F. K., Hosseini, A. R., Khorshid, A. R., Mahvi, A. H., 2013. Decolorisation of Reactive Red 120 Dye by Using Single-Walled Carbon Nanotubes in Aqueous Solutions. *J. Chem.*, 8.

199. Moradi, O., 2013. Adsorption Behavior of Basic Red 46 by Single-Walled Carbon Nanotubes Surfaces. *Fullerenes Nanotubes Carbon Nanostruct.* 21, 286

200. Szlachta, M., Wojtowicz, P., 2013. Adsorption of methylene blue and Congo red from aqueous solution by activated carbon and carbon nanotubes. *Water Sci. Technol.* 68, 2240–2248.

201. Zhao, D., Zhang, W., Chen, C., Wang, X., 2013. Adsorption of Methyl Orange Dye Onto Multiwalled Carbon Nanotubes. *Procedia Environ. Sci.* 18, 890.

202. Prola, L. D., Machado, F. M., Bergmann, C. P., de Souza, F. E., Gally, C. R., Lima, E. C., Calvete, T., 2013. Adsorption of Direct Blue 53 dye from aqueous solutions by multi-walled carbon nanotubes and activated carbon. *J. Environ. Manage.* 130, 166–175

203. Chen, J., Sheng, Y., Song, Y., Chang, M., Zhang, X., Cui, L., Zou, H., 2018. Multimorphology Mesoporous Silica Nanoparticles for Dye Adsorption and Multicolor Luminescence Applications. *Sustain. Chem. Eng.* 6, 3533

204. Suzimara, R., Jonnatan, J. S., Paola, C., Denise, A. F., 2018. Highly Pure Silica Nanoparticles with High Adsorption Capacity Obtained from Sugarcane Waste Ash. *ACS Omega* 3, 2618–2627

205. Qin, P., Yang, Y., Zhang, X., Niu, J., Yang, H., Tian, S., Lu, M., 2017. Highly Efficient, Rapid, and Simultaneous Removal of Cationic Dyes from Aqueous Solution Using Monodispersed Mesoporous Silica Nanoparticles as the Adsorbent. *Nanomaterials*, 8, 4.

206. Shinde, P., Gupta, S. S., Singh, B., Polshettiwar, V., Prasad, B. L., 2017. Amphifunctional mesoporous silica nanoparticles for dye separation. *J. Mater. Chem. A.* 5, 14914.

207. Tavassolia, N., Ansaria, R., Mosayebzadeh, Z., 2017. Synthesis and Application of Iron Oxide/Silica Gel Nanocomposite for Removal of Sulfur Dyes from Aqueous Solutions. *Arch. Hyg. Sci.*, 6, 214–220.

208. Salimi, F., Tahmasobi, K., Karami, C., Jahangiri, A., 2017. Preparation of Modified nano-SiO2 by Bismuth and Iron as a novel Remover of Methylene Blue from Water Solution. *J. Mex. Chem. Soc.* 61, 250–259

209. Zhang, L., Zhang, G., Wang, S., Peng, J., Cui, W., 2016. Cation-functionalized silica nanoparticle as an adsorbent to selectively adsorb anionic dye from aqueous solutions. *Environ. Prog. Sustain. Energy*, 35, 1070.

210. Mahmoodi, N. M., Maghsoodi, A., 2015. Kinetics and isotherm of cationic dye removal from multicomponent system using the synthesized silica nanoparticle. *Desalin. Water Treat.* 54, 562–571.

211. Li, Y., Zhou, Y., Nie, W., Song, L., Chen, P., 2015. Highly efficient methylene blue dyes removal from aqueous systems by chitosan coated magnetic mesoporous silica nanoparticles. *J. Porous Mater.* 22, 1383

212. Araghi, S. H., Entezari, M. H., 2015. Amino-functionalized silica magnetite nanoparticles for the simultaneous removal of pollutants from aqueous solution. *Appl. Surf. Sci.* 333, 68

213. Mahmoodi, N. M., Maghsoudi, A., Najafi, F., Jalili, M., Kharrati, H., 2014. Primary–secondary amino silica nanoparticle: synthesis and dye removal from binary system. *Desalin. Water Treat.* 52, 7784–7796

214. Zarezadeh-Mehrizi, M., Badiei, A., 2014. Highly efficient removal of basic blue 41 with nanoporous silica. *Water Resour. Ind.* 5, 49.

215. Chawla, S., Uppal, H., Yadav, M., Bahadur, N., Singh, N., 2017. Zinc peroxide nanomaterial as an adsorbent for removal of Congo red dye from waste water. *Ecotoxicol. Environ. Saf.* 135, 68–74.

216. Heidarizad, M., Şengor, S. S., 2016. Synthesis of graphene oxide / magnesium oxide nanocomposites with high-rate adsorption of methylene blue. *J. Mol. Liq.* 224, 607–617

217. Atta, A. M., Al-Hodan, H. A., Al-Hussain, S. A., Ezzat, A. O., Tawfik, A. M., El-Dosary, Y. A. 2016. Preparation of magnetite and manganese oxide ionic polymer nanocomposite for adsorption of a textile dye in aqueous solutions. *Dig. J. Nanomater. Bios.* 11(3), 909–919.

218. Salem, A. N. M., Ahmed, M. A., El-Shahat, M. F., 2016. Selective adsorption of amaranth dye on Fe3O4/MgO nanoparticles, *J. Mol. Liq.*, 219, 780–788.

219. Istratie, R., Stoia, M., Pǎcurariu, C., Locovei, C., 2016. Single and simultaneous adsorption of methyl orange and phenol onto magnetic iron oxide/carbon nanocomposites. *Arabian J. Chem.*.

220. Chang, J., Ma, J., Ma, Q., Zhang, D., Qiao, N., Hu, M., Ma, H., 2016. Adsorption of methylene blue onto Fe3O4/activated montmorillonite nanocomposite. *Appl. Clay Sci.* 119, 132–140.

221. Wang, P., Wang, X., Yu, S., Zou, Y., Wang, J., Chen, Z., Wang, X., 2016. Silica coated Fe_3O_4 magnetic nanospheres for high removal of organic pollutants from wastewater *Chem. Eng. J.* 306, 280–288

222. Satheesh, R., Vignesh, K., Rajarajan, M., Suganthi, A., Sreekantan, S., Kang, M., Kwak, B. S., 2016. Removal of congo red from water using quercetin modified α-Fe2O3 nanoparticles as effective nanoadsorbent *Mater. Chem. Phys.*.

223. Alqadami, A., Naushad, M., Abdalla, M. A., Khan, M. R., Alothman, Z. A., 2016. Adsorptive Removal of Toxic Dye Using Fe_3O_4–TSC Nanocomposite: Equilibrium, Kinetic, and Thermodynamic Studies. *J. Chem. Eng. Data.* 61(11), 3806–3813

224. Abdel Ghafar, H. H., Ali, G. A., Fouad, O. A., Makhlouf, S. A., 2015. Enhancement of adsorption efficiency of methylene blue on Co3O4/SiO2 nanocomposite. *Desalin. Water Treat.* 53(11), 2980–2989

225. Nekouei, F., Nekouei, S., Tyagi, I., Gupta, V. K., 2015. Kinetic, thermodynamic and isotherm studies for acid blue 129 removal from liquids using copper oxide nanoparticle-modified activated carbon as a novel adsorbent. *J. Mol. Liq.* 201, 124–133.

226. Hao, T., Yang, C., Rao, X., Wang, J., Niu, C., Su, X., 2014. Facile additive-free synthesis of iron oxide nanoparticles for efficient adsorptive removal of Congo red and Cr (VI) *Appl. Surf. Sci.* 292, 174–180

227. Wu, R., Liu, J. H., Zhao, L., Zhang, X., Xie, J., Yu, B., Liu, Y., 2014. Hydrothermal preparation of magnetic Fe_3O_4@C nanoparticles for dye adsorption. *J. Environ. Chem. Eng.* 2(2), 907–913

228. Zhang, Y. R., Shen, S. L., Wang, S. Q., Huang, J., Su, P., Wang, Q. R., Zhao, B. X., 2014. A dual function magnetic nanomaterial modified with lysine for removal of organic dyes from water solution. *Chem. Eng. J.* 239, 250–256

229. Shariati-Rad, M., Irandoust, M., Amri, S., Feyzi, M., Jafari, F., 2014. Magnetic solid phase adsorption, preconcentration and determination of methyl orange in water samples using silica coated magnetic nanoparticles and central composite design. *Int. Nano Lett* 4, 91–101

230. Mohammadi, H., Daemi, M., 2014. Fast removal of malachite green dye using novel superparamagnetic sodium alginate-coated Fe3O4 nanoparticles. *Int. J. Biol. Macromol.* 69, 447–455.

231. Che, H. X., Yeap, S. P., Ahmad, A. L., Lim, J., 2014. Layer-by-layer assembly of iron oxide magnetic nanoparticles decorated silica colloid for water remediation. *Chem. Eng. J.* 243, 68–78.

232. Wu, J., Wang, J., Li, H., Du, Y., Huang, K., Liu, B., 2013. Designed synthesis of hematite-based nanosorbents for dye removal. *J. Mater. Chem. A* 1(34), 9837–9847

233. Kumar, K. Y., Muralidhara, H. B., Nayaka, Y. A., Balasubramanyam, J., Hanumanthappa, H., 2013. Low-cost synthesis of metal oxide nanoparticles and their application in adsorption of commercial dye and heavy metal ion in aqueous solution. *Powder Technol.* 246, 125–136

234. Mahapatra, B. G., Mishra, G. H., 2013. Adsorptive removal of Congo red dye from wastewater by mixed iron oxide–alumina nanocomposites *Ceram. Int.* 39(5), 5443–5451.

235. Li, D. P., Zhang, Y. R., Zhao, X. X., Zhao, B. X., 2013. Magnetic nanoparticles coated by aminoguanidine for selective adsorption of acid dyes from aqueous solution. *Chem. Eng. J.* 232, 425–433

236. Zhang, X., Zhang, P., Wu, Z., Zhang, L., Zeng, G., Zhou, C., 2013. Adsorption of methylene blue onto humic acid-coated Fe_3O_4 nanoparticles. *Colloids Surf: A Physicochem. Eng. Asp.* 435, 85–90.

237. Rahmi, Ismaturrahmi, Mustafa, 2019. Methylene blue removal from water using H_2SO_4 crosslinked magnetic chitosan nanocomposite beads. *Microchem. J.* 144, 397–402

238. Massoudinejad, M., Rasoulzadeh, H., Ghaderpoori, M., 2019. Magnetic chitosan nanocomposite: Fabrication, properties, and optimization for adsorptive removal of crystal violet from aqueous solutions. *Carbohydr. Polym.* 206, 844–853

239. Rezaei, H., Razavi, A., Shahbazi, A., 2017. Removal of Congo red from aqueous solutions using nano-chitosan. *Env. Resour. Res.* 5, 25.

240. Wang, Y., Zhang, Y., Hou, C., Liu, M., 2016. Mussel-inspired synthesis of magnetic polydopamine–chitosan nanoparticles as biosorbent for dyes and metals removal. *J. Taiwan. Inst. Chem. Eng.* 61, 292.

241. Tanhaei, A., Ayati, M., Lahtinen, M. S., 2015. Preparation and characterization of a novel chitosan/Al2O3/magnetite nanoparticles composite adsorbent for kinetic, thermodynamic and isotherm studies of Methyl Orange adsorption. *Chem. Eng. J.* 259, 1.

242. Kadam, A., Lee, D. S., 2015. Glutaraldehyde cross-linked magnetic chitosan nanocomposites: Reduction precipitation synthesis, characterization, and application for removal of hazardous textile dyes. *Bioresour. Technol.* 193, 563.

243. Mahmoodian, H., Moradi, O., Shariatzadeha, B., Salehf, T. A., Tyagi, I., Maity, A., 2015. Enhanced removal of methyl orange from aqueous solutions by poly HEMA–chitosan-MWCNT nano-composite. *J. Mol. Liq.* 202, 189

244. Zhou, Z., Lin, S., Yue, T., Lee, T. C., 2014. Adsorption of food dyes from aqueous solution by glutaraldehyde cross-linked magnetic chitosan nanoparticles. *J. Food Eng.* 126, 133–141.

245. Cao, L., Xiao, C., Chen, X., Shi, Q., Cao, L.G., 2014. In situ preparation of magnetic Fe_3O_4/chitosan nanoparticles via a novel reduction–precipitation method and their application in adsorption of reactive azo dye. *Powder Technol.* 260, 90–97.

246. Sarkheil, H., Noormohammadi, F., Rezaei, A. R., Borujeni, M. K., *Dye Pollution Removal from Mining and Industrial Wastewaters using Chitson Nanoparticles*, *International Conference on Agriculture, Environment and Biological Sciences (ICFAE'14)*, Antalya (Turkey), 2014

247. Zhu, H., Fu, Y., Jiang, R., Yao, J., Liu, L., Chen, Y., Zeng, G., 2013. Preparation, characterization and adsorption properties of chitosan modified magnetic graphitized multi-walled carbon nanotubes for highly effective removal of a carcinogenic dye from aqueous solution. *Appl. Surf. Sci.* 285, 865–873.

248. Fan, L., Luo, C., Sun, M., Qiu, H., Li, X., 2013. Synthesis of magnetic β-cyclodextrin-chitosan/graphene oxide as nanoadsorbent and its application in dye adsorption and removal. *Colloids Surf. B* 103, 601–607.

249. Soltani, R. D. C., Khataee, A. R., Safari, M., Joo, S. W., 2013. Preparation of bio-silica/chitosan nanocomposite for adsorption of a textile dye in aqueous solutions. *Int. Biodeterior. Biodegrad.* 85, 383–391

250. Dimapilis, E.A.S., Hsu, C.S., Mendoza, R.M.O., Lu, M.C., 2018. Zinc oxide nanoparticles for water disinfection. *Sustain. Environ.* 28, 47–56.
251. Sultana, S.R., Khan, M.Z., Umar, K., Ahmed, A.S., Shahadat, M., 2015. SnO2-SrO based nanocomposites and their photocatalytic activity for the treatment of organic pollutants. *J. Mol. Struct.* 1098, 393–399
252. Faisal, M., Tariq, M.A., Khan, A., Umar, K., Muneer, M., 2011. Photochemical reactions of 2, 4-dichloroaniline and 4-nitroanisole in aqueous suspension of titanium dioxide. *Sci. Adv. Mater.* 3, 269–275.
253. He, D., Sun, Y., Xin, L., Feng, J., 2014. Aqueous tetracycline degradation by non-Thermal plasma combined with nano-TiO2. *Chem. Eng. J.* 258, 18–25.
254. Mir, N.A., Khan, A., Umar, K., Muneer, M., 2013. Photocatalytic Study of a Xanthene Dye Derivative, Phloxine B in Aqueous Suspension of TiO2: Adsorption Isotherm and Decolourization Kinetics. *Energy Environ. Focus* 2, 208–216.
255. Bushra, R., Shahadat, M., Ahmad, A., Nabi, S.A., Umar, K., Muneer, M., Raeissia, A.S., Owais, M., 2014. Synthesis, characterization, antimicrobial activity and applications of composite adsorbent for the analysis of organic and inorganic pollutants. *J. Hazard. Mater.* 264, 481–489.

8 Application of Biogenic Nanoparticles for a Clean Environment

Jishma Panichikkal and
Radhakrishnan Edayileveetil Krishnankutty
Mahatma Gandhi University, India

CONTENTS

8.1 INTRODUCTION

The most discussing problems facing by the world today due to the urbanization and industrialization are environmental degradation and contamination. Contamination from both inorganic and organic contaminants has an adverse effect on the quality of air, drinking water, aquatic ecosystems, food, soil, agricultural systems, urban systems, and natural habitats (Thompson and Darwish, 2019). A contaminant can be any chemical compound (heavy metals, radionuclides, organophosphorus compounds, gases) or geochemical substance (dust and sediment), biological organism or product, or physical substance (heat, radiation, sound wave) that is liberated intentionally or accidentally by humans into the environment with damaging effects (Rai, 2016). Chemical pollutants can persist in the environment without degradation for long periods and cause detrimental toxicity. Toxic chemicals discharged from industries and vehicles, agricultural runoff and wastewater discharge and algal and bacterial blooms pollute the air, soil, and water bodies and are life-threatening to all living forms (Enault et al., 2015).

Pollutants accumulated in water bodies decrease oxygen levels, which can kill fish and alter food chain composition, reduce species biodiversity, and foster invasion by new thermophilic species (Laws, 2017). Environmental pollutants have various adverse health effects, some of the most important being perinatal disorders, infant mortality, respiratory disorders, allergy, malignancies, cardiovascular disorders, increase in stress oxidatives, endothelial dysfunction, mental disorders, and various other harmful effects (Kelishadi, 2012). Hence it is important to remove such toxic contaminants from the environment and conserve a healthy ecosystem.

And the bioremediation for these problems is also getting noticed as the existing methods have limitations. In this chapter, nanotechnological applications for the detoxification and removal of toxic contaminants from the environment are highlighted.

8.2 POLLUTANTS AND ENVIRONMENTAL CONTAMINATION: A GLOBAL ISSUE

Environmental contamination is one of the major global issues affecting biota and the ecosystem adversely due to the accumulation of toxic chemicals and gases in air, soil, and water (Hader, 2013). Pollution is a serious problem, as it has harmful effect on the essentials that are necessary for life to exist on earth, like air, soil, and water and in contaminated environment, plants, animals, humans, and microbiota could not stay alive. There are various types of pollution on the earth such as air pollution, water pollution, and soil pollution (Figure 8.1).

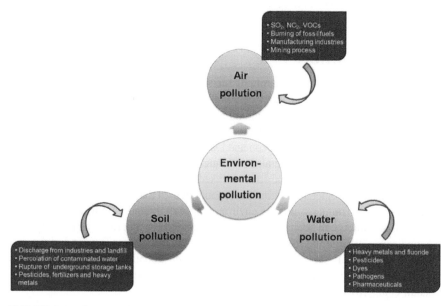

FIGURE 8.1   Sources of environmental contamination.

8.2.1 Air Pollution

Air is contaminated by high concentration of toxic substances released due to human activities and from industries for a prolonged time, resulting in unfavorable effects on all living forms and as well as on ecosystems. Major air pollutants include particulate matters (PMs), ground-level ozone (O_3), sulfur dioxide (SO_2), nitrogen dioxides (NO_2), and volatile organic compounds (VOCs) (Wei et al., 2017). A physical, biological, or chemical variation to the air in the atmosphere can cause air contamination. It happens when toxic gases, dust, or smoke come into the atmosphere and makes it unfavorable for plants, animals, and humans to survive. Air pollution can further be classified into visible air pollution and invisible air pollution (Florentina IIo, 2011).

Primary and secondary pollutants are the major cause of air pollution. The sulfur dioxide emitted from factories is a primary pollutant which is a direct result of the process. Secondary pollutants are the compounds which are caused by the reactions and combination of primary pollutants like smog (Isaksson, 2010). Other important causes of air pollution are burning of fossil fuels like coal, petroleum, and other factory combustibles; use of insecticides, pesticides, and fertilizers in agricultural fields release harmful chemicals into the air and can also cause water pollution. Manufacturing industries liberate a large amount of carbon monoxide, hydrocarbons, organic compounds, and other toxic chemicals into the air and thereby reduce the quality of air. Petroleum refineries also release hydrocarbons and various other chemicals that pollute the air and also cause land pollution. In addition to this, during the mining process dust and chemicals are released in the air, causing air pollution, which is responsible for worsening health conditions of workers and nearby inhabitants; household cleaning products and painting supplies emit toxic chemicals in the air, also causing pollution (Wake, 2005).

8.2.2 Soil Pollution

Soil pollution is due to the presence and persistence of toxic compounds, chemicals, salts, radioactive materials, or disease-causing agents, which have adverse effects on plant growth and animal health. Discharge from landfills and industries into the soil, oozing out of contaminated water into the soil, cracks in underground storage tanks, excess application of pesticides, herbicides or fertilizer, and solid waste leakage lead to the contamination of soil (Okrent, 1999). Further examples include heavy metals such as copper, mercury, cadmium, lead, nickel, arsenic. Through industrial waste water, sewage and mine washing and fungicides containing copper and mercury get accumulated in the soil and smoke from automobiles containing lead is adsorbed by soil particles which increase the soil pollution, are toxic to plant vegetation and animals (Aqeel et al., 2014).

The presence of heavy metals in soil is a serious issue due to its entry into food chains, thus destroying the entire ecosystem. As much as organic pollutants can be biodegradable, their biodegradation rate is reduced by the presence of heavy metals in the environment and enhances the effect of environmental pollution. There are various ways through which heavy metals risk to humans, animals, plants, and ecosystems as a whole, like direct ingestion, absorption by plants, food chains, consumption of contaminated water, and alteration of soil pH, porosity, color, and its natural chemistry, which in turn impact on soil quality (Musilova et al., 2016).

8.2.3 WATER POLLUTION

Water pollution occurs when harmful substances such as chemicals or microorganisms contaminate a stream, river, lake, ocean, aquifer, or other water body, degrading water quality and rendering it toxic to humans or the environment. Rainwater from roofs and roads in urban areas and pesticides and fertilizers applied in agricultural fields are the major factors for water pollution. Groundwater gets polluted when contaminants from pesticides and fertilizers to waste are leached from landfills (Srivastava et al., 2018). Half of our rivers and streams, and more than one-third of our lakes are polluted and unfit for swimming, fishing, and drinking. Nutrient pollution, which includes nitrates and phosphates, is the leading type of contamination in these freshwater sources. While plants and animals need these nutrients to grow, they have become a major pollutant due to farm waste and fertilizer runoff. Municipal and industrial waste discharges also lead to massive water contamination. Contamination originates from a single source is called point source pollution, which is due to the legal or illegal wastewater discharge by a manufacturer, oil refinery, or wastewater treatment facility and contamination from leaking septic systems, chemical and oil spills, and illegal dumping (Oplatowska et al., 2011).

When water pollution causes an algal bloom in a lake or marine environment, the proliferation of newly introduced nutrients stimulates plant and algae growth, which in turn reduces oxygen levels in the water, which cause eutrophication, suffocates plants and animals and can cause "dead zones," where water is essentially devoid of life. In certain cases, these harmful algal blooms can also produce neurotoxins that affect wildlife, from whales to sea turtles. Also, toxic heavy metals released into the environment via different routes such as industries, mining activities, and agricultural activities make the water contaminated (Cao et al., 2017). Bioavailable metals present in the soil may be absorbed by plants, resulting in serious plant metabolism dysfunction. High heavy metal ion concentrations are also known to damage the cell membrane, and affect enzymes involved in chlorophyll production, thus reducing photosynthetic rates as well as affecting plant reproduction via a decrease in pollen and seed viability. Humans and animals can be exposed to heavy metal toxicity through the food web, direct consumption of water containing metal, or via inhalation which causes bioaccumulation in vegetables and enters man and animal through the food chain and leads to serious health issues and their existence (Xun et al., 2017).

8.3 NANOTECHNOLOGY: NEW EMERGING TECHNOLOGY

Nanotechnology is the novel, promising and exciting branch of science which deals with particles with nanoscale dimension (Heiligtag and Niederberger, 2013). Nanotechnology has been explored to quantify, visualize, operate, and manufacture things on an atomic or molecular scale, commonly between 1 to 100 nm size

(Rajasekhar and Kanchi, 2018). As the nanoparticles possess a large surface area to volume ratio, noble metal nanoparticles like gold, silver, platinum, palladium, and nonmetallic, inorganic oxides like zinc oxide, and titanium oxide have been extensively used in biotechnology, optics, mechanics, electronics, microbiology, environmental remediation, medicine, many engineering fields, and material science (Luechinger et al., 2010).

Nanotechnology as a techno-economic model is still in the early phase of its research, development, and innovation. Nanotechnology is a new, emerging technology, which is quite different from sustaining technologies with distinct features (Fraceto et al., 2016). Nanoparticles are composed of three layers: (a) the surface layer, which may be functionalized with a variety of small molecules, metal ions, surfactants, and polymers, (b) the shell layer, which is chemically different material from the core in all aspects, and (c) the core, which is essentially the central portion of the nanoparticle (Shin et al., 2016).

There are different types of nanoparticles categorized depending on their morphology, size and chemical properties and they are used for energy conversion and energy storage, for water purification, environmental remediation, and also in biomedical and electronic industries (Lee et al., 2010; Zhang and Li, 2012). Different classes of nanoparticles are carbon-based nanoparticles, metal nanoparticles, ceramics nanoparticles, semiconductor nanoparticles, polymeric nanoparticles, and lipid-based nanoparticles (Khan et al., 2019). Two important types of carbon-based nanoparticles are fullerenes and carbon nanotubes. Nanomaterial made of globular hollow cage, such as allotropic forms of carbon with high strength, structure, electron affinity, and versatility due to its electrical conductivity, is present in fullerenes. Carbon nanotubes are an elongated, tubular structure resembling graphite rolled sheets which are used for commercial applications, such as fillers; efficient gas adsorbents for environmental remediation; and support medium for different inorganic and organic catalysts (Elliott et al., 2013; Astefanei et al., 2015). Metal nanoparticles are made up of pure metal precursors, which possess opto-electrical properties due to localized surface plasmon resonance (LSPR) (Dreaden et al., 2012). Ceramic nanoparticles are inorganic nonmetallic solids, found in amorphous, polycrystalline, dense, porous, or hollow forms and are used for catalysis, photocatalysis, photodegradation of dyes, and imaging applications (Thomas et al., 2015b). Semiconductor materials showed properties between metals and nonmetals with wide band gaps and used in photocatalysis, photo optics and electronic devices (Ali et al., 2017). Polymeric nanoparticles are mostly nanospheres or nanocapsular shaped and, like polymeric nanoparticles, lipid nanoparticles possess a solid core made of lipid and a matrix containing soluble lipophilic molecules with a spherical shape. These are used in biomedical applications such as drug carriers and delivery and RNA release in cancer therapy (Rawat et al., 2011; Gujrati et al., 2014; Mansha et al., 2017).

Particle size and surface area of nanoparticles have a major role in the interaction of materials with the biological system. The smaller size of the materials leads to an exponential increase in surface area relative to volume, thus making the nanomaterial surface more reactive on itself and to its adjacent environment. The size and surface area are important factors in determining the activity of nanoparticles and other factors, such as the chemical nature of the constituents, may also contribute to the functioning of the nanoparticles (Lherm et al., 1992; Chen et al., 2006). Shape-dependent toxicity has been reported for numerous nanoparticles including carbon nanotubes, silica, allotropies, nickel, gold, and titanium nanomaterials (Hamilton et al., 2009; Petersen and Nelson, 2010; Verma and Stellacci, 2010; Kim et al., 2012). The stability and potential to resist agglomeration of nanoparticles influence their functioning. Mostly, the stability of nanoparticles depends on size, surface charge, and composition, among others (Yang et al., 2008).

8.4   GREEN SYNTHESIS OF NANOPARTICLES AND ITS PROPERTIES

Various methods can be applied for the fabrication of nanoparticles, including physical, chemical, and biological methods (Figure 8.2). The synthesis of nanoparticles through conventional physical and chemical procedures results in toxic by-products that are environmental hazards. Additionally, these particles cannot be used in medicine due to health-related issues, especially in clinical fields (Dhand et al., 2015). Physical and chemical methods are not cost-effective because of the expensive vacuum components, high temperature process, and toxic and corrosive gases, particularly in the case of chemical vapor deposition. Because of the limitations of both methods, the biological method is more attractive for the synthesis of nanoparticles.

The biological method for the synthesis of nanoparticles includes using microorganisms like bacteria, fungi, and actinomycetes and using plant extracts, agricultural waste materials, and enzymes. The application of microorganisms as eco-friendly and cost-effective means avoids the use of toxic chemicals and the high energy demand required for physiochemical synthesis. Microorganisms have the ability to accumulate and detoxify heavy metals due to various reductase enzymes, which are able to reduce metal salts to metal nanoparticles with a narrow size distribution and,

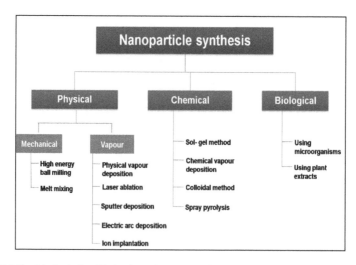

FIGURE 8.2   Methods for fabrication of nanoparticles.

as a result, less polydispersity (Singh et al., 2016). Previous studies by Roshmi et al. have reported the biological synthesis of silver nanoparticles using different bacteria such as *Bacillus* spp., *Ochrobactrum* sp., and *Pseudomonas* sp., and these biogenic nanoparticles were found to have potential antimicrobial activity, which can be used for clinical applications (Thomas et al., 2012; Das et al., 2013; Thomas et al., 2014a). Also, they have reported the ability of silver nanoparticles producing *Bacillus amyloliquefaciens* and *Bacillus subtilis* to fabricate gold nanoparticles from aurochloric acid with the potential for photocatalytic and agricultural applications (Roshmi et al., 2015; Panichikkal et al., 2019).

Recently, phytonanotechnology has been applied as a novel approach for the synthesis of nanoparticles and is an eco-friendly, simple, rapid, stable, and cost-effective method. Phytonanotechnology has many advantages, including biocompatibility, scalability, and the medical applicability of synthesizing nanoparticles using the universal solvent, water, as a reducing medium. Thus, plant-derived nanoparticles produced by easily available plant materials and nontoxic plants are suitable for fulfilling the high demand for nanoparticles with applications in biomedical and environmental areas (Noruzi, 2014). There are many previous studies on the biofabrication of nanoparticles, especially silver nanoparticles using plant extracts (Pirtarighat et al., 2018; Aritonang et al., 2019). Mathew et al. have been reported the synthesis of silver nanoparticles with antibacterial activity using ginger rhizome extract under sunlight (Mathew et al., 2018).

Shape- and size-controlled synthesis of nanoparticles is possible through biological methods by modulating the pH or the temperature of the reaction mixture. The rate of reduction of metal ions using biological agents is found to be more rapid at ambient temperature and pressure conditions. Gericke and Pinches (2006) have reported the fabrication of different shaped nanoparticles (triangle, hexagons, spheres, and rods) by modulating the pH of reaction mixture to 3, 5, 7 and 9. Riddin et al. (2006) also reported that reduced yield of nanoparticles was obtained when synthesized at 65 °C, whereas yield was enhanced at 35 °C (Gericke and Pinches, 2006; Riddin et al., 2006). The bioactive molecules present in biological agents serve as metal reducing agents and capping agents, which facilitates high stability to the biogenic nanoparticles and protects it from agglomeration. Agglomeration affects the functioning and active properties of nanoparticles (Janardhanan et al., 2013). This green synthesis method avoids the use of additional capping agents and toxic chemicals which makes the method more attractive, non-hazardous, cost-effective, and eco-friendly.

8.5 EXPLORATION OF NANOPARTICLES IN DIFFERENT FIELDS

Nanotechnology is an integration of different fields of science with potential applications in the pharmaceutical industry, medicine, agriculture, and environmental remediation (Figure 8.3). The nanoparticles can be used as antimicrobial agents, electrochemicals and biosensors, surface coating agent on medical devices, carrier for drug delivery, photocatalysts, detoxifying agents for environmental remediation, and delivery system for agricultural applications like nanofertilizers and nanopesticides (G. Ingale, 2013; Khan et al., 2019).

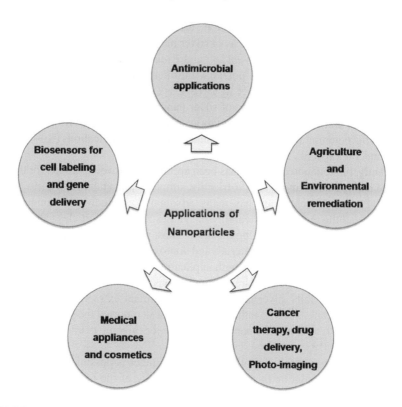

FIGURE 8.3 Application of nanoparticles in various fields.

Due to the antimicrobial property of nanomaterials, nanotechnology can be executed as preventives, diagnostics, drug carriers, and synergetics in antibacterial therapies. Nanoparticles can be designed as targeted, environmentally responsive, combinatorial systems for antibacterial drug delivery systems and as a vaccine that contains nanoparticles as adjuvants or delivery vehicles, which provoke immune responses against bacterial infection (Karaman et al., 2017). This nanoscopic delivery system has the following advantages: improved serum solubility, prolonged systemic circulation lifetime of the drug, targeted delivery of the drug to the site of infection, sustained and controlled release of the drug, and also combinatorial drug delivery to the site of interest (Zhang et al., 2007; Rosenholm et al., 2012). Previous studies have reported the antimicrobial potential of bacteriogenic silver nanoparticles against *Staphylococcus aureus*, *Salmonella* Typhi, *Salmonella* Paratyphi, and *Vibrio cholera* (Thomas et al., 2014b; Roshmi et al., 2016). In addition to this, the surface fabrication of medical devices to inhibit staphylococcal biofilm formation was also demonstrated using biogenic nanoparticles which revealed its potential in medical applications (Thomas et al., 2015a; Thomas et al., 2017).

Oxide, metal, and semiconductor nanoparticles have been utilized for constructing electrochemical sensors and biosensors, and these nanoparticles have various roles in different sensing systems. The significant functions provided by nanoparticles comprise the immobilization of biomolecules, the catalysis of electrochemical reactions, the improvement of electron transfer among electrode surfaces and proteins, and the labeling of biomolecules due to their unique physical and chemical properties. Gold nanoparticles are most frequently used for the immobilization of proteins and also DNA, which can be used for the creation of electrochemical DNA sensors (Cai et al., 2001; Liu et al., 2003).

Nanotechnology is also integrated in the field of agriculture for the enhanced productivity of agroproducts, where nanoparticles are used as nanofertilizers and nanopesticides to enhance the growth and health of crops without causing contamination of soils and water. Biogenic silver nanoparticles have been reported for their potential to enhance plant growth and diosgenin content in Fenugreek (*Trigonella foenum-graecum* L.). The effect of silver metal nanoparticles on the growth and antioxidant status of *Brassica juncea* seedlings has also been reported previously which indicate the role of silver nanoparticles in agricultural field (Sharma et al., 2012; Jasim et al., 2017). Additionally, the gold nanoparticles synthesized by *Bacillus subtilis* enhanced the production of phytohormone indole acetic acid (IAA) by plant growth promoting rhizobacteria *Pseudomonas monteilii*, which readily augmented the growth in *Vigna unguiculata* seedlings (Panichikkal et al., 2019).

Nanotechnology is also used as novel approach to protect the environment from pollution by the cleanup of hazardous waste sites. Environmental cleanup using nanomaterials includes removal of heavy metals and other toxic contaminants from the environment mainly from air, soil and water by a process known as nanoremediation (Ingle et al., 2014).

8.6 EXPLORATION OF BIOGENIC NANOPARTICLES FOR CLEAN ENVIRONMENT: NANOREMEDIATION

Nanoremediation process involves the application of reactive nanomaterials for the transformation and detoxification of pollutants for water and air purification, sewage treatment, environmental remediation, and waste management. These nanomaterials have properties that facilitate both chemical reduction and catalysis to transform the pollutants to nontoxic form (Karn et al., 2009). Many different nanoscale materials have been explored for remediation, such as nanoscale zeolites, metal oxides, carbon nanotubes, and fibers, and various noble metals such as bimetallic nanoparticles (BNPs) and titanium dioxide. A previous study showed that nanoscale iron particles are very effective for the transformation and detoxification of a wide variety of common environmental contaminants, such as chlorinated organic solvents, organochlorine pesticides, and polychlorinated biphenyls (PCBs) (Zhang, 2003). Furthermore, highly reactive iron oxide nanoparticles have the potential to enhance the efficiency of remediating the pollutant and the application of microorganisms also facilitates a faster effect on environmental remediation. Another study reported that *Geobacter metallireducens*, when coupled with macroparticulate ferrous oxide, were reported to reduce 4-nitroacetophane, a highly contaminating organic compound (Braunschweig et al., 2013).

Biogenic metal nanoparticles such as palladium, gold, and iron are used for environmental remediation from soil and water based on their catalytic properties. The waste water discharged from the textile industries contains toxic dyes which cause much environmental concern as these exist in the environment for prolonged times without degradation. Previous studies reported the detoxification of such dyes using biogenic nanomaterials. The reduction of two azo dyes, such as methyl orange (MO) and Evan's blue (EB), was achieved by using reduced palladium nanomaterial (Watson et al., 1999). Also, silver nanoparticles formed by *Bacillus amyloliquefaciens* were explored for the degradation of Victoria blue B and methyl violet dyes (Jishma et al., 2016; Jishma et al., 2018b). Victoria blue dye can also be detoxified by gold nanoparticles with biogenic origin as demonstrated previously (Jishma et al., 2018a). The degradation of textiles dyes is made possible by nanoparticles due to its photocatalytic property. The superiority of photocatalytic degradation is due to its low cost, high stability, and quick oxidation process (Lu et al., 2016).

Biogenic nanomaterials remediate environmental pollutants by three possible mechanisms: adsorption, transformation, and catalysis. Various biologically produced inorganic and organic nanoparticles have reduced pollution due to toxic metals and organic compounds by adsorption, transformation, photocatalysis, and catalytic reduction (Das et al., 2018). Adsorption is the surface phenomenon involving the binding of contaminants on the surface of an adsorbent. Several biologically synthesized nanoadsorbents, such as metal oxide nanoparticles, bimetallic nanoparticles, modified nanoparticles, and multifunctional nanocomposites, are currently used to trap environmental pollutants from polluted sites. Biologically produced Fe_3O_4 nanosorbent from tea extract was effective and economical in nature for the rapid removal and recovery of metal ions from wastewater effluents (Bansal et al., 2014). Metal oxide nanoparticles and bimetallic nanoparticles usually show very high reactivity, surface energy, and magnetic properties, and therefore they are very easily oxidized by reacting with dissolved air in water and can be used for water purification (Martínez-Cabanas et al., 2016). Also, the cluster of hybrid silver, copper, and zerovalent iron nanoparticles was previously demonstrated to treat different industrial effluents from tannery, textile, and pharmaceutical wastes and Alsabagh et al. have reported the efficacy of multifunctional nanocomposite of chitosan/silver nanoparticle/copper nanoparticle/carbon nanotubes for the treatment of water containing toxic metals such as copper, cadmium, and lead (Suman et al., 2014; Alsabagh et al., 2015). The transformations of heavy metals, such as cadmium, chromium, nickel, zinc, and lead, have been extensively studied using zerovalent iron nanoparticles. The reduction of chromium by zerovalent iron nanoparticles has been investigated in various field-scale studies, which serve as an electron donor for the reduction of chromium (Singh et al., 2011). Hence these biogenic nanoparticles have attracted great attention for the removal of environmental contaminants through nanoremediation.

The global demand for nanoscale materials, tools, and devices has been increasing due to the extended advances of nanotechnology in various fields, especially for environmental protection and remediation. The green aspects of nanotechnology in the process of its synthesis and elimination of contaminants from the environment make it a more attractive, cost-effective, and ecologically safe approach.

REFERENCES

Ali S, Khan I, Khan SA, Sohail M, Ahmed R, Rehman Au, Ansari MS, Morsy MA. Electrocatalytic performance of Ni-Pt core–shell nanoparticles supported on carbon nanotubes for methanol oxidation reaction. *Journal of Electroanalytical Chemistry* 2017;795:17–25.

Alsabagh AM, Fathy M, Morsi RE. Preparation and characterization of chitosan/silver nanoparticle/copper nanoparticle/carbon nanotube multifunctional nano-composite for water treatment: heavy metals removal; kinetics, isotherms and competitive studies. *RSC Advances* 2015;5(69):55774–55783.

Aqeel M, Jamil M, Yusoff I. Soil Contamination, Risk Assessment and Remediation. Environmental Risk Assessment of Soil Contamination, 2014.

Aritonang HF, Koleangan H, Wuntu AD. Synthesis of silver nanoparticles using aqueous extract of medicinal plants' (Impatiens balsamina and Lantana camara) fresh leaves and analysis of antimicrobial activity. *International Journal of Microbiology* 2019;2019:1–8.

Astefanei A, Núñez O, Galceran MT. Characterisation and determination of fullerenes: A critical review. *Analytica Chimica Acta* 2015;882:1–21.

Bansal V, Salvadori MR, Ando RA, Oller do Nascimento CA, Corrêa B. Intracellular biosynthesis and removal of copper nanoparticles by dead biomass of yeast isolated from the wastewater of a mine in the Brazilian Amazonia. *Plos One* 2014;9(1):e87968.

Braunschweig J, Bosch J, Meckenstock RU. Iron oxide nanoparticles in geomicrobiology: from biogeochemistry to bioremediation. *New Biotechnology* 2013;30(6):793–802.

Cai H, Xu C, He P, Fang Y. Colloid Au-enhanced DNA immobilization for the electrochemical detection of sequence-specific DNA. *Journal of Electroanalytical Chemistry* 2001;510(1-2):78–85.

Cao Y, Zhang S, Wang G, Li T, Xu X, Deng O, Zhang Y, Pu Y. Enhancing the soil heavy metals removal efficiency by adding HPMA and PBTCA along with plant washing agents. *Journal of Hazardous Materials* 2017;339:33–42.

Chen Z, Meng H, Xing G, Chen C, Zhao Y, Jia G, Wang T, Yuan H, Ye C, Zhao F, Chai Z, Zhu C, Fang X, Ma B, Wan L. Acute toxicological effects of copper nanoparticles in vivo. *Toxicology Letters* 2006;163(2):109–120.

Das VL, Thomas R, Varghese RT, Soniya EV, Mathew J, Radhakrishnan EK. Extracellular synthesis of silver nanoparticles by the Bacillus strain CS 11 isolated from industrialized area. *3 Biotech* 2013;4(2):121–126.

Das S, Chakraborty J, Chatterjee S, Kumar H. Prospects of biosynthesized nanomaterials for the remediation of organic and inorganic environmental contaminants. *Environmental Science: Nano* 2018;5(12):2784–2808.

Dhand C, Dwivedi N, Loh XJ, Jie Ying AN, Verma NK, Beuerman RW, Lakshminarayanan R, Ramakrishna S. Methods and strategies for the synthesis of diverse nanoparticles and their applications: a comprehensive overview. *RSC Advances* 2015;5(127):105003–105037.

Dreaden EC, Alkilany AM, Huang X, Murphy CJ, El-Sayed MA. The golden age: gold nanoparticles for biomedicine. *Chem Soc Rev* 2012;41(7):2740–2779.

Elliott JA, Shibuta Y, Amara H, Bichara C, Neyts EC. Atomistic modelling of CVD synthesis of carbon nanotubes and graphene. *Nanoscale* 2013;5(15):6662.

Enault J, Robert S, Schlosser O, de The C, Loret JF. Drinking water, diet, indoor air: Comparison of the contribution to environmental micropollutants exposure. *International Journal of Hygiene and Environmental Health* 2015;218(8):723–730.

Florentina Ilo B. The effects of air pollutants on vegetation and the role of vegetation in reducing atmospheric pollution. In: Khallaf MK editor. *The Impact of Air Pollution on Health, Economy, Environment and Agricultural Sources*: IntechOpen, 2011.

Fraceto LF, Grillo R, de Medeiros GA, Scognamiglio V, Rea G, Bartolucci C. Nanotechnology in agriculture: Which innovation potential does it have? *Frontiers in Environmental Science* 2016;4.

Gericke M, Pinches A. Microbial production of gold nanoparticles. *Gold Bulletin* 2006;39(1):22–28.

Gujrati M, Malamas A, Shin T, Jin E, Sun Y, Lu Z-R. Multifunctional cationic lipid-based nanoparticles facilitate endosomal escape and reduction-triggered cytosolic siRNA release. *Molecular Pharmaceutics* 2014;11(8):2734–2744.

Hader D-P. A grand challenge for environmental toxicity: what are the permissive limits of toxic pollutants in the environment? *Frontiers in Environmental Science* 2013;1.

Hamilton RF, Wu N, Porter D, Buford M, Wolfarth M, Holian A. Particle length-dependent titanium dioxide nanomaterials toxicity and bioactivity. *Particle and Fibre Toxicology* 2009;6(1).

Heiligtag FJ, Niederberger M. The fascinating world of nanoparticle research. *Materials Today* 2013;16(7-8):262–271.

Ingale, G. Biogenic synthesis of nanoparticles and potential applications: An eco-friendly approach. *Journal of Nanomedicine & Nanotechnology* 2013;04(02).

Ingle AP, Seabra AB, Duran N, Rai M. *Nanoremediation: . Microbial Biodegradation and Bioremediation*. Elsevier, 2014. p. 233–250.

Isaksson C. Pollution and its impact on wild animals: A meta-analysis on oxidative stress. *EcoHealth* 2010;7(3):342–350.

Janardhanan A, Roshmi T, Varghese RT, Soniya EV, Mathew J, Radhakrishnan EK. Biosynthesis of silver nanoparticles by a Bacillus sp. of marine origin. *Materials Science-Poland* 2013;31(2):173–179.

Jasim B, Thomas R, Mathew J, Radhakrishnan EK. Plant growth and diosgenin enhancement effect of silver nanoparticles in Fenugreek (Trigonella foenum-graecum L.). *Saudi Pharmaceutical Journal* 2017;25(3):443–447.

Jishma P, Thomas R, Narayanan R, Radhakrishnan EK. Exploration of photocatalytic properties of microbially designed silver nanoparticles on Victoria blue B. *Bioprocess and Biosystems Engineering* 2016;39(7):1033–1040.

Jishma P, Roshmi T, Snigdha S, Radhakrishnan EK. Kinetic study of gold nanoparticle mediated photocatalytic degradation of Victoria blue. *3 Biotech* 2018a;8(2).

Jishma P, Narayanan R, Snigdha S, Thomas R, Radhakrishnan EK. Rapid degradative effect of microbially synthesized silver nanoparticles on textile dye in presence of sunlight. *Biocatalysis and Agricultural Biotechnology* 2018b;14:410–417.

Karaman DŞ, Manner S, Fallarero A, Rosenholm JM. *Current Approaches for Exploration of Nanoparticles as Antibacterial Agents*. IntechOpen; 2017.

Karn B, Kuiken T, Otto M. Nanotechnology and in situ remediation: a review of the benefits and potential risks. *Environmental Health Perspectives* 2009;117(12):1813–1831.

Kelishadi R. Environmental pollution: Health effects and operational implications for pollutants removal. *Journal of Environmental and Public Health* 2012;2012:1–2.

Khan I, Saeed K, Khan I. Nanoparticles: Properties, applications and toxicities. *Arabian Journal of Chemistry* 2019;12(7):908–931.

Kim ST, Chompoosor A, Yeh Y-C, Agasti SS, Solfiell DJ, Rotello VM. Dendronized gold nanoparticles for siRNA delivery. *Small* 2012;8(21):3253–3256.

Laws EA. 2017. *Aquatic Pollution: An Introductory Text*. Wiley

Lee J, Mahendra S, Alvarez PJJ. Nanomaterials in the construction industry: A review of their applications and environmental health and safety considerations. *ACS Nano* 2010;4(7):3580–3590.

Lherm C, Müller RH, Puisieux F, Couvreur P. Alkylcyanoacrylate drug carriers: II. Cytotoxicity of cyanoacrylate nanoparticles with different alkyl chain length. *International Journal of Pharmaceutics* 1992;84(1):13–22.

Liu S, Leech D, Ju H. Application of colloidal gold in protein immobilization, electron transfer, and biosensing. *Analytical Letters* 2003;36(1):1–19.

Lu H, Wang J, Stoller M, Wang T, Bao Y, Hao H. An overview of nanomaterials for water and wastewater treatment. *Advances in Materials Science and Engineering* 2016;2016:1–10.

Luechinger NA, Grass RN, Athanassiou EK, Stark WJ. Bottom-up fabrication of metal/metal nanocomposites from nanoparticles of immiscible metals. *Chemistry of Materials* 2010;22(1):155–160.

Mansha M, Khan I, Ullah N, Qurashi A. Synthesis, characterization and visible-light-driven photoelectrochemical hydrogen evolution reaction of carbazole-containing conjugated polymers. *International Journal of Hydrogen Energy* 2017;42(16):10952–10961.

Martínez-Cabanas M, López-García M, Barriada JL, Herrero R. Sastre de Vicente ME. Green synthesis of iron oxide nanoparticles. Development of magnetic hybrid materials for efficient As(V) removal. *Chemical Engineering Journal* 2016;301:83–91.

Mathew S, Prakash A, Radhakrishnan EK. Sunlight mediated rapid synthesis of small size range silver nanoparticles using Zingiber officinale rhizome extract and its antibacterial activity analysis. *Inorganic and Nano-Metal Chemistry* 2018;48(2):139–145.

Musilova J, Arvay J, Vollmannova A, Toth T, Tomas J. Environmental contamination by heavy metals in region with previous mining activity. *Bulletin of Environmental Contamination and Toxicology* 2016;97(4):569–575.

Noruzi M. Biosynthesis of gold nanoparticles using plant extracts. *Bioprocess and Biosystems Engineering* 2014;38(1):1–14.

Okrent D. On intergenerational equity and its clash with intragenerational equity and on the need for policies to guide the regulation of disposal of wastes and other activities posing very long-term risks. *Risk Analysis* 1999;19(5):877–901.

Oplatowska M, Donnelly RF, Majithiya RJ, Glenn Kennedy D, Elliott CT. The potential for human exposure, direct and indirect, to the suspected carcinogenic triphenylmethane dye Brilliant Green from green paper towels. *Food and Chemical Toxicology* 2011;49(8):1870–1876.

Panichikkal J, Thomas R, John JC, Radhakrishnan EK. Biogenic gold nanoparticle supplementation to plant beneficial *Pseudomonas monteilii* was found to enhance its plant probiotic effect. *Current Microbiology* 2019;76(4):503–509.

Petersen EJ, Nelson BC. Mechanisms and measurements of nanomaterial-induced oxidative damage to DNA. *Analytical and Bioanalytical Chemistry* 2010;398(2):613–650.

Pirtarighat S, Ghannadnia M, Baghshahi S. Green synthesis of silver nanoparticles using the plant extract of Salvia spinosa grown in vitro and their antibacterial activity assessment. *Journal of Nanostructure in Chemistry* 2018;9(1):1–9.

Rai PK. Particulate Matter and Its Size Fractionation. *Biomagnetic Monitoring of Particulate Matter*, Elsevier, 2016, 1–13.

Rajasekhar C, Kanchi S. Green Nanomaterials for Clean Environment. In: Martínez LM et al. (eds.) editor. *Handbook of Ecomaterials*, Springer International Publishing AG, 2018. p. 1–18.

Rawat MK, Jain A, Singh S. Studies on binary lipid matrix based solid lipid nanoparticles of Repaglinide: in Vitro and in Vivo evaluation. *Journal of Pharmaceutical Sciences* 2011;100(6):2366–2378.

Riddin TL, Gericke M, Whiteley CG. Analysis of the inter- and extracellular formation of platinum nanoparticles byFusarium oxysporumf. sp.lycopersiciusing response surface methodology. *Nanotechnology* 2006;17(14):3482–3489.

Rosenholm JM, Mamaeva V, Sahlgren C, Lindén M. Nanoparticles in targeted cancer therapy: mesoporous silica nanoparticles entering preclinical development stage. *Nanomedicine* 2012;7(1):111–120.

Roshmi T, Soumya KR, Jyothis M, Radhakrishnan EK. Effect of biofabricated gold nanoparticle-based antibiotic conjugates on minimum inhibitory concentration of bacterial isolates of clinical origin. *Gold Bulletin* 2015;48(1-2):63–71.

Roshmi T, Jishma P, Radhakrishnan EK. Photocatalytic and antibacterial effects of silver nanoparticles fabricated byBacillus subtilisSJ 15. *Inorganic and Nano-Metal Chemistry* 2016;47(6):901–908.

Sharma P, Bhatt D, Zaidi MGH, Saradhi PP, Khanna PK, Arora S. Silver nanoparticle-mediated enhancement in growth and antioxidant status of *Brassica juncea*. *Applied Biochemistry and Biotechnology* 2012;167(8):2225–2233.

Shin W-K, Cho J, Kannan AG, Lee Y-S, Kim D-W. Cross-linked composite gel polymer electrolyte using mesoporous methacrylate-functionalized SiO2 nanoparticles for lithium-ion polymer batteries. *Scientific Reports* 2016;6(1).

Singh R, Misra V, Singh RP. Removal of hexavalent chromium from contaminated ground water using zero-valent iron nanoparticles. *Environmental Monitoring and Assessment* 2011;184(6):3643–3651.

Singh P, Kim Y-J, Zhang D, Yang D-C. Biological synthesis of nanoparticles from plants and microorganisms. *Trends in Biotechnology* 2016;34(7):588–599.

Srivastava A. Jangid N, Srivastava M. Pesticides as Water Pollutants, 2018.

Suman, KA, Gera M, Jain VK. A novel reusable nanocomposite for complete removal of dyes, heavy metals and microbial load from water based on nanocellulose and silver nano-embedded pebbles. *Environmental Technology* 2014;36(6):706–714.

Thomas R, Viswan A, Ek R. Evaluation of antibacterial activity of silver nanoparticles synthesized by a novel strain of marine Pseudomonas sp. *Nano Biomedicine and Engineering* 2012;4(3).

Thomas R, Janardhanan A, Varghese RT, Soniya EV, Mathew J, Radhakrishnan EK. Antibacterial properties of silver nanoparticles synthesized by marine Ochrobactrum sp. *Brazilian Journal of Microbiology* 2014a;45(4):1221–1227.

Thomas R, Janardhanan A, Varghese RT, Soniya EV, Mathew J, Radhakrishnan EK. Antibacterial properties of silver nanoparticles synthesized by marine Ochrobactrum sp. *Brazilian journal of microbiology : [publication of the Brazilian Society for Microbiology]* 2014b;45(4):1221–1227.

Thomas R, Soumya KR, Mathew J, Radhakrishnan EK. Inhibitory effect of silver nanoparticle fabricated urinary catheter on colonization efficiency of Coagulase Negative Staphylococci. *Journal of Photochemistry and Photobiology B: Biology* 2015a;149:68–77.

Thomas S, Harshita BSP, Mishra P, Talegaonkar S. Ceramic nanoparticles: Fabrication methods and applications in drug delivery. *Current Pharmaceutical Design* 2015b;21(42):6165–6188.

Thomas R, Mathew S, Nayana AR, Mathews J, Radhakrishnan EK. Microbially and phytofabricated AgNPs with different mode of bactericidal action were identified to have comparable potential for surface fabrication of central venous catheters to combat Staphylococcus aureus biofilm. *Journal of Photochemistry and Photobiology B: Biology* 2017;171:96–103.

Thompson LA, Darwish WS. Environmental chemical contaminants in food: Review of a global problem. *Journal of Toxicology* 2019;2019:1–14.

Verma A, Stellacci F. Effect of Surface Properties on Nanoparticleâ Cell Interactions. *Small* 2010;6(1):12–21.

Wake H. Oil refineries: a review of their ecological impacts on the aquatic environment. *Estuarine, Coastal and Shelf Science* 2005;62(1-2):131–140.

Watson JHP, Ellwood DC, Soper AK, Charnock J. Nanosized strongly-magnetic bacterially-produced iron sulfide materials. *Journal of Magnetism and Magnetic Materials* 1999;203(1-3):69–72.

Wei X, Lyu S, Yu Y, Wang Z, Liu H, Pan D, Chen J. Phylloremediation of air pollutants: Exploiting the potential of plant leaves and leaf-associated microbes. *Frontiers in Plant Science* 2017;8.

Zhang W-X. Nanoscale iron particles for environmental remediation: An overview. *Journal of Nanoparticle Research* 2003;5(3/4):323–332.

Xun E, Zhang Y, Zhao J, Guo J. Translocation of heavy metals from soils into floral organs and rewards of Cucurbita pepo: Implications for plant reproductive fitness. *Ecotoxicology and Environmental Safety* 2017;145:235–243.

Yang S-T, Wang X, Jia G, Gu Y, Wang T, Nie H, Ge C, Wang H, Liu Y. Long-term accumulation and low toxicity of single-walled carbon nanotubes in intravenously exposed mice. *Toxicology Letters* 2008;181(3):182–189.

Zhang J, Li CM. Nanoporous metals: fabrication strategies and advanced electrochemical applications in catalysis, sensing and energy systems. *Chemical Society Reviews* 2012;41(21):7016.

Zhang L, Gu FX, Chan JM, Wang AZ, Langer RS, Farokhzad OC. Nanoparticles in medicine: Therapeutic applications and developments. *Clinical Pharmacology & Therapeutics* 2007;83(5):761–769.

9 Phycoremediation of Heavy Metals in Wastewater

Strategy and Developments

Velusamy Priya
SNS College of Engineering, India

Sivakumar Vivek and Muthulingam Seenuvasan
Hindusthan College of Engineering and Technology, India

Carlin Geor Malar
Rajalakshmi Engineering College, India

CONTENTS

9.1 INTRODUCTION

In general, wastewater is of poor quality, consisting of pollutants and microorganisms. The discharge of wastewater into nearby water bodies causes serious effects on the environment and health problems for human beings. Wastewater treatment is an important measure to reduce pollutants and other contaminants present in wastewater (Rawat et al. 2010; Balaji et al. 2016; Kumar et al. 2017a). Industrialization and human activities add huge waste materials into different water resources such as running water, lakes, rivers and seas. These pollutants are primarily from the discharge of inadequately treated industrial, agricultural and municipal wastewater (Lim et al. 2010). It contains high levels of inorganic and organic pollutants, heavy metals, pesticides, and pathogens and causes diffusion limitations (Malar et al. 2018). Many conventional methods are being adopted to treat the wastewater containing these heavy metal contaminants. However, these methods are more expensive and less effective in removing all types of contaminants and also produce more sludge, contributing additional pollution (Kumar et al. 2017b, 2017c). These limitations made it possible to explore one of the alternative methods of wastewater treatment known as phycoremediation. Phycoremediation is an effective and economic algal wastewater treatment for the removal of inorganic compounds containing nitrogen and phosphorus, coliform bacteria, heavy metals and for the reduction of chemical oxygen demand (COD) and biological oxygen demand (BOD) (Rawat et al. 2010; Abdel-Raouf et al. 2012; Cai et al. 2013).

9.1.1 Wastewater and Its Characteristics

Wastewater is defined as any water that has been negatively affected in quality by humans. Wastewater is composed of liquid and solid waste that is discharged from domestic residences, commercial properties, industrial plants, and agriculture facilities or land. Wastewater contains a wide range of contaminants at various concentrations. The three main characteristics of wastewater include physical, chemical, and biological characteristics by the presence of respective contaminants.

9.1.2 Heavy Metals and Its Effects in Wastewater

Heavy metals are one of the most persistent pollutants in wastewater. They are also referred to as trace elements and are the metallic elements of the periodic table (Salem et al. 2000). Some of the most common toxic heavy metals in wastewater include arsenic, lead, mercury, cadmium, chromium, copper, nickel, silver, and zinc (Balaji et al. 2014). The release of high amounts of heavy metals into water bodies creates serious health and environmental problems (Table 9.1).

TABLE 9.1

Effects of Chief Heavy Metals on Human Health

Metals	Effects on Human Health	References
Arsenic (As)	Skin cancer	Manju and Mahurpawar (2015)
Lead (Pb)	Reduction in hemoglobin formation cause anemia, mental retardation, abnormalities in fertility	Manju and Mahurpawar (2015)
Nickel (Ni)	Carcinogenic causes dramatitis, cyanosis, and weakness	Manju and Mahurpawar (2015)
Cadmium (Cd)	Visual disturbance, mental deterioration, chromosomal disorder and convulsion	Manju and Mahurpawar (2015)
Copper (Cu)	Hypertension, nausea and vomiting	Singh Jiwan et al. (2011)

9.2 TREATMENT OF HEAVY METALS IN WASTEWATER

The presence of heavy metals in water results in serious effects so that many conventional methods are being adopted to treat wastewater containing heavy metals. A method for removing heavy metals from wastewater generally consists of physical, chemical, and biological technology which comes under primary, secondary and tertiary treatment processes. Some of the conventional methods for eliminating the heavy metals from wastewater involve electrocoagulation, chemical precipitation, ion exchange and reverse osmosis (Ali and Gupta 2006).

9.2.1 ELECTROCOAGULATION

Electrocoagulation (EC) is a process of removing heavy metals in wastewater by using electrical currents. It is also effective in removing suspended solids, dissolved metals and dyes. The electrical charges from the current will maintain the contaminants present in wastewater. The ions and other charged particles will get neutralized with the ions of opposite charges provided by the electrocoagulation system and they become destabilized and precipitate in a stable form. It has been shown that EC is able to eliminate a variety of pollutants from wastewaters, such as, for example, metals, arsenic strontium and cesium, phosphate sulfide, sulfate and sulfite (Apaydin et al. 2009). Some of the limitations of electrochemical coagulation which makes us seek alternative options of the treatment process are as follows (Mollah et al., 2004):

- The high cost of electricity can result in an increase in operational cost of EC process;
- In case of the removal of organic compounds, from effluent containing chlorides, there is a possibility of formation of toxic chlorinated organic compounds;
- The sacrificial anodes need to be replaced periodically.

9.2.2 CHEMICAL PRECIPITATION

Chemical precipitation is one of the methods for heavy metal removal from inorganic effluents. In this process, the dissolved metal ions get precipitated by the chemical reagents and result in the formation of metal hydroxides, sulfides, carbonates, and phosphates that can be separated by sedimentation or filtration.

9.2.3 ION EXCHANGE

Ion exchange is based on the reversible exchange of ions between solid and liquid phases. An ion exchanger is a solid resin capable of exchanging both cations and anions from an electrolytic solution and releases counter-ions of similar charge in a chemically equivalent amount.

9.2.4 REVERSE OSMOSIS

Reverse Osmosis is a pressure-driven membrane separation process that forces the solution to pass through a semi-permeable membrane for the removal of heavy metals from various industries. Reverse osmosis was used for the removal of Cu (II), Ni (II), and Zn (II) by using a polyamide thin-film composite membrane.

9.2.5 NATURAL RESOURCES IN THE TREATMENT OF HEAVY METALS

The purpose of wastewater treatment is to remove the solids present in it like heavy metals and some other impurities, although few are not harmful to humans and do not have significant drawbacks (Bina et al. 2010). Many water treatment methods have been used, and most of them still require high investments (Ghebremichael et al. 2005; Sumathi and Alagumuthu 2014). Many natural resources like rice husk, peanut, bean, Moringa oleifera have been used for treating the heavy metals in wastewater (Eman et al. 2014). Since the investment made on natural resources is much less, nowadays natural resources are used in the aspect of removing heavy metals instead of chemical processes. Some of the heavy metals like arsenic (As), chromium (Cr), mercury (Hg), cadmium (Cd), Lead (Pd) are more toxic which pose threats to both environment and human health. A new idea is emerging using algae in phycoremediation of heavy metals, due to several benefits like more availability, more economical, and excellent metal removal efficiency (Leong Yoong Kit and Jo-Shu Chang 2020).

9.2.6 MICRO AND MACRO ALGAE

Macro and microalgae are used in a variety of commercial products with many more in development. The microalgae like Spirulina and Chlorella have been widely marketed as nutritional supplements for both humans and animals. Many microalgae containing high nutritional value and energy are grown commercially as aquaculture feed. Hydrocolloids, including carrageenan, alginates, and agars, are the major processed products from macro algae that are used as gelling agents in a variety of foods and healthcare products. B-carotene, astaxanthin, and phyco-biliproteins are some of the pigments extracted from algae. These are generally used as food colorants, as additives in animal feed or as nutraceuticals for their antioxidant properties (Radmer 1996). The bioremediation process and CO_2 sequestration also have the applications of algae and in producing many interesting bioactive compounds.

9.3 PHYCOREMEDIATION

Phycoremediation is also an effective ecofriendly treatment process of removing heavy metals from wastewater. It is the use of plants, including algae or lower plants, for the removal of contaminants containing nutrients, heavy metals, and so on from wastewater. It is a highly recommended choice as a cost-effective and best alternative to the currently available physiochemical contaminant remediation methods (Olguín et al. 2003). Remediation by algae is taken as a feasible choice for the removal of metal ions, leading to the improvement of water quality and the sustainable development of aquatic systems (Kumar et al. 2007).

9.3.1 CULTIVATION METHOD

There are three main types of system for cultivation of algae. They are the open system, closed system, and immobilized system. Cultivation of algae using the natural open and closed pond system is technologically simple to conduct and is cheaper (Atkinson and Mavituna, 1983). However, an open pond system is exposed to the environment and its effects due to factors such as temperature and light, but the closed system allows condition control for cultivation. The third is an immobilized system where algae are trapped in a solid medium (Borowitzka 1998).

9.3.1.1 Open Pond Cultivation

Open pond cultivation is a low-cost technique and can be done on a large scale so it is the most preferred of all methods. Moreover, it is easy to manage and more durable than the closed system cultivation. Open system cultivation can be carried out in natural or artificial lakes and ponds. There are many types of ponds that have been designed and experimented with before for the optimum cultivation of microalgae. The three major designs of open systems are raceway ponds, inclined systems, and circular ponds. Normally we prefer all types of ponds but due to high cell growth, raceway ponds are used exclusively.

9.3.1.2 Closed Photobioreactor

Photobioreactors are much more universal devices that may be applied under various climatic conditions. The closed character of bioreactors restricts evaporation, eliminates the problem of parasites and predators, whilst artificial lighting assures optimal conditions for photosynthesis. Such conditions afford the possibility for running cultures of specific algae species, for example those with a high concentration of oil in biomass (Cuaresma et al. 2011). The first closed photo bioreactor in use was a big gags photobioreactor (Baynes et al. 1979). It consisted of large sterile plastic bags ca. 0.5 m in diameter with an adjusted aeration system. Most of those systems have been designed to operate in batch mode; however, semi-continuous systems do happen as well. This variant of the system was developed and its modification consisted in using bags with a smaller diameter. The key problem linked with its exploitation is the necessity of running the process indoors, for there is no possibility of controlling temperature. In addition, a relatively large diameter of the bags poses problems with culture lighting, which in turn constitutes a factor diminishing the system's productivity.

9.3.1.3   Immobilized Cultivation

The accumulation of a biocatalyst (cell or enzyme) either on surfaces or within particles is called Immobilization. Travieso et al. (2002) designed a bio-alga reactor, which consisted of a pilot scale model that was operated with synthetic wastewater with an initial concentration of 3000 µG/L of cobalt ion. Scenedesmusobliquus was immobilized in the reactor, which was operated in batch mode. They recorded that the maximum removal of cobalt ion of 94.5% was reached after 10 days. The unicellular green microalga, Chlorella sorokiniana, was immobilized on a loofa sponge and successfully used as a new bio-sorption system for the removal of lead (II) ions from aqueous solutions (Akhtar et al. 2004). Moreno-Garrido et al. (2005) used the marine microalga Tetraselmischui, to perform a short-term heavy metal accumulation experiment. The binding of metals to algal surface occurs in living and non-living algae (Greene and Bedell, 1990), cell surface area being a major parameter in the uptake of metals by microalgae (Khoshmanesh et al. 1997). Packed-bed columns containing immobilized cells seem to be very efficient in the removal of metals from aquatic media (Moreno-Garrido et al. 2002).

9.3.2   Algae Harvesting Technologies

Algae are typically in a dilute concentration in water, and biomass recovery from a dilute medium accounts for 20%–30% of the total production cost. Algae can be harvested using: 1) Sedimentation (Gravity settling), 2) Membrane Separation (Micro/Ultrafiltration), 3) Flocculation, 4) Flotation, and 5) Centrifugation.

9.3.2.1   Sedimentation

Sedimentation is an easy technique and the initial phase of separating the algae from water. Once agitation is completed, the algae are allowed to settle and densify. However, other methods most likely will also be required to achieve complete separation.

9.3.2.2   Membrane Separation

Membrane separation is a form of filtration process and is typically done on a small scale. In a lab, the content of algae is allowed to pour into the filter on the funnel which is attached to a vacuum flask, so that the algae are allowed to dry on the filter as the vacuum continues to be pulled. Those methods are used to collect microalgae with low density. The main disadvantage is membrane fouling.

9.3.2.3   Flocculation

Flocculation is another technique where some flocculants like alum and ferric chloride are added to the mixture of water and algae that causes the algae to clump together to form colloids. Chitosan is a biological flocculant but has a fairly high cost. Rather than adding chemicals, CO_2 can be introduced to an algal system to cause algae to flocculate on their own.

9.3.2.4   Froth Floatation

Froth floatation is a technique used for harvesting and separating algae from water. Here the air bubbles are incorporated into the unit. For algal systems, the algae will accumulate with the froth of bubbles at the top, and there is some way to collect or scrape the froth and algae from the top to separate it from the water. It is an expensive technology to use commercially.

9.4   PHYCOREMEDIATION MECHANISM OF HEAVY METAL REMOVAL

The mechanism of phycoremediation involves the removal of contaminants present in a water body using micro and macro algae. Algae fix the carbon dioxide by the process of photosynthesis and remove the excess nutrients or contaminants effectively at minimal cost. It removes pathogens and toxic materials as well as heavy metals from wastewater. Xenobiotic, chemicals, and heavy metals are known to be detoxified, transformed, accumulated, or volatilized by algal metabolism. For example, algae like Chlorella Vulgaris will oxidize the heavy metals like chromium and zinc. This process is more advantageous over conventional methods of remediation by its effectiveness, efficiency, and ecofriendly nature (Table 9.2).

TABLE 9.2
Removal of Heavy Metals Using Different Algae

S. No	Algae	Heavy Metals Removed	Metal Concentration	References
1	Oscillatoriaanguistissima	copper	90%	Ahuja et al. (1997)
2	Spirogyra	lead (Pb)	85%	Gupta and Rastogi (2008)
3	Spirogyra hyalina	lead (Pb) and cobalt (Co)	80%	Harneeth Kaur et al. (2019)
4	Dunaliella	cadmium (Cd), lead (Pb) and mercury (Hg)	72% 65% 67%	Harneeth Kaur et al. (2019)
5	Chlorella pyrenoidosa and Scenedesmusobliquus	zinc (Zn) and copper (Cu)	100%	Zhou et al. (2012)
6	Nostocmuscorum	lead (Pb) and cadmium (Cd)	93.3% and 85.2%	Dixit and Singh (2013)

9.4.1 Phycoremediation Strategy in Heavy Metal Removal

9.4.1.1 Facultative Stabilization Pond

Facultative ponds are the simplest variant of the stabilization pond systems. Here, the natural organic matter stabilization process takes place due to the retention of wastewater for a long period. The mechanism occurs in three zones of the pond, that is, anaerobic, aerobic, and facultative zones.

In an anaerobic zone, the suspended organic matter tends to settle the sludge at the bottom. The anaerobic microorganisms will decompose the sludge and slowly convert it into carbon dioxide, methane, and others. After a certain period only the non-biodegradable particles remain in the bottom layer. The hydrogen sulphide generated will get oxidized by chemical and biochemical processes.

In an aerobic zone, dissolved organic matter and the suspended organic matter does not settle and remains dispersed in the liquid mass. Oxygen is required to oxidize the organic matter in this zone. Hence, the oxygen is made available through the photosynthesis of algae and creates a balance between the consumption, production of oxygen, and carbon dioxide as depicted in Figure 9.1.

The presence or the absence of oxygen can occur, called a facultative zone. This condition also gives its name to the ponds (facultative ponds). The process of facultative ponds is essentially natural and does not need any equipment. Figure 9.2 reveals the most efficient biological working process of the Facultative Stabilization Pond. Here the stabilization of organic matter takes place at slow rates, implying the need for a high detention time in the pond.

9.4.1.2 High Rate Algae Ponds

A high rate algae pond (HRAP) achieves two purposes namely secondary wastewater treatment and algal biomass production. This pond is a combination of oxidation pond and an algal reactor. In this process algae supplies oxygen for bacterial degradation of organic matter and heavy metals present in the water and the bacteria excrete mineral compounds that provide the algae with nutrition. High rate algae ponds are greatly effective in removing the heavy metals by oxidizing those using suitable algae. They are shallow, open raceway ponds and provide much more efficient

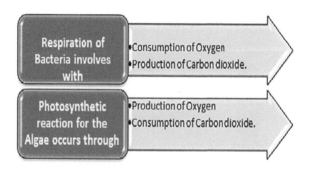

FIGURE 9.1 Algal symbiosis relationship.

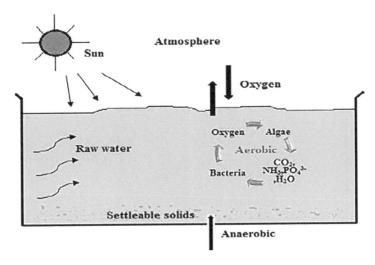

FIGURE 9.2   Facultative waste stabilization pond.

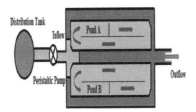

FIGURE 9.3   Constructional details of high rate algal pond.

treatment of wastewater than the conventional oxidation ponds with the support of Dual Stabilization units represented in the Figure 9.3. High rate algal ponds have been studied for many years as a means of wastewater treatment and it enables resource recovery in the form of protein rich microalgaebiomass (Oswald et al. 1957). High rate algae ponds are more active in nutrient and heavy metal removal especially in the removal of phosphate and heavy metals like zinc, iron, etc. HRAPs are much more cost-effective and high efficiency than energy intensive mechanical wastewater treatment systems providing similar wastewater treatment.

9.4.1.3   Algae Settling Pond

Algae settling ponds helps in natural settling of the algal biomass and provides storage for the periodic recovery of the settled algae. This pond is designed to promote the efficient gravity settling using lamella plates and secondary thickening of settled algae to 1%–3% oxidized solids which is outlined in the given Figure 9.4. Settled algal biomass is removed continuously or daily to avoid deterioration of the harvested algae.

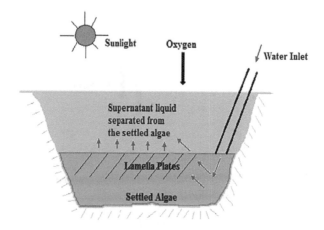

FIGURE 9.4　Settling process of algal settling pond.

9.4.1.4 Maturation Pond

Maturation ponds are the low-cost polishing ponds that generally follow either the primary or secondary facultative pond which is primarily designed for tertiary treatment, that is, the removal of nutrients and possibly algae. This pond is very shallow and it allows light to penetrate it and the process is carried under aerobic conditions and the relevant systematic working process of the Maturation Pond is shown in Figure 9.5. The loading on the maturation pond is calculated based on the assumption that 80% removal of BOD and heavy metals occur in the treatment. If anaerobic and secondary facultative ponds are used, this will produce an effluent suitable only for restricted irrigation. Therefore, additional maturation ponds will only be needed if a higher quality effluent is required for unrestricted irrigation.

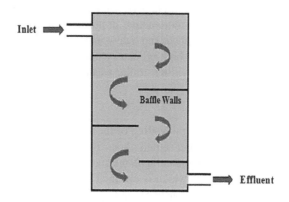

FIGURE 9.5　Working process of maturation pond.

9.5 CHALLENGES AND REMEDIAL MEASURES

Phycoremediation is a time-consuming process and it takes several growing seasons to grow the algae. Phycoremediation is also limited by the growth rate of the plants. The growth of algae will be affected due to the various climatic conditions. The effect of temperature on micro-algal growth makes it one of the most important environmental factors influencing growth rate and biochemical composition of algae. Since algal growth is expensive due to the requirement of large land areas, revolving algal biofilm systems could be involved. This facilitates efficient space utilization, reduced light limitation and enhances algal productivity.

9.5.1 RECENT DEVELOPMENTS IN PHYCOREMEDIATION OF HEAVY METALS

Phycoremediation has involved some of the recent developments in removing heavy metals from wastewater such as using transgenic algae which is a complex but fast-growing technique. A microalgae in wastewater treatment at cold climate is a process of genetic manipulation to improve the detoxification ability. By observing the properties and ability of the microalgae in the uptake of carbon, nitrogen, phosphorus and heavy metal, microalgae show a potential in wastewater treatment for various types of effluent. It is recommended for starting a microalgae wastewater treatment by inoculating a large variety of algae because this will create a mixture of algae where the best suited species will strive and grow faster and dominate the treatment steps. This approach requires less supervision and operation than if a particular algal is chosen to be cultivated for any purpose.

9.6 SUMMARY

Phycoremediation is a novel technique that uses algae to clean up polluted water. It takes advantage of the algal natural ability to take up, accumulate and degrade the constituents that are present in their growth environment. Treatment of heavy metals in wastewater systems offers more simple and economical technology as compared to the other environmental protection systems. Some of the microalgae used in the phycoremediation process will reduce the nutrient load of polluted water samples collected from industries. To safeguard the health of living organisms and for environmental sustainability, a variety of biological treatment processes are employed for the removal of heavy metals from wastewater effluents. Biological removal of heavy metals in wastewater is a selective technique that utilizes the operational flexibility of microorganisms for the elimination of pollutants from wastewater.

REFERENCES

Abdel-Raouf, N., Al-Homaidan A.A., Ibraheem A.A. 2012. Microalgae and wastewater treatment. *Saudi J. Biol. Sci.* 19: 257–275.

Ahuja, P., Gupta, R., Saxena, R.K. 1997. Oscillatoriaanguistissima: A promising Cu2+ biosorbent. *Curr. Microbiol.* 35: 151–154.

Akhtar, Y., Isman, B. 2004. *Feeding Responses of Specialist Herbivores to Plant Extracts and Pure Allelochemicals: Effects of Prolonged Exposure.* Wiley Online Library.

Ali, I., Gupta, V. 2006. Advances in water treatment by adsorption technology. *Nat Proto.* 6: 2661–2667.

Apaydin, K.U., Gonullu, M.T. 2009. An investigation on the treatment of tannery wastewater by electrocoagulation. *Global Nest J.* 11: 546–555.

Atkinson, B., Mavituna, F. 1983. *Biochemical Engineering and Biotechnology Handbook.* MacMillan, Surrey, 1117.

Balaji, N., Kumar, K.S., Seenuvasan, M., Kumar, M.A. 2014. Efficacy of laccase to mineralize the aromatic amines released during the oxido-reductive cleavage of sulphonated reactive azo dye. *Int. J. Appl. Eng. Res.* 9: 9116–9118.

Balaji, N., Kumar, K.S., Seenuvasan, M., Vinodhini, G., Kumar, M.A. 2016. Immobilization of laccase onto micro-emulsified magnetic nanoparticles for enhanced degradation of a textile recalcitrant. *J. Environ. Biol.* 37: 1489.

Baynes, S.M., Emerson, L., Scott, A.P. 1979. Production of algae for use in the rearing of larval fish. *Fish. Res. Tech. Report.* 53: 13–18.

Bina, B., Mehdinejad, M.H., Gunnel, D. et al. 2010. Effectiveness of *Moringaoleifera* coagulant protein as natural coagulant aid in removal of turbidity and bacteria from turbid waters. *World Acad. Sci. Eng. Technol.* 4: 7–28

Borowitzka, M.A. 1998. Limits to growth in Wastewater treatment with algae. *Springer Verlag* 1: 203–226.

Cai, Y., Zheng, H., Ding, S., Kropachev, K., Schwaid, A.G., Tang, Y., Mu, H., Wang, S., Geacintov, N.E., Zhang, Y., Broyde, S. 2013. Free energy profiles of base flipping in intercalative polycyclic aromatic hydrocarbon-damaged DNA duplexes: energetic and structural relationships to nucleotide excision repair susceptibility. *Chem. Res. Toxicol.* 26(7): 1115–1125.

Cuaresma, M., Janssen, M., Vi Lchez, C., Wijffels, R.H. 2011. Horizontal or vertical photobioreactors? How to improve microalgae photosynthetic efficiency. *Bioresourc. Technol.* 102: 5129–5137.

Dixit, S., Singh, D.P. 2013. Phycoremediation of lead and cadmium by employing Nostocmuscorum as biosorbent and optimization of its biosorption potential. *Int. J. Phytorem.* 15: 801–813.

Eman, N.N., Tan, C.S., Makky, E.A. 2014. Impact of Moringaoleifera Cake Residue Application on Waste Water Treatment: A Case Study. *J. Water Resourc. Protect..* 6: 677–687.

Ghebremichael, K.A., Gunaratna, K.R., Henriksson, H. et al. 2005. A simple purification activity assay of the coagulant protein from *Moringaoleifera* seed. *Water Res.* 39: 2338–2344.

Greene, B., Bedell, G.W. 1990. Algal gels or immobilized algae for metal recovery. *Introduction Appl. Phycol.*, 109–136.

Gupta, V.K., Rastogi, A. 2008. Biosorption of lead from aqueous solutions by green algae Spirogyra species: Kinetics and equilibrium studies. *J. Hazard. Mater.* 152: 407–414.

HarneetKaur, A.R., Kaleka, A.S. 2019. Role of phycoremediation to remove heavy metals from sewage water: Review article. *J. Environ. Sci. Technol.* 12: 1–9.

Khoshmanesh, A., Lawson, F., Prince, I.G. 1997. Cell surface area as a major parameter in the uptake of cadmium by unicellular green microalgae. *Chem. Eng. J.* 65: 13–19.

Kit, L.Y., Chang, J.S. 2020. Bioremediation of heavy metals using microalgae: Recent advances and mechanisms. *Bioresourc. Technol.* 303:122886.

Kumar, K.S., Ganesan, K., Rao, P.V.S. 2007. Phycoremediation of heavy metals by the three-color forms of Kappaphycusalvarezii. *J. Hazard Mater.* 143: 590–593.

Kumar, M.A., Vigneshwaran, G., Priya, M.E., Seenuvasan, M., Kumar, V.V., Anuradha, D., Sivanesan, S. 2017a. Concocted bacterial consortium for the detoxification and mineralization of azoic-cum-sulfonic textile mill effluent. *J. Water Process. Eng.* 16: 199–205.

Kumar, M.A., Poonam, S., Kumar, V.V., Baskar, G., Seenuvasan, M., Anuradha, D., Sivanesan, S. 2017b. Mineralization of aromatic amines liberated during the degradation of a sulfonated textile colorant using Klebsiella pneumoniae strain AHM. *Process Biochem.* 57: 181–189.

Kumar, M.A., Harthy, K.D., Kumar, V.V., Balashri, G.K., Seenuvasan, M., Anuradha, D., Sivanesan, S. 2017c. Detoxification of a triphenylmethane textile colorant using acclimated cells of *Bacillus mannanilyticus* strain AVS. *Environ. Progress Sustain. Energy* 36: 394–403.

Lim, S., Chu, W., Phang, S. 2010. Use of *Chlorella vulgaris* for bioremediation of textile wastewater. *J. Bioresour. Technol.* 10: 7314–7322.

Malar, C.G., Seenuvasan, M., Kumar, K.S. 2018. Prominent study on surface properties and diffusion coefficient of urease-conjugated magnetite nanoparticles. *Appl. Biochem. Biotechnol.* 186: 174–185.

Manju, M.2015. Effects of heavy metals on human health. Social Issues and Environmental Problems. ISSN: 2350-0530(O); ISSN: 2394-3629.

Mollah, M.Y.A., Morkovsky, P., Gomes, J.A.G. Kesmez, M., Parga, J., Cocke, D.L.2004. Fundamentals, present and future perspectives of electrocoagulation. *J. Hazard Mater.* 114(1–3): 199–210.

Moreno-Garrido, I., Codd, G.A., Gadd, G.A., Lubian, L.M. 2002. Cu and Zn accumulation by calcium alginate immobilized marine microalgal cells of Nannochloropsisgaditana (Eustigmatophyceae). *Ciencias Marinas.* 28(1): 107–119.

Moreno-Garrido, I., Campana, O., Lubian, L.M., Blasco, J. 2005.Calcium alginate immobilized marine microalgae: experiments on growth and short-term heavy metal accumulation . *Mar. Pollut. Bull.* 51: 823–829.

Olguín, E.J. 2003. Phycoremediation: Key issues for cost–effective nutrient removal processes. *Biotechnol. Adv.* 22: 81–91.

Oswald, W.J., Gotaas, H.B., Golueke, C.G., Kellen, W.R. 1957. Algae in waste treatment. *Sewage Ind. Wastes* 29: 437–457.

Radmer, R.J.1996. Algal diversity and commercial algal products. *BioScience.* 46(4): 263–270.

Rawat, I.R., Ranjith, K., Mutanda, T., Bux, F. 2010. Dual role of microalgae: Phycoremediation of domestic wastewater and biomass production for sustainable biofuels production. *Appl. Energy.* doi:10.1016/j.apenergy.2010.11.025

Salem, H.M., Eweida, E.A., Farag, A. 2000. Heavy metals in drinking water and their environmental impact on human health. *ICEHM* 2000: 542–556.

Singh, J., Kalamdhad Ajay, S. 2011. Effects of heavy metals on soil, plants, human health and aquatic life. *Int. J. Res. Chem. Environ.* 1: 15–21.

Sumathi, T., Alagumuthu, G. 2014. Adsorption studies for arsenic removal using activated *Moringaoleifera. Int. J. ChemEng* 1–6.

Travieso, T., Pellon, A., Benitez, F., Sanchez, E., Borja, R., Farrill, N.O., Weiland, P. 2002. BIOALGA reactor: Preliminary studies for heavy metals removal. *Biochem. Eng. J.* 12(2): 87–91.

Zhou, G.J., Peng, F.Q., Zhang, L.J., Ying, G.G. 2012. Biosorption of zinc and copper from aqueous solutions by two freshwater green microalgae *Chlorella pyrenoidosa* and *Scenedes musobliquus. Environ. Sci. Pollut. Res.* 19: 2918–2929.

10 An Economic Perspective of Bio-waste Valorization for Extended Sustainability

Suganya Subburaj and Madhava Anil Kumar

CSIR-Central Salt & Marine Chemicals Research Institute,
India; Academy of Scientific and Innovative Research, India

CONTENTS

10.1 INTRODUCTION

Due to rapid urbanization and a growing population, global waste is expected to grow by 70% by 2050 [1]. The World Bank urges high-income countries, Sub-Saharan Africa, and the East Asia and Pacific region to take "critical action" to control their waste stream. Solid waste management is still challenging, especially, and plastic is problematic if not managed properly. It will contaminate waterways and ecosystems which is critical for communities and cities. Therefore, human sustainability is often overlooked in low-income countries. The high-income countries, on the other hand, focus on recovering waste and recycling them for a variety of by-products. From the volume of waste generated in low-income countries, only 4% undergoes recycling based on composition and carbon-dioxide-equivalence. By increasing the rate of recycling, global carbon-dioxide emissions can be diminished. Climate change, also, is a factor that is harmful to human health and the local environment for the waste mismanagement.

Unfortunately, the practice of waste management is often the poorest, for example, the most toxic everyday waste ends up in landfill that carries toxic substances as they subsequently break down, leach into groundwater and earth over time, and often contribute to global warming. Rather, it is time to build a circular economy from good waste management practice, where waste must be designed, categorized, and optimized for reuse and recycling. National and local governments need to promote or embrace the circular economy in smart and sustainable ways to make ensure economic growth whilst simultaneously reducing environmental impact [2]. The cost of addressing uncollected waste, poorly disposed waste and its impact is many times higher than the cost of developing and operating simple, adequate waste management systems. With the right management policies, system, financing, and planning decision, it could be the key to successful recycling. The fastest-growing countries, the most in need, can avail monetary support to develop state-of-the-art waste management. The rapid reduction in consumption of solid waste, for example plastics and marine litter, through comprehensive waste reuse and recycling programs is possible if consumer education is made with coordinated organics and food waste management. Importantly, since 2000, the World Bank has committed to over 340 solid waste management programs in countries across the globe to build an essential circular economy [3].

10.1.1 Disposal and Management

Bio-waste, agricultural, municipal and food solid waste, sludge, and wastewater are no longer considered as low-value materials because they are the resources to develop by-products such as high levels of cellulose, hemicellulose, starch, protein, and lipid extracted from agricultural waste. A breakthrough in technology prevails s an inexpensive conversion of waste ground candidates into liquid biofuel. The end-product is tailored with functional groups accordingly and reactivates with the ever-growing need for world resource depletion.

The Waste Framework Directive 7 (revised) defines the large-scale generation of bio-waste (garden and park waste; food and kitchen waste from households, restaurants, caterers, retail premises, and comparable waste from food processing plants) are eligible candidates to produce biofuel. At the same time, the Landfill Directive (1999/31/EC) defines bio-waste as not including forestry or agricultural residue, thus, is further called "biodegradable waste". For instance, wood, paper, cardboard, sewage sludge, and natural textiles are called biodegradable materials [4]. This section describes new methods of different waste disposal in detail.

Other than resources, bio-waste can cause infectious diseases in humans in non-infectious ways since they contain chemicals, medical, and industrial waste. They are classified into the following sections.

10.1.1.1 Medical Bio-waste

Pharmaceutical waster, blood-soaked materials and other sharp objects like needles, syringes, lancets, and blood vials (which can lacerate the skin) are subject to eWaste disposal by a licensed or insured agency. They are required to provide a complete scheme for the value chain of waste, including the transportation of biohazard materials, treatment, and disposal. This is followed by a minimum knowledge about low-level radio-active elements, off-specification chemicals, and various hazardous waste streams like formaldehyde is required before treatment [5].

Because biohazardous materials or infectious agents (bacteria, fungi, virus, parasite, allergen, and cultured human, and animal cells) are biological, capable of self-replication, and can produce deleterious effects upon biological organisms, the segregation of chemicals from med biological waste, for instance, ethidium bromide contaminated agarose gel (i.e. arsenic, chromium), needs further inspection. When both biological and chemical waste exists, biological waste has to be first treated or deactivated using an autoclave or chemical disinfection. The chemical waste that remains is considered a hazardous material. Similarly, biologically/chemically contaminated needles are segregated and labeled as hazardous materials in a sealed container and in-tact.

10.1.1.2 Liquid Waste

Chemical disinfection is recommended for liquid biological waste before disposal: for instance melted agarose cannot be poured down the drain, instead, allow it to cool and solidify, then dispose of it as solid waste in biohazardous waste bags. Solidified agarose gels might undergo chemical disinfection. It is to be noted that category I biological waste as infectious waste can be treated with suitable agents like bleach. Category II – non-infectious waste containing animal tissues, fluids, cell cultures – requires cautious transport to a disposal site. A landfill is a good option for non-infectious waste and waste containing no liquid [6]. Animal carcasses require incineration using chemical or inorganic material.

10.1.1.3 Biodegradable Waste

Biodegradable waste including food, kitchen waste from households, restaurants, and comparable waste from food processing plants are considered not waste. Forestry or agricultural residues, manure, other natural textiles, paper or processed wood or the by-products of food production [7] are also excluded. An environmental threat lies in decomposing the above-mentioned waste such as the production of methane from landfills that account for 3% of total greenhouse gas emissions. The landfill of this waste can certainly reduce the amount of diverted waste through specific treatment options. The advantages of bio-waste management are as follows:

- To avoid greenhouse gas emission,
- To produce biogas and compost,
- To improve soil fertility and resource efficiency.

Owing to the production of compost or biodiesel, researchers opt for the easiest strategies such as incineration or landfill and disregard the actual environmental benefits and costs.

10.2 THE TRANSITION FROM LINEAR TO A CIRCULAR ECONOMY

An increase in the generation of solid waste blatantly affects the nation's linear economy. Unquestionably, the landfill is not suitable for bio-waste management and that needs to be diverted to environmentally favorable solutions. Recycle of waste, in contrast, is the key to energy conversion via a "circular economy" that develops closed-loop technical and biological cycles. Cradle-to-cradle defines a closed-loop system whereas the recycling of bio-waste retains the original properties and ensures no harm to the natural ecosystem [8]. But, in the circular economy, a cradle to grave approach aims to eradicate urban waste generation or poor conversion of abandoned natural sources. For a sustainable global economy, resource scarcity or trouble in recovering natural sources must be mitigated. The sustainable approach must secure the conversion rate of waste fractions into industrial production systems. This also addresses industrial ecology by coupling bio-waste and its transformation to resources.

For example, organic waste from municipal solid waste contains high moisture and salt content, which leads to rapid decomposition and unpleasant odors as well as greenhouse gas emissions, leachate, and sanitary problems. Agriculture and food processing industries require biomass feedstock to produce renewable energy sources such as organic fertilizers, biopesticides, and bioplastics and biorefinery streams. However, by identifying the most optimal process and output, the concept of "circular economy" can contribute to functional elements on future bio-based "cradle-to-cradle" regenerative business. This chapter attributes to existing "recycle strategies of bio-waste" and their valorization pathways, efficiency, and potential revenue. It could consummate by identifying hotspots in steps, methodologies, production, and conversion efficiency (bio-waste is returned to the ecosystem) and analysis of revenue for future circular economic systems [9].

10.2.1 THE VALUE CHAIN OF BIOMASS WASTE

"Waste hierarchy" is a tool used in the evaluation of processes throughout the value chain in waste treatment schemes. The hierarchy establishes types of waste, a preferred program of treatment, protection of the environment, and which resources and energy are most favorable for the least favorable actions.

The priorities are set based on sustainability. This deals with the transformation of biomass into a broad plethora of end-use products through physical, chemical, or

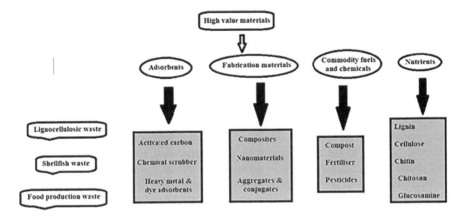

FIGURE 10.1 The value chain of biomass waste.

biological methods [10]. In a value chain analysis, the end-use product values are ensured with tailored properties rather than the existence of complexities in market. Such a radical transformation obeys the correlation between the end-products and their applications (Figure 10.1) across the technological cost more as a value chain than a hierarchy, where biomass waste is the input feedstock.

10.2.2 WASTE HIERARCHY

Waste management requires a treatment priority that relies on human health and environmental protection. The Waste Framework Directive exemplifies recycling targets for household waste and quality criteria for the end-of-waste by permitting bio-waste treatment installations of a capacity exceeding 50 tonnes/day. It also diverts from the poorest landfill practice to waste incineration that regulates health rules via "Animal by-products Regulation" for compost and biogas plants [11].

Recycle of bio-waste offers good quality compost at local, regional, or supra-regional level. Additionally, anaerobic digestion in vogue is plausible for biogas production if the biomass is rich in organic content. Bio-based chemicals or green chemicals are novel, established in conjunction with compost, thermal degradation, microbial decomposition, and anaerobic digestion, particularly heterogenous bio-waste for bioplastic. The organic residues are turned to digestive products; thus, the biological cycle is closed for improved soil health.

The hierarchy in bio-waste management begins from the environmental balance of waste that depends on numerous local factors, inter alia collection systems, quality, and composition, climatic conditions, the potential of the use of various waste-derived products such as electricity, heat, methane-rich gas, or compost. Hence, transparency in a structured and comprehensive approach is mandatory. Life Cycle Thinking (LCT) or decision-making to set guidelines to technological and business models for a circular economy is necessary [12]. Industrial partners and researchers are fascinated to extract the chemical richness of the waste contributing to valorization for effective waste management or reduction of impact on the environment and economic profit.

10.2.3 LIFE CYCLE ASSESSMENT OF BIO-WASTE

The management of bio-waste relies on the environmental balance of local factors, inter alia collection systems, waste composition and quality, climatic conditions, and the potential use of waste-derived products such as electricity, heat, methane-rich gas compost, waste composition, and quality climatic conditions. Therefore, national strategies for the management of bio-waste demands transparency on a structural and comprehensive approach such as Life Cycle Analysis (LCA). Based on LCA, decision-making falls in line with the waste hierarchy with a set of guidelines. Such intrinsic benefits to the environment by developing new technologies, livelihoods, and waste valorization can help in adopting a circular economy. This section provides an overview of present trends and future potential in the conversion of residues from different waste sectors into valuable bio-based materials.

Cradle-to-cradle analysis of bio-waste describes when material or waste is recycled into a new product at the end of its life, where the by-products can be reutilized through valorization. The value chain or production chain promotes the rate of conversion of bio-waste to valuable secondary materials [13]. Different typologies and methodologies will recuperate the supply chain of biomaterials valorized with distinct characteristics.

10.3 VALORIZATION OF BIO-WASTE AND THEIR BY-PRODUCT

Bioeconomy, whether linear or circular, is an important area of research to unravel the potentiality of unused biomaterials or bioenergy without compromising the positive effects. The global validation of bioeconomy makes the highest turnover in the manufacturing of liquid biofuels, the production of bio-based chemicals, pharmaceuticals, plastics, and rubber. Considering technical and non-technical barriers for the use of bio-waste to produce biorefineries is still a niche market, though an uncertainty of high-value compounds limits the application and diffusion of their cascading use. Indeed, bioeconomy to circular economy depends upon the practicability of valorization strategies.

The holistic approach of bio-waste valorization includes the production of energy fuels (ethanol, methane, and hydrogen), organic acids extracted from volatile fatty acids. The established biorefineries, biopesticides, enzymes, and plastic with conversion efficiency (e.g. kg output product/kg of bio-waste input) are highly welcome by the vendors [14]. The market price of the end-product depends on the identified product per dry weight of input feedstock and valorization efficiency (kg output/kg input). The market demand for eco-labeled end-use products can strengthen the national economy with continuous reclamation of waste.

10.3.1 WASTE TO WEALTH CONCEPT

Waste valorization is the key to the "waste to wealth concept" with an intrinsic benefit to a circular economy that consumes biological sustainability. Most importantly, fisheries and aquaculture consume residues to recognize the primary role of recovery of waste. The recovery of waste from marine fisheries and aquaculture by anthropogenic activity ensures producing value-end-products from the waste biological resources. The potential molecules such as oils, proteins, pigments, bioactive peptides, amino acids, collagen, chitin, gelatin, and so on, are extracted from fish waste, that is, skin, blood, bones, gonads, guts, muscle tissue, and so on. The impressive growth of by-products with controlled odor is a result of extremely defined degradation strategies of crustacean waste and the addition of recalcitrants [15].

The valorization chain of crustacean waste deals with innovation programs of cleaner production despite transport, storage, and delivery. The output is supplied to biorefining operations to overcome the scarcity of biopolymers, activated carbon for adsorption, chitosan, and the recovery of carotenoids. The production of high-value proteins from shrimp shell waste meets the environmental, social, and economic problems associated with stinky valuable substances. Importantly, feed quality is not sacrificed in the value chain of crustacean waste processing.

Fly ash is also used to produce pulverized coal from combustion, unburned carbon, and inorganic oxides including hematite, lime, silica, alumina, potassium oxide, titania, and magnesium oxide. The process of co-combustion of fly ash for cereal production and oil extraction generates less emission in comparison with the conventional process. However, low heating value and corrosivity of bio-residues, since salt content is low, is a pressing challenge, along with the production of coal bio ash from burning wood in the presence of a low-alkali solvent which is widely accepted [16]. It is been considered as an alternative to cement. The retained physicochemical characteristics such as size, the shape of particles, porosity, texture, high surface area, and low bulk density account for water remediation whereas the soil and water contaminants are recovered. The end-product is also used as a catalyst and precursor in developing composites.

In addition, lignocellulosic waste from agricultural industries is the most abundant precursor to produce lignocellulose. It is obtained through microbial decomposition and transformations to afford CO_2 and water during the natural carbon cycle, because lignocellulosic waste contains phenols, polysaccharides, and proteins and creates the interaction between lignin, cellulose, and hemicellulose. The tailored characteristics of these enzymes as a resistant and recalcitrant depends on the lignin content, crystallinity of cellulose, and particle size. The commonly available methods to fractionate lignocellulose is as follows: as thermochemical conversions such as enzymatic hydrolysis, hydrothermal liquefaction, acid and alkaline hydrolysis, ammonia fiber expansion, steam explosion, and mechanical milling [17]. Recently, ionic liquid solvents are used.

Lignocellulosic waste-based bioethanol is highly welcomed by biochemists due to being green and sustainable transportation fuel. The major constituents of lignocellulose, that is, cellulose, hemicellulose, and lignin, are used to develop soluble pulp for the production of textile fibers and chemicals and carbon products.

Further, pig bristles are low-value bio-residues consisting of 90% valuable protein such as keratin. The digestibility of this is challenging due to its complex protein arrangement cross-linked by multiple disulfide bonds that confer exceptional stability towards common proteolytic enzymes such as pepsin, trypsin, and papain [18]. The degradation of keratin by tailor-made microbial consortia produces protein-rich products which are to be expected as an alternate high-value feed for fish.

Next in order, cattle manure enriched in nitrogen content is used to generate soil fertilizer. The existing inappropriate treatment of animal debris or cattle manure leads to the contamination of water and soil resources and increased greenhouse gas (GHG) emissions. The anaerobic digestion of manure involves the production of bio-methane, however, burning animal debris to generate power and CO_2 extremely mitigates GHG concentrations due to methane from unmanaged livestock residues. The organic matter of low carbon to nitrogen ratio (C/N) in animal manure is essential to produce about a moderate yield of biogas, whereas, the overall productivity relies on the introduction of carbon-rich co-substrates into the anaerobic digester. A break-through in co-digestion that has a mix of manure containing agro/lignocellulosic wastes, energy crops, and bulking agents bring hydrothermal liquefaction to produce bio-oil regardless of the initial composition of residues [19]. Therefore, manure valorization meets an industrial demand.

Further, household waste with a high heterogeneous composition (C-rich organic materials) is subjected to chemical and mechanical treatments, anaerobic and aerobic digestion, oxidation ditch, and biochemical recycling processes to produce high-value protein biomass. The composition of household waste is influenced by various factors like climate, economic development, and culture, for instance, low- and middle-income countries are 40%–80% rich in organic matter.

Biofuel is produced from the valorization of food waste or the fraction of discards as single waste via any of the following steps: i) transesterification of oil; ii) fermentation of carbohydrates to bioethanol or bio-butanol; iii) anaerobic digestion to bio-methane; iv) dark fermentation to produce hydrogen, pyrolysis, v) gasification to obtain oil and syngas, and vi) hydrothermal carbonization to get hydrochar. The green technology of using chemo-enzymatic and biotechnological treatments is emerging to obtain bio-monomers for polylactates and polyhydroxyalkanoates, succinic acid, furfural and furans, and phenolic compounds [20]. The consolidative approach of food waste valorization brings alternatives and additional revenues to the sector involved in energy recovery. Overall, the production of biodegradable residues is the basis for the concept of bio-waste valorization.

10.4 SUSTAINABILITY OF BIO-VALORIZATION

Bio-waste valorization is successful on a large-scale perspective in dealing with the extraction, purification, processing, and degradation of natural polymers to fragmented usable chemicals. Despite rising concerns in the disposal of waste, bio-valorization is a promising approach to formulate sustainability of area-specific or regional characteristics of different compositions of waste, land availability with the consideration of staggering practices, and so forth. It helps to contemplate the adoption of a strategy on waste disposal.

The sustainable approach of bio-valorization contributes green production strategies from municipal waste into the production of chemicals, materials, and fuels, highlighting the recent legislation on the management of wastes worldwide. The Environmental Protection Agency (EPA) incorporates valorization strategies to build a circular economy by treating domestic waste into industrially important materials, commercial establishments, and occupational sources. The highlight of a sustainable source begins from unsuitable human consumption which is further characterized by a high diversity and variability, a high proportion of organic matter, and high moisture content, volatile solids to total solids ratio, and carbon to nitrogen ratio (mentioned above). A significant challenge exists to control the growth of pathogens, odor due to high microbial activity, the release of foul-smelling fatty acids, and rapid autoxidation due to high lipid content from uneaten food or residential waste. The decomposition of food waste causes methane, a GHG, rather than carbon dioxide under anaerobic environment conditions. A high concentration of methane can cause an explosion; similarly, a high concentration of inorganic salts, heavy metals, or organic compounds from ground water contamination can also lead to gas explosion/seepages. Therefore, valorization practices in the present and foreseeable future must be encouraged to address utility of energy-efficient feedstocks rather eliminating waste and to produce biofuel.

The stable polymer of lignin deconstruction approach using a heterogeneous catalyst to produce simple aromatics and depolymerization of lignin using hydrogen donating solvents owing to hydrogenolysis reactions serve the best methods to recover high-value chemicals [21]. The generation of renewable gasoline, phenolic compounds with potential antioxidant properties using an organic solvent-extraction step with a polystyrene resin can be derived from food waste. To commercialize a valorization product to an industrial upscale requires advanced protocols based on diversified feedstock.

Similarly, the numerous options of greener valorization to produce biodiesel from waste oil at optimum reaction conditions exists with a novel solid acid catalyst. The carbon-based catalysts (e.g. porous starch) for waste valorization other than metal functionalized acids is appreciable and ease to be recycled. It can be used in biodiesel production having activity 2–10 times greater than common microporous carbonaceous catalysts. Particularly, carbonaceous residues are subjected to gasification to attain maximum ester conversion yields from fatty acids to methyl esters. These are so-called "designer catalysts" used in bio valorization.

10.4.1   BIO-WASTE TO BIOMATERIALS

Biowastes are engineered, mimicking the function of the human body as a biomaterial with stringent requisites of biocompatibility, pharmacological acceptability (nontoxicity, non-allergenicity, non-immunogenicity, etc.), mechanical strength, suitable weight and density, and cost-effectiveness due to their direct interaction with living organisms. This section describes the development of biomaterial using valorization strategies from natural waste sources.

10.4.1.1   Biopolymers

Chitin is a natural amino polysaccharide which is similar to cellulose. It is widely extracted from the crustacean waste of the fishing industry such as shrimp, crab, lobster, prawn, and krill shells. They usually contain 20%–30% of chitin, 30%–40% of protein, inorganic salts, mainly 30%–50% of calcium carbonate and phosphate, and 5%–14% of lipids. Therefore, isolation of chitin generally requires consecutive steps of deproteinization, demineralization, and discoloration. Chemical treatment with alkaline solutions or biological fermentative treatments at the acid condition or enzymatic treatments is recommended. Again, chitin is analogous having crystalline microfibrils that form structural components. It has also been extracted from fungal and yeast cell walls [22]. However, the extensive use of crystalline chitin as a biomaterial displays biocompatibility, biodegradability, antimicrobial activity, and mainly eco-safety. With the right choice of extraction strategy, hydrogen-bonded structure can be disturbed. Consequently, fabrication techniques including self-assembly, microcontact printing, and electrospinning to maintain the fiber orientation are used to mimic the multi-layer maple structure and are applied in dye adsorption.

The N-deacetylated derivative of chitin, a copolymer, that is chitosan consisting of glucosamine and N-acetylglucosamine with a varying degree of acetylation, is extracted from a significant part of the wall of fungal cells. The shrimp waste is subjected to biological treatment in a reactor or acid/base chemical treatment. Crystallinity index and acetylation degree are potential factors to determine a block copolymer structure inherited from the parent crystalline chitin. Discoloration or bleaching is advisable in the end using ethanol or other solvents. In contrast to parent chitin, chitosan is readily soluble in dilute acids due to the easy protonation of free amino groups [23]. Hence, chitosan is far more accessible than chitin with antifungal and antibacterial activities with a low molecular weight of 15–200 kDa.

Animal and human bodies constitute a fibrous protein, collagen, that is made up of three α-chains intertwined in the collagen triple-helix. The stability of the structure arises from intra- and inter-chain hydrogen bonding comprised of 15–26 amino acid residues. This polymer exhibits remarkable bioactive properties, which is inspired by its extensive use as a biomaterial in medical applications. Among existing methods such as acidic, alkaline, or neutral solubilization or enzymatic treatments, the moderate yield of extraction is achieved through terminal cleaving using an appropriate solvent/solid ratio. An enzymatic digestion to extract collagen from fish residues exhibit solubility, high activity coefficient, triple helical structure, and a high fibril-forming ability.

10.4.1.2 Hydroxyapatite

The invention of bio-waste valorization deals with the extraction of significant natural hydroxyapatite $Ca_{10}(PO_4)6(OH)_2$, a compound of bone filler. The chemical composition of synthetic hydroxyapatite resembles the mineral component of animal bones, fish scales, eggshells, seashells, and algae. Hence, these natural residues are treated under thermal decomposition and subcritical water and alkaline hydrothermal processes. After extraction, the synthetic hydroxyapatite is used to fabricate a nano-composite scaffold of chitosan to produce a super-paramagnetic material, potentially suitable for bone healing therapies [24]. Most recently, eggshells and seashells are reported as a natural source to use in catalytic applications.

10.4.1.3 Bioplastics

Bioplastics are confined within the biomass, in particular, the emergence of biodegradable plastics from the activity of naturally occurring microorganisms including algae, fungi, and bacteria such as the breakdown of organic matter and gases. On the other hand, all bio-based plastics are not necessarily biodegradable. The European

Bioplastic reports that non-biodegradable polymer of bio-polyethylene terephthalate made from bio ethylene glycol accounts for 30% [25]. However, it can be improved with a barrier and thermal properties for packing applications.

10.4.1.4 Silica and Silicates

Silica and silicate salts are prepared from bio-waste, especially, the biogeochemical cycle of silicon with the uptake of silicic acid (H_4SiO_4) by plants from soil water. The polymerization of silica forms hydrated amorphous silica that has been extracted from bio-waste like terrestrial and marine plants, algae, and rice husks. The biogenic silica extraction requires acidic pre-treatment followed by pyrolytic procedures in the range of 500°C–700°C. A well-defined semi-crystalline silica porous material is fabricated for the extensive use of lithium batteries. And the silica-based composites are welcome in the automobile industry. A 3D crystalline porous silica nanopowder is extracted from an essential agricultural residue of sugarcane bagasse and bamboo culm [26].

Another breakthrough is the extraction of silicate from rice husks and eggshells as a source of calcium with the right choice of dissolving agents such as nitric acid, magnesium nitrate, citric acid, and tetraethyl orthosilicate. These silicate particles are used in hard tissue regeneration.

10.4.1.5 Biopesticides

Biopesticides from the strain of *Bacillus thuringiensis* through semi-solid fermentation attains higher valorization efficiency than solid-state fermentation. The growth of microorganisms on the solid-substrate, otherwise, without water, has constrained or fails to yield higher fermentation productivity and product stability. Higher sugar content in the substrate can cause low mass-transfer efficiency that leads to substrate inhibition [27]. Henceforth, the water content of the substrate is suggested to increase to attain a semi-solid-state fermentation for high valorization efficiency.

10.4.1.6 Enzymes

Enzymes are catalysts that induce biochemical reactions, such as the hydrolysis of cellulose to glucose. The use of enzymes in a biorefinery is fundamental for the production of biofuels and bulk chemicals since it determines the efficiency and velocity of the reactions involved in the process. Despite the expensiveness of enzymes, the substrate extracted from mixed organic waste can be useful through semi-solid fermentation. For instance, microorganisms like fungus can be grown in an optimized media for enhanced production of the complete cellulase system. The rate of release of sugars from mixed organic waste seems compatible with commercial enzyme preparation.

10.5 BIO-BASED CIRCULAR ECONOMY

The growing consumption of bio-waste facilitates the valorization strategy through "limit the waste growth" and "utilize". The circular economy steps forward to design business models for valorized products from the business community; thereby the government conducts audit, validates valorized products, makes sure no harmful impact to the environment, and finally issues licences or eco-labels. It is believed to create awareness in civil society to identify eco-friendly products and use them wisely. An empirical principle of the circular economy towards the "bio-sector" is to implement policy development for a better understanding of business perception. This section delivers the concept of bioeconomy and how that can remain circular with the principles, scopes, and business models for the extent of innovation potentials.

The zero-waste hierarchy has been established for the transition of modern waste to resource management, limited solely on an environmental standpoint. Social, economic, and logistical factors are excluded to preserve the economy for new generations. The current waste hierarchy pivots around the extraction of energy from waste, recycling, value products, and energy preservation. In sustainability aspects, the goals are defined to change consumption habits, rethink business models, and make them waste-free by design and tools. A new set of legislation regarding valorization practices, for instance incineration of bio-waste at high temperature to recover energy, resulting in the emission of persistent organic pollutants as well as a lower destruction efficiency, is to be considered carefully from a reuse perspective.

Since 2014, the introduction of a circular economy strategy has pushed the dominant linear economic strategy of "take, make and dispose of" towards more circular resource management as well eliminating landfill by 2030 [28]. The regional and global level of "circular economy strategy" requires changes in the whole value chain; business models; consumer behavior; awareness or habits of civil society. The homogenous feedstock from the agricultural industry to produce petroleum refinery, protein from mixed organic waste, bioplastic, probiotic in animal feed, graphic carbon, and plasticizers are an integrated approach.

10.5.1   Revenue of Bio-valorization

The potential revenue of mixed waste-based production of products for valorization efficiency is noted worldwide. The bioplastics have shown a less significant value, assuming that enzymatic activity is not altered when transforming and transporting crude products to the market. The adequate design of production and transformation phases to preserve enzymatic and entomotoxic activities could guarantee high revenues per unit of treated mixed organic waste. The industrial implementation of the mixed organic waste valorization is a critical point for the large-scale development of organic waste biorefineries. Moreover, operational costs and environmental impacts and benefits from substituting fossil-based products have to be considered as well. The economic and environmental impacts of developed products can be characterized. Accordingly, further research and development are needed before a complete assessment of the economic and environmental costs and benefits of the cascading utilization of organic mixed waste and associated bioproducts can be characterized. This is left as a key limitation, but such a research objective is short.

10.6   CONCLUSION

This chapter has paved the way for a circular economy, a state-of-the-art concept for the application of mixed organic waste as a feedstock to produce high-valued materials. The valorization efficiency or conversion rate of waste to biopolymer or bioplastic seems high that hit the market demand for the end-use of bioproducts. It reveals that developing bio-valorized products such as biorefineries, biopesticides, and enzymes from bio-waste feedstock can generate high revenues. However, challenges in upscaling exist due to the complexity of biologically-produced products. Using mixed waste as an inexpensive material for the production of enzymes may contribute to the reduction of the operational cost of bioplastics and other high-value products. It is evidencing that bio-waste is a potential resource and not a waste. For this reason, it is recommended to concentrate further efforts in technology development. Also, operational cost and environmental benefits and the impact of these emerging technologies should be addressed in future research.

ACKNOWLEDGMENT

The manuscript has been assigned as CSIR-CSMCRI: 56/2020 under the CSIR-Research Associateship of No. 31/28(0260)/2019-EMR-I. Authors are thankful to the Director of CSIR-CSMCRI for his support and encouragement.

REFERENCES

1. What a Waste 2.0: A Global Snapshot of Solid Waste Management to 2050, http://hdl. handle.net/10986/30317
2. Blikra, E, Romeo D, Thomsen M (2018) Biowaste valorisation in a future circular bio-economy. *Procedia CIRP* 69:591–596.
3. Xu C, Nasrollahzadeh M, Selva M, Issaabadib Z, Luque R (2019) Waste-to-wealth: biowaste valorization into valuable bio(nano)materials. *Chem. Soc. Rev.* 10.1039/c8cs00543e
4. Abdel-Shafy, M (2018). Solid waste issue: Sources, composition, disposal, recycling, and valorization. *Egypt. J. Petrol.* https://doi.org/10.1016/j.ejpe.2018.07.003
5. Corrado S, Sala S (2018) Bio-Economy contribution to circular economy. *Design. Sustain. Technol. Products Policies*, https://doi.org/10.1007/978-3-319-66981-6_6
6. Crenna E, Sozzo S, Sala S (2017) Natural biotic resources: towards an impact assessment framework for sustainable supply chain management, *J. Cleaner Product.*, 172:3669–3684.
7. Saraiva AB (2017) System boundary setting in life cycle assessment of biorefineries: A review. *Int. J. Environ. Sci. Technol.* 14(2):435–452.
8. Pham TPT, Kaushik R, Parshetti GK, Mahmood R, Balasubramanian R (2015) Food waste-to-energy conversion technologies: Current status and future directions. *Waste Manage.* 38:399–408.
9. Sette P, Fernandez A, Soria J, Rodriguez R, Salvatori D (2020) Integral valorization of fruit waste from wine and cider industries. *J. Cleaner Product.* 242:118486.
10. Ayala-Cortes A, Lobato-Peralta D, Arreola-Ramos CE, Martínez-Casillas DC, Pacheco-Catalan DE, Cuentas-Gallegos AK, Arancibia-Bulnes CA, Villafan-Vidales HI (2019). Exploring the influence of solar pyrolysis operation parameters on characteristics of carbon materials. *J. Anal. Appl. Pyrolysis* 140, 290–298. https://doi.org/10.1016/j.jaap.2019.04.006.
11. Di Blasi C (2009) Combustion and gasification rates of lignocellulosic chars. *Prog. Energy Combust. Sci.* 35 (2), 121–140. https://doi.org/10.1016/j.pecs.2008.08.001.

12. Eshtaya MK, Rahman AA, Hassan MA (2013) Bioconversion of restaurant waste into Polyhydroxybutyrate (PHB) by recombinant E. coli through anaerobic digestion. *Int. J. Environ. Waste Manag.* 11: 27.

13. Zhang W, Zou H, Jiang L, Yao LJ, Wang Q (2015) Semi-solid-state fermentation of food waste for production of Bacillus thuringiensis biopesticide. *Biotechnol. Bioprocess Eng.* 20:1123–1132.

14. Pleissner D, Lau KY, Zhang C, Lin CSK (2015) Plasticizer and surfactant formation from food-waste- and algal biomass-derived lipids. *ChemSusChem.* 8:1686–1691.

15. Zhao B, O'Connor D, Zhang J, Peng T, Shen Z, Tsang D, Hou D (2018) Effect of pyrolysis temperature, heating rate, and residence time on rapeseed stem derived biochar Author links open overlay panel. *J. Clean. Prod.* 174, 977–987. https://doi.org/10.1080/09064710.2016.1214745.

16. Phan M, Paterson J, Bucknall M, Arcot J (2017) Interactions between phytochemicals from fruits and vegetables: effects on bioactives and bioavailability. *Crit. Rev. Food Sci. Nutr.* 58 (8), 1310–1329. https://doi.org/10.1080/10408398.2016.1254595.

17. Gua Y, Zhang X, Deala B, Hanc L (2019) Biological systems for treatment and valorization of wastewater generated from hydrothermal liquefaction of biomass and systems thinking: A review *Bioresourc. Technol.* 278, 329–345.

18. López Barreiro D, Bauer M, Hornung U, Posten C, Kruse A, Prins W (2015) Cultivation of microalgae with recovered nutrients after hydrothermal liquefaction. *Algal Res.* 9, 99–106.

19. Shanmugam SR, Adhikari S, Wang Z, Shakya R (2017) Treatment of aqueous phase of bio-oil by granular activated carbon and evaluation of biogas production. *Bioresour. Technol.* 223, 115–120.

20. Zhou Y, Schideman L, Zhang Y, Yu G (2011) Environment-enhancing energy: a novel wastewater treatment system that maximizes algal biofuel production and minimizes greenhouse gas emissions. *Proc. Water Environ. Fed.* 2011, 7268–7282.

21. Khoshnevisan B, Tabatabaei M, Tsapekos P, Rafiee S, Aghbashlo M, Lindeneg S, Angelidaki I (2020) Environmental life cycle assessment of different biorefinery platforms valorizing municipal solid waste to bioenergy, microbial protein, lactic and succinic acid. *Renew. Sustain. Energy Rev.* 117: 109493.

22. Petkowicz C, Vriesmann L, Williams P (2017) Pectins from food waste: extraction, characterization and properties of watermelon rind pectin. *Food Hydrocolloids* 65:57–67.

23. Dreschke G, Probst M, Walter A, Pümpel T, Walde J, Insam H (2015) Lactic acid and methane: improved exploitation of biowaste potential. *Bioresour Technol.* 176:47–55.

24. Uranga J, Etxabide A, Cabezudo S (2018) Valorization of marine-derived biowaste to develop chitin/fish gelatin products as bioactive carriers and moisture scavengers. *Sci. Total Environ.* https://doi.org/10.1016/j.scitotenv.2019.135747

25. Yu Z, Lau D (2017) Flexibility of backbone fibrils in α-chitin crystals with different degree of acetylation. *Carbohydr. Polym.* 174, 941–947.

26. Zhang C, Zhao F, Li R, Wu Y, Liu S, Liang Q (2019) Purification, characterization, antioxidant and moisture-preserving activities of polysaccharides from Rosa rugosa petals. *Int. J. Biol. Macromol.* 124, 938–945.

27. Mihai F-C, Ingrao C (2016) Assessment of biowaste losses through unsound waste management practices in rural areas and the role of home composting, *J. Cleaner Product.* doi: 10.1016/j.jclepro.2016.10.163.

28. Nisticò R, Cesano F, Franzoso F, Magnacca G, Scarano D, Funes IG, Carlos L, Parolo ME (2018) From biowaste to magnet-responsive materials for water remediation from polycyclic aromatic hydrocarbons, *Chemosphere.* 10.1016/j.chemosphere.2018.03.153

11 Recovery of Energy from Plastic Wastes by Pyrolysis Process for Sustainable Waste Management

Riasha Pal, Nilanjan Paul, Deep Bhattacharya,
Rajupalepu S. Monish and Samuel Jacob
SRM Institute of Science and Technology, India

CONTENTS

11.1 INTRODUCTION

Waste management refers to the collection, transportation, recovery, and neutralization of waste in a manner that does not pose a threat to the environment. Constant
growth in waste quantity has led to increased research and development in waste
management strategies (Qu et al., 2013; Shao et al., 2013). Plastic waste production
is a growing impediment in today's scenario. Although plastics can be recycled and
reused, in the end, it is converted into a non-degradable waste that is harmful to the
environment (Janusz, 2015; Wilk and Hofbauer, 2013). This waste contains plenty of
potential energy inside them as carbon, linked with high energy bonds. This energy
can be transformed and converted into useful forms such as bio-oils or synthetic
pyro-oils through different thermochemical processing methods as represented in
Figure 11.1. The first law of thermodynamics, which states that "energy can neither
be created nor destroyed, only transformed," supports this theory too (Khan et al.,
2016; Environment and Plastics Industry Council, 2004).

Recently, with the pandemic occurrence of COVID-19, many environmentalists
have raised their concern about the rise in demand and the use of plastic products
such as PPE, gloves, syringes, and single-use plastic bags to reduce cross-contami-
nation. Furthermore, many plastic companies have made use of this fearful and
uncertain atmosphere to rebuke many environment-protection laws against the use of
recalcitrant products (Tenenbaum, 2020). As the pandemic continues, plastic waste is
going to increase. Hence, it is crucial to invest in an alternative strategy that is economical, beneficial, and safe to dispose of contaminated personal protective equipment (PPE).

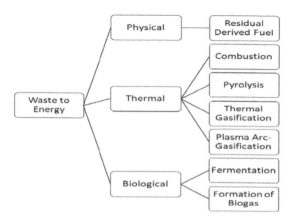

FIGURE 11.1 Methods of waste disposal for the recovery of energy. (Arthur et al., 2014)

One of the most prominent physical methodologies employed in waste management is the use of thermal treatment. The use of heat energy to process the waste has the following advantages:

- Conversion of waste into its safer subsidiaries;
- Notable reduction in weight and volume of the waste;
- Ample possibility of recovery of energy in the form of heat (Bujak, 2015).

Similar to other organic compounds, plastics, too, have a range of calorific value that can be made use of in thermal treatment methods (Paradela et al., 2009; Environment and Plastics Industry Council, 2004). The calorific value of some plastics is shown in Table 11.1 followed by the calorific value of certain waste fractions in Table 11.2.

The above tables are indicative of a higher calorific value presented by plastics, yet underutilized. This data ensures the viability of using plastic as an alternate source of fuel. Moreover, it makes the thermal treatment of plastics an economic and lucrative solution for plastic waste disposal. Energy can be recovered with the help of electrical or thermal energy generated through pyrolysis, gasification, or incineration (Buekens, 1978; Al-Salem et al., 2009). From our earlier discussion, it is evident that plastic has become a necessary evil in our daily lives. Now, more than ever, it is important to develop technologies that would help us manage this plastic solid waste

TABLE 11.1
Calorific Value of Plastics (Environment and Plastics Industry Council, 2004)

Material	Btu	Kilojoules per kg
Plastics		
Polyethylene	20,000	45,500
Polypropylene	19,300	45,000
Polystyrene	17,900	41,600
PET	9,290	21,600
PVC	8,170	19,000
Fuel oil	20,900	48,500
Coal	11,500	27,000
Wood	6,700	15,500

TABLE 11.2
Calorific Value of Waste Fractions (Al-Salem et al., 2009)

Material	Calorific Value (GJ/t)
Food	2
Garden	5
Paper	15
Wood	15
Textile	16
Plastics	30

more efficiently. As interpreted in Tables 11.1 and 11.2, plastics have the highest calorific value amongst other waste fractions. This enables us to look at energy recovery technologies as a feasible solution for plastic solid waste (PSW) management which has been discussed in detail in this chapter.

11.2 THERMOCHEMICAL TREATMENT

Amongst other methods, chemical recycling is most suitable for the conversion of polymers into useful monomers or petrochemicals (Mastellone, 1999). Advanced thermochemical processes have proven to result in a higher energy yield with minimum waste (Smolders and Baeyens, 2004). The thermal treatment has enabled monomer recovery by up to 60% and has led to the production of valuable petrochemicals such as gases, tars, and char (Al-Salem et al., 2009). Thermochemical treatment can be subdivided amongst gasification (conversion of organic matter in the presence of thermal energy and limited oxygen for the formation of synthetic gas), hydrogenation (hydrogen cracking), and pyrolysis (conversion of matter in an inert environment) (Bridgwater, 2003).

11.2.1 GASIFICATION

This process deals with the reaction of PSW with gasification aiding agents like steam, oxygen, or air at a temperature range of 500°C–1300°C to produce a valuable by-product known as synthetic gas or syngas (Tsiamis and Castaldi, 2018) (Figure 11.2). This syngas has proven to be very useful in making fuel cells and fuel for the generation of electricity (Saebea et al., 2019; Brems et al., 2013).

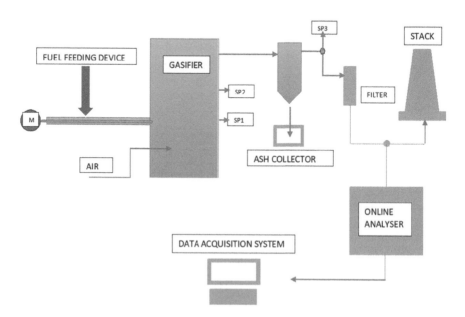

FIGURE 11.2   Pre-Pilot bubbling fluidized bed gasifier (Zaccariello et al., 2015).

11.2.2 HYDROGENATION

This process is employed in polyvinyl chlorides (PVC) to obtain hydrochloric acid, halogenated solid residue, and gas as the by-products. Hydrogenation causes dechlorination in PVC under 300°C–400°C (Awasthi et al., 2013).

11.2.3 PYROLYSIS

Pyrolysis is a process wherein PSW is heated in the absence of oxygen until it gets decomposed into gases and oil with high calorific value. Pyrolysis is carried out at high temperatures up to 600°C or higher to produce gas molecules. At a lower temperature <400°C, a viscous liquid is produced which is termed as pyro-oil or bio-oil (Ioannis and Guohong, 2007). This is the most suitable method of thermochemical treatment as it reduces carbon dioxide emissions, landfilling, and encourages the development of bio-oils and gas fuels that can help with the energy crisis (Chen et al., 2014; Awasthi et al., 2013).

Pyrolysis can be divided into three types (Figure 11.3):

- **Slow/conventional pyrolysis** – Adequate heating rates of approximately 10°C min⁻¹ and maximum temperature of around 300°C, giving an almost equal distribution of oils, chars, and gases, is referred to as slow or conventional pyrolysis (Wu et al., 2014; Chowdhury and Sarkar, 2012).
- **Intermediate pyrolysis** – This process includes a reaction temperature of 400°C–500°C, a relatively slow heating rate between 1°C–1000°C/s, and a residence time of 5–10 min. Solid residues (biochar) produced during this process can be used for many other purposes of energy recovery (Li et al., 2011; Li et al., 2013). The latest advances in intermediate pyrolysis have solved the problems of fast pyrolysis, including waste conditioning and low moisture content (Hornung, 2013).
- **Fast pyrolysis** – This thermochemical process is responsible for producing bio-oil from various wastes. Traditional fast pyrolysis conditions include rapid

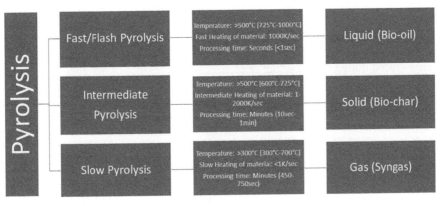

FIGURE 11.3 Different types of pyrolysis, optimized conditions and products.

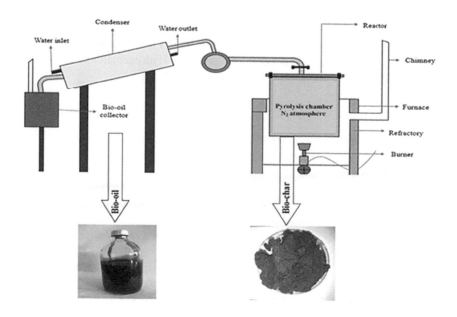

FIGURE 11.4 The overview of the pyrolysis process. (From the lab of the authors).

heating levels >100°C/min, temperatures ranging between 400°C–600°C, and a brief residence time of waste substances and high-temperature pyrolysis vapors for 0.5 s to 2 s (Mašek, 2016).

- **Flash pyrolysis** – Also known as very fast pyrolysis, it includes high reaction temperatures of 900°C–1300°C, a short residence time of <0.5 s (less than that of fast pyrolysis), and a prompt heating rate of more than 1000°C/s. Flash Pyrolysis showcases low water content yield, high liquid biofuel production, and conversion efficiency of up to 70% (Li et al., 2013).

Organic vapors that form are condensed into an oil as a liquid fuel. It takes around 2–4 h. to turn waste into usable energy products (Figure 11.4). Approximately 80% of plastic can be converted into liquid oil by pyrolysis (Miandad et al., 2016; Demirbas and Arin, 2002).

Plastics are often made up of different types of polymers. One of the major challenges faced during the recycling of plastics is the presence of heterogeneous polymers in them. This makes their recovery complex and economically draining (Beyene, 2014). Table 11.3 represents some of the indicator tests for the identification of types of plastics.

11.2.3.1 Pyrolysis of Different Types of Plastics

11.2.3.1.1 Polyethylene Terephthalate (PET)

PET is a thermoplastic polymer with a semi-crystalline structure used for manufacturing several products for different purposes such as fibers and filaments, sheets, and bottles of soft drink based on their physical characteristics. The manufacture of PET

TABLE 11.3

Identification of Different Types of Plastics (Jefferson and Robert, 2009)

Test	PE	PP	PS	PVC
Water	Floats	Floats	Sinks	Sinks
Burning	Blue flame with yellow tip	Yellow flame with blue base	Yellow sooty flame	Yellow sooty, Smoke
Smell after burn	Like candle wax	Like candle wax but less strong	Sweet	Hydrochloric acid
Scratch	Yes	No	No	No

fibers started in the 1960s, and since then global demand has risen steadily at a rate of 8.3% per year. As a result of the versatility of its applications in different consumer goods, a large quantity of PET waste is also produced, which includes both wastes from polymer production and used goods. With an indispensable need to keep the environment safe, recycling PET waste in an eco-friendly manner is the only option (Shukla and Harad, 2006). Liquid oil production from PET pyrolysis is 23%–40% by wt, while the gaseous output is 52%–77% by wt. According to this, polyethylene terephthalate (PET) is considered the most appropriate polymer used in the thermo-chemical process of pyrolysis when the production of gas fuel or syngas is required, for generating heat or thermal energy (Sharuddin et al., 2016; Sharuddin et al., 2017).

11.2.3.1.2 High-density Polyethylene (HDPE)

HDPE is a long, linear chain of polymer made from the monomer ethylene that falls under the category of thermoplastics. HDPE has less branching and therefore a high degree of crystallinity and durability (Sharuddin et al., 2016). When used for tubing, HDPE is often referred to as "alkathene" or "polythene". Due to an extraordinary ratio of strength-to-density, it is used in the manufacture of plastic lumber, corrosion-resistant tubing, geo-membranes, and plastic bottles. The resin identification code (RIC) for HDPE is #2. Due to its extensive applications, HDPE is the third-largest type of plastic wastes, contributing approximately 17.6% to this category. The global demand for HDPE reached more than 30 million tons in 2007. HDPE pyrolysis is performed in the presence of nitrogen as an inert carrier gas medium between temperatures of 500°C–800°C for the processing of bio-oil, syngas, and biochar. The optimal temperature of 550°C is required for achieving the highest efficiency of 70% for the production of bio-oil (Al-Salem, 2019). Temperatures beyond this tend to reduce the yield of bio-oil and increase the production of syngas as the cycle reaches the highest level of thermal degradation.

11.2.3.1.3 Polyvinyl Chloride (PVC)

PVC, also known as vinyl, is a highly durable thermoplastic that is commonly purposeful in manufacturing medical equipment, tubing, wire, cable insulation, and so on. It is a thin and brittle material, obtained in the form of granules and powder. PVC production takes place from vinyl chloride monomer (VCM) through polymerization, forming rigid (RPVC) and flexible PVCs. It is positioned as the third-largest polyethylene and polypropylene thermoplastic by volume in the world. Tacticity is the property of a polymer wherein the conformation of the main chain determines its flexibility. The relative stereochemistry of the chloride moiety in a PVC is random, thereby having an atactic (non-regular) stereochemistry. PVC has a limited percentage of crystallinity that affects the properties of the material due to the syndiotactic property of the polymer. Chlorine comprises approximately 57% of PVC by mass. Products made of PVC showcase flexible properties like durability, lightweight, easy to process, and low cost because of which they are known to replace building materials like concrete, wood, ceramics, metal, rubber, and so on, used for several purposes. PVC pyrolysis takes place in a batch reactor with the rate of thermal heating at approximately 10°C/min and temperatures ranging from 225°C to 520°C, under the influence of vacuum and a minimum pressure of 2 kPa. Bio-oil obtained ranges between 0.45%–12.79% by wt. with a rise in temperature. Biochar accumulation is higher than the bio-oil produced with an ultimate yield of to 19.6% by wt. (Czajczyńska et al., 2017). Therefore, pyrolysis of PVC is not preferred as the yield of bio-oil is moderately less whereas the discharge of hazardous products such as hydrochloric acid and other chlorinated compounds such as chlorobenzene present in the end liquid product of pyrolysis may be detrimental to the environment (Aracil et al., 2005). To resolve this, the dechlorination of PVC is essential to decrease the level of chlorine in the residual bio-oil. PVC dechlorination is carried out through pyrolysis with adsorbents applied to the PVC sample, catalytic pyrolysis, or stepwise pyrolysis. Pyrolysis of PVC, therefore, requires an additional dechlorination step which is one of its disadvantages (Sharuddin et al., 2016; Sharuddin et al., 2017).

11.2.3.1.4   Low-density Polyethylene (LDPE)

LDPE is a translucent polymer with a semi-rigid structure due to its increased side-chain branching. LDPE processing takes place at a high pressure of about 1000–3000bar and a temperature of about 80°C–300°C through the polymerization cycle of a free radical. LDPE consists of 4,000–40,000 carbon atoms with several small branches. The production of LDPE takes place in two processes, a stirred autoclave or a tubular route. Production of LDPE is favored in the tubular routes over the autoclave one due to its higher rates of conversion of ethylene. Low-density polyethylene is primarily used for manufacturing containers, bottles, molded laboratory equipment, tubes, and, most importantly, in the production of plastic bags (https://omnexus.spe-cialchem.com/selection-guide/polyvinyl-chloride-pvc-plastic).

LDPE pyrolysis takes place between temperatures ranging from 500°C to 700°C which results in the formation of aliphatic compounds consisting of a sequence of alkanes, alkenes, and alkynes. The concentration of these aliphatic compounds decreases with a rise in temperature. The key gases produced during this process include hydrogen, methane, ethane, ethene, propane, propene, butane, and butene, where its yield drastically rises with a rise in pyrolysis temperature. Bio-oil obtained as a residue contains an increased aromatic composition with increasing temperature. A significant concentration of polycyclic aromatic and single-ring aromatic compounds is detected at a temperature of 700°C. Oil and wax derived from pyrolysis have immense purpose to be reused in the petrochemical industry (Williams and Williams, 1999).

11.2.3.1.5 Polypropylene (PP)

PP is characterized as thermoplastics belonging to the polyolefin group (Aguado et al., 2002). It is made using Zieglere-Natta catalysts from propylene monomers in a process of chain-growth polymerization. Polypropylene is partially crystalline and non-polar and can be found in three forms, isotactic, syndiotactic, and atactic (Begum et al., 2020). PP exhibits high stiffness, hardness, low specific gravity, excellent tensile strength, and improved chemical, heat, and stain resistance due to its higher level of crystallinity. PP is the second largest manufactured generic plastic (after polyethylene) and mostly used for packaging and labeling purposes due to its high ratio of strength-to-weight which makes it an effective choice. The density of polypropylene (PP) is between 0.895 and 0.92 g/mL, the melting point occurs in a range, isotactic PP is 171°C (340°F). The thermal expansion of polypropylene is lesser than that of polyethylene (Whiteley et al., 2000). PP used for industrial purposes is primarily isotactic and with an intermediate degree of crystallinity between LDPE and HDPE. Isotactic and atactic PP has a high solubility in para-xylene at a temperature of 140°C. The atactic component of PP remains soluble when cooled to a temperature of 25°C while the isotactic form precipitates at this temperature. At a melting point of 130°C (266°F), syndiotactic polypropylene has a crystallinity of 30%. PP becomes brittle below 0°C. At a temperature of 300°C, PP pyrolysis produces maximum bio-oil at around 69.82% by wt. with a total conversion rate of 98.66% and at 400°C, char formation increases from 1.34% by wt. to 5.7% by wt. and the total conversion rate drops to 94.3%. This indicates that biochar formation increases at higher temperatures and reduces the amount of bio-oil produced (Ahmad et al., 2014).

11.2.3.1.6 Polystyrene (PS)

PS is one of the most commonly used plastics for the manufacture of molded pots, lids, bottles, radios, and so on, with a production volume of several million tons per year (Maul et al., 2007). It is a synthetic polymer consisting of linear molecules composed of monomeric styrene which is chemically inert. Commercial PS is a rigid, amorphous, inexpensive polymer with a glassy texture. General-purpose PS (GPPS) or crystal PS (CPS) is hard, clear, and brittle with a glittering appearance because it is unfilled. The addition of butadiene or rubber copolymer increases the strength and impact of PS producing high-impact polystyrene (HIPS) (Begum et al., 2020).

The traditional pyrolysis process for polystyrene takes place in a semi-batch reactor. Pyrolytic gases produced can be condensed and stored as pyrolysis oil with a reactor temperature of 500°C and further cooled to 20°C to collect the leftover residue as char (Prathiba et al., 2018).

11.3 HAZARDOUS PLASTICS

Plastics are mainly differentiated based on their type marks called RIC, which differentiates the types of resins used for the manufacturing of different plastic products. Figure 11.5 shows the different RIC for their respective types of plastics. Out of the seven different RICs, RIC #3, #6 and #7 are considered as hazardous plastics.

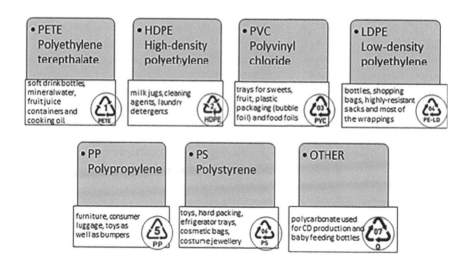

FIGURE 11.5 Plastic coding system guide for resin types (https://polychem-usa.com/plastic-coding-system/).

RIC #3 contains polyvinyl chloride (PVC). It is the most toxic and hazardous type of plastic. PVC contains mercury, bisphenol-A (BPA), lead, cadmium, dioxins, and phthalates, which can be carcinogenic and lead to different forms of cancer, such as breast cancer and can also cause ADHD, endocrine disturbances, allergies, and asthma in children. When burnt, PVC releases dioxins, which are organic pollutants and one of the most toxic chemicals known and potentially harmful to human health. RIC #6 contains polystyrene (PS). PS can leach styrene, a chemical found in exhaled cigarette smoke. Styrene is a probable carcinogen and can also affect the nervous system. Experimental studies with animal models have shown that RIC #6 plastics can damage the liver, lungs, cause mutation in genes, and have adverse effects on the immune system.

RIC #7 contains all the other types of hazardous plastics (O), containing polycarbonates (PC) like acrylonitrile butadiene styrene (ABS), nylons, polyamide (PA), polybutylene terephthalate (PBT), or polyethylene (PE). Polycarbonates may leach bisphenol-A (BPA). BPA is an endocrine disruptor and affects estrogen levels as well as sperm production in males. It causes chromosomal defects in women and leads to puberty disturbances, behavioral changes, disrupts neurological functions, causes adverse impacts on immunity, and even deteriorates cardiovascular health in people. BPA is carcinogenic and may cause breast and prostate cancer. It is also responsible for obesity, infertility, metabolic disorders, diabetes, and cause resistance to chemotherapy being administered as part of treatment measures (TNN, 2018). The plastic coding system followed for different types of chemical (Resin) nature has been represented in Figure 11.5.

11.3.1 Treatment Process

The pre-requisites for successful pyrolysis to proceed are the appropriate choice of input materials and the determination of optimal conditions. Due to these reasons, it is necessary to verify the suitability of the plastic waste several times before the pyrolysis process proceeds along with an assessment of the quantity and quality of the individual pyrolysis products as shown in Figure 11.6.

The crucial steps required in the pyrolysis of plastic wastes are:

- Adequate heating of the plastic within a temperature range;
- Expulsion of oxygen out of the pyrolysis reactor containing the biochar before it starts acting as a thermal insulator and reduces the heat transfer to the plastic wastes;
- Monitored fractionation and condensation of the gaseous product to produce a good quality distillate of the required consistency.

11.3.1.1 Types of Pyrolysis Reactors Used

A significant volume of heat transfer must take place within the reactor walls for pyrolysis treatment, which ensures the utmost importance to the type of reactor suitable for particular material degradation. Various types of reactors are designed to attain sustainable waste management techniques, which include fluidized-bed or fixed-bed reactors, batch or semi-batch reactors, microwave-assisted reactors, rotary kiln reactors and a few advanced revolutionary technologies like solar and plasma reactors (Aishwarya and Sindhu, 2016). Recent developments show that these

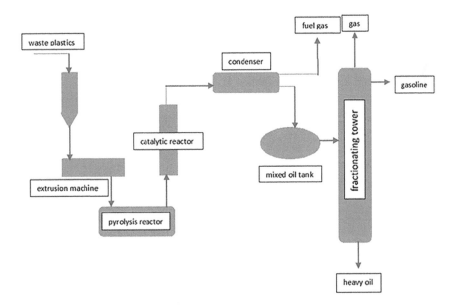

FIGURE 11.6 Pyrolysis-catalytic upgrade technique of plastic wastes (Aishwarya and Sindhu, 2016).

reactors are economically viable when used on an industrial scale which is further elaborated in subsequent sections.

11.3.1.2   Fixed-bed Reactors (FBR)

These reactors are very efficient and have a comparatively lucid design. In this reactor type, the feedstock (often stainless steel) is allocated in the reactor, having an external heating source. An electric furnace is the most common solution. Before the experiment, inert gases like N_2 or Ar are flushed into the reactor, and the gas flow is monitored throughout the entire cycle to maintain an anaerobic surrounding. During pyrolysis, the gaseous emissions are released from the reactor. This process also produces char which is separated after the process and treated for energy recovery. This reactor has a fixed bed with a low rate of thermal heating. Besides, the input feed remains stagnant throughout the cycle on an industrial scale and it is very difficult to assume a consistent distribution of heat of a significant portion of plastic waste. Fixed-bed reactors could be used on a larger scale for tackling and modifying the problem of uniform heating.

11.3.1.3   Batch Reactors

Batch reactors are characterized as closed system reactors that forbid the flow of reactants or products when the reaction is in process. This method results in a higher degree of conversion of feed into their products. On the contrary, semi-batch reactors require the flow of reactants into the reactor and removal of products, while undergoing the process. Maintaining the scale of the reaction is difficult because the feed varies from batch to batch. Long residence time along with difficulty in removing the char adds to some of the disadvantages of batch reactors.

This type of reactor (Figure 11.7) is used to treat polyvinyl chloride (PVC) plastics, so pyrolysis is conducted at a temperature of 225°C–520°C. Hydrogen chloride is the main product obtained which is corrosive and may harm the process equipment. HCl in the presence of chlorinated compounds like chlorobenzene can be very harmful and toxic to the environment hence a dechlorination step was essentially carried out after pyrolysis. This dechlorination reduces the chlorine content in the oil, which is obtained after the process. Hence this step adds a disadvantage to this process in terms of the high cost.

11.3.1.4   Fluidized-bed Reactors (FBR)

Fluidized-bed reactors support uniform mixing of the feed and are designed to have a high heating rate. Rapid pyrolysis of waste can be studied through this reactor which is followed by secondary oil cracking as it has a long residence time. Therefore, the impact of the synchronization of temperature and the time of residence on pyrolysis is typical of a fluidized-bed reactor (Xiao et al., 2007). This reactor is useful for energy recovery from plastic waste through pyrolysis. In other words, plastic waste pyrolysis in a fluidized-bed reactor can provide exceptional benefits over the processes. As polymers have a high viscosity and very low thermal conductivity, efficient and uniform heat transfers cannot arise for cracking of polymers (Wang et al., 2009).

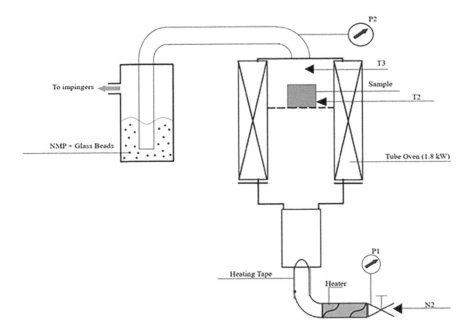

FIGURE 11.7 Batch reactor layout for pyrolysis.

There are certain limitations for this reactor:

- The feed must be crushed into small pieces so that it floats in the liquid;
- Segregating the biochar from the fluidized bed becomes tedious work;
- The strenuous task of feed preparation creates issues related to viability and industrial scale-up.

However, the preparation of pyrolytic oil or bio-oil of a higher-quality from plastic wastes can be extracted significantly from a fluidized-bed reactor (Ding et al., 2016).

11.3.1.5 Conical Spouted Bed Reactors (CSBR)

CSBR shown in Figure 11.8 ensures uniform mixing with the potential of distributing large particle size and particle density variances. This reactor solves the drawback of turning the feed into smaller particles by a fluidized-bed reactor. Due to their rotational movement and high temperature, the melting of waste plastics in this type of reactor provides a regular coating around the particles of sand of the conical spouted bed. According to the latest research, CSBR can be used for catalytic cracking of plastic materials to recover energy (Olazar et al., 2009). There are certain advantages of CSBR over conventional fluidized bed:

- It has low bed segregation and attrition;
- It allows high heat transfer between phases;
- It has very few de-fluidization problems while handling sticky feedstock.

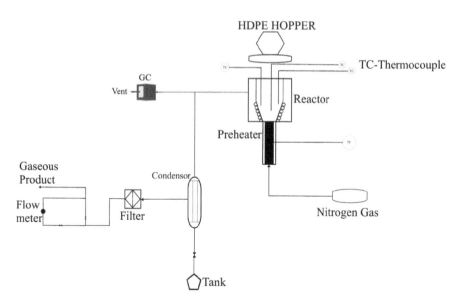

FIGURE 11.8   CSBR in HDPE pyrolysis using zeolite catalyst (Elordi et al., 2009).

CSBR performs HDPE pyrolysis at a temperature of 500°C with a zeolite catalyst that produces 68.7% by wt. of gasoline (C5–C10) from waste plastic (Elordi et al., 2009). There was a high yield of wax when the sticky feed was pyrolyzed at 450°C–500°C (Arabiourrutia et al., 2012). In particular, the design of the spouted bed was suitable for the production of wax at low-temperature pyrolysis.

11.3.1.6   Rotary Kiln Reactors

Rotary kiln reactor shown in Figure 11.9 is useful for the slow pyrolysis of plastic wastes in the industry. This process is performed at an approximate temperature of 500°C with a time of residence as short as 1 hr.

Rotary kiln reactors are the sole reactors that have successfully enacted in various scales as a functional solution for industries to date. Sometimes the feedstock requires a pre-treatment ahead of pyrolysis. It is essential to sort out the wastes appropriately to get rid of unwanted materials. The sorting is followed by shredding to minimize the particle size. Though this process is quite simple compared to other reactors, rotary kiln reactors provide more consistent heat transfer to the feed than fixed-bed reactors, and the operations are less complex than fluidized-bed reactors (Li et al., 2013; Li et al., 2011). The time of residence of the feed within the reactor is the principle parameter during pyrolysis, as it determines the energy provided by the charge at a given heating rate. Usually, the residence time in rotary kiln reactors depends on the mean volumetric flow and the rotational speed of the kiln (Fantozzi et al., 2007).

The rotary kiln has certain advantages like the slow rotational speed of the tilted kiln which allows for reasonable mixing of waste, and uniform heating which yields more renewable pyrolytic products.

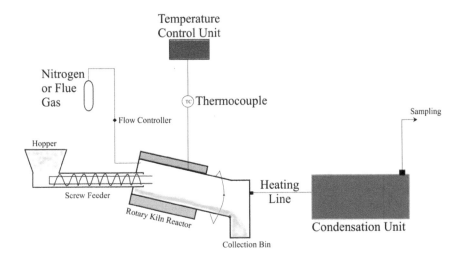

FIGURE 11.9   Rotary kiln reactor for pyrolysis (Undri et al., 2011).

11.3.1.7   Microwave-assisted Reactors

The combination of microwave and pyrolysis has been gaining significant interest due to the complexity and various advantages of the application of microwave for heating. The main advantages of microwaves are efficient and fast internal heating of large biomass particles, prompt response to quick start-up and shutdown, and efficient energy consumption. Agitation and controllability are redundant. A microwave, having a wavelength (1 mm to 1 m) that falls between radio frequencies and infrared rays in the electromagnetic spectrum, is a sustainable energy recovery process of pyrolysis which was successfully studied (Lahm and Chase, 2012). Besides the clarity of this method, the drawbacks that hinder the emergence of industrial microwave heating applications stem from insufficient knowledge of the operations in a microwave and technical skills involved in microwave design. Microwave-assisted pyrolysis was used with absorbers like tires and carbon to treat polyolefin waste (Undri et al., 2014). Feedstock particles have to be granular, and organic vapors must be removed immediately from the reactor to prevent reactions of secondary cracking. Furthermore, account must be taken of the high maintenance costs associated with high electrical power usage.

   Radiation from microwaves has various advantages over the traditional methods of pyrolysis, such as quick heating, higher reaction rate, and lower production costs. The energy emissions from a microwave are provided directly to the material by molecular contact with the electromagnetic field as opposed to traditional methods, therefore less time is taken to heat the surrounding environment. In spite of the benefits of heating using a microwave, there exists a big drawback preventing widespread use of this technology. The comparison between microwave and thermal heating is shown in Table 11.4. When explored on an industrial scale, it lacks sufficient data for which the dielectric characteristics of the waste stream cannot be

TABLE 11.4

Comparison between Microwave and Thermal Heating (Lahm and Chase, 2012)

Microwave Dielectric Heating	Conventional Thermal Heating
Conversion of energy and selective heating	Transfer of energy and non-selective heating
Rapid and efficient	Slow and inefficient
Dependent on the material's properties	Less dependent on the material's properties
Precise and controlled heating	Less controllable

quantified. In other words, using a carbon mixture as a microwave absorber will increase the absorbed energy during pyrolysis, so that it can be converted to heat in a shorter period of time since plastic waste has a reduced dielectric constant. Thus, heating efficiency will vary from one material to another which is a major challenge for the industry. The difference and comparison of microwaves with thermal heating of waste plastic has been tabulated in Table 11.4.

11.3.1.8 Plasma Reactors

Plasma is known to be an ionized gas, which is a mixture of gases of electrons and cations. When the carbonaceous particles are extracted from the waste and injected into the plasma, it undergoes rapid heating of the plasma. It releases volatile matter and after cracking it gives rise to light hydrocarbons like acetylene and methane. Plasma pyrolysis is becoming increasingly important because:

- Quite easily manageable;
- Enables fast and uniform heating;
- Function efficiently under relatively low power consumption.

Organic wastes undergo thermal plasma pyrolysis which produces only two streams: gaseous fuel and solid residue, which are both quite useful and simple to treat. Gaseous output ranges from 50%–98% by wt. The fuel gas is composed of CO, H_2, CH_4, C_2H_2, and C_2H_4, and has a heating value of 4–9 MJ/Nm^3. This can also be used in various energy applications as direct power sources, such as diesel operated engines, boiler systems, and gas turbines (Guddeti et al., 2000).

This type of reactor is used primarily for polypropylene pyrolysis that contains nearly 99.5% carbon. Advanced structures of carbon are found which suggests the potential of various high-value applications of this carbon residue, such as the development of carbon adsorbents, high-surface catalysts, or supercapacitors.

11.3.1.9 Solar Reactors

Solar reactors formulate an innovative way of heating the pyrolysis reactor without much power expenditure in a way of renewable energy supply. Experiments on pyrolysis are conducted under an argon surge in a transparent Pyrex globular reactor. The pellet made of wood was placed into an enclosed graphite sink with black spray in a vertical axis in the middle of a solar furnace of 1.5 kilowatts. Such architecture

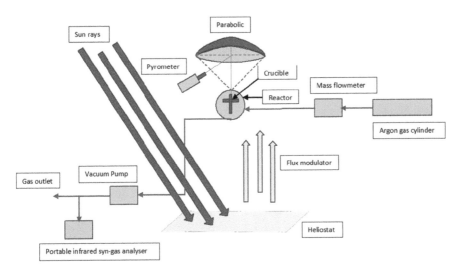

FIGURE 11.10   Pyrolytic heating process using solar reactor (Zeng et al., 2015).

enables the device to hit temperatures ranging from 600°C to 2000°C in the absence of external heating sources. This analysis aims to monitor the effects of the reactor's heating and temperature to initiate pyrolysis. This reactor mechanism enables the usage of renewable energy sources to provide energy for pyrolysis, which in turn generates eco-friendly products (Zeng et al., 2015). The reactor process is elaborated in Figure 11.10.

11.4   VALUE-ADDED PRODUCTS OF PYROLYSIS

With the ever-increasing need for clean and sustainable energy, being able to use thermal energy to convert plastics into fuels (Biochar and Syngas) and bio-oils seems like a novel approach to solving the current energy crisis (Fivga and Ioanna, 2018). Thermal cracking or pyrolysis is the only method that degrades plastic polymers to give carbonized char and a volatile fraction that is later converted into condensable hydrocarbon oil or a non-condensable high calorific gas (Fakhrhoseini and Dastanian, 2013; Zhang and Zhu, 2006). Thus, pyrolysis helps in the making of solid (Biochar), liquid (Bio-oil), and gas (syngas) fuel.

11.4.1   BIO-OIL

Bio-oil can be defined as plastic-derived liquid hydrocarbon at normal temperature and pressure. Three kinds of plastics, namely, PE, PP, and PS, can be used in the feedstock to obtain such a fuel source. This, however, cannot be used as it is. Bio-oil may contain numerous contaminants such as amines, water, and alcohol that can affect its yield drastically. Moreover, to be able to use bio-oil in traditional engines, it must conform to pre-existing viscosity and ash concentration parameters.

11.4.2 PRODUCTION OF BIO-OIL

To convert PSW to bio-oil, pyrolysis is employed exclusively. The process begins with the thermo-lysis of plastics in a bioreactor (300°C–550°C) for decomposition to begin. Depending on the kind of plastic there is a carbon-rich matter that develops as a deposit. This is removed in later stages when pyrolysis is over (Zhang and Zhu, 2006). Figure 11.11 is a schematic representation of a liquid fuel production plant (Beyene, 2014). The mixture of hydrocarbons results in oil-like consistency and is distilled out immediately. The evaporated oil is cracked with the help of a catalyst (microporous zeolites). This is where modifications like an attachment of fractional distillation equipment are installed to meet the specific requirements. After the distillation is over there are certain oils with high boiling points such as kerosene and diesel which are condensed using a water condenser.

11.4.3 BIOCHAR

Biochar is produced from heating (300°C–1000°C) waste in the absence of oxygen and is a co-product of the pyrolysis and gasification processes used to manufacture liquid and gaseous fuels (Arutselvy et al., 2020; Kumar et al., 2010).

Biochar produced through pyrolysis of plastics can be used as a fuel for combustion and gasification and also serve as a raw material for producing activated carbon and carbon nano-tubes. The production of biochar from PSW has been explained in subsequent sections.

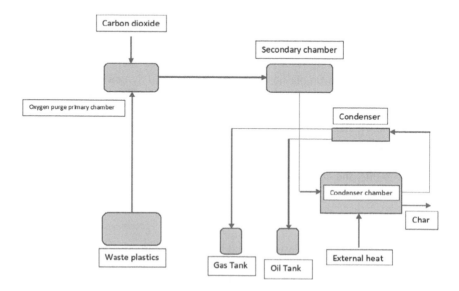

FIGURE 11.11 Schematic diagram of a pyrolysis plant for thermal degradation of plastic wastes (Kalargaris et al., 2017).

11.4.3.1 Production of Biochar from Plastic Wastes

Pyrolysis of plastic wastes with HDPE as its major constituent through fast pyrolysis results in the formation of biochar at a temperature of 400°C–450°C. The biochar formed post-pyrolysis contains a sufficient amount of fixed carbon and volatile matter, a small amount of moisture, and a negligible amount of ash. The amount of hydrogen, carbon, sulfur, and nitrogen constituent in the biochar was found to be 3.06%, 42.65%, 1.80%, and 0.43% respectively. Thermal heating of biochar produced from the plastic wastes pyrolysis increases the number of pores on its surface to a considerable extent to form a mass with extremely high adsorption capacity and form-activated carbon (Figure 11.12). The char derived can be used as a fuel for cooking purposes. Compared to fuel from wood, biochar derived from plastic pyrolysis requires an extended time to ignite and obtain a stable flame and releases an unwanted amount of smoke and smell but can be reused several times due to the continuous process of melting and solidification of the plastic (Jamradloedluk and Lertsatitthanakorn, 2014). Therefore, biochar produced from pyrolysis of plastic wastes cannot be used as a substitute to fuel but can be successfully used for surfacing of roads and in water treatment due to its high porosity and adsorption capacity.

Biochar can be used as a raw material for the molding of materials. The addition of biochar to a considerable extent will significantly enhance the ductility, tensile, and impact properties of the materials (Zhang et al., 2017). Biochar can, therefore, be used as a sound substitute for conventional (carbon) fillers in both thermoset and thermoplastic composite materials, due to their high carbon content and porosity, for the manufacture of reinforced plastics (Bartoli et al., 2020).

11.4.4 Gas Fuel

This refers to the highly volatile gas obtained from the decomposition of PSW. They may be broadly classified into gaseous hydrocarbons that are present in the gaseous state under normal temperature and pressure. The other is syngas, which is a mixture

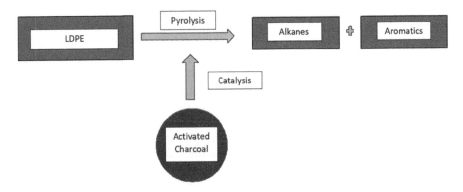

FIGURE 11.12 Activated carbon produced from plastic wastes (Huo et al., 2020).

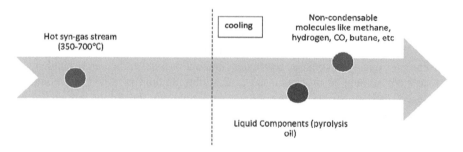

FIGURE 11.13 Syngas production stream (http://www.biogreen-energy.com/wp-content/uploads/2017/08/Biogreen-syngas-repartition.png).

of harmful gases like carbon dioxide, carbon monoxide, and hydrogen with trace amounts of hydrocarbons. Both are equally beneficial as a fuel source (Figure 11.13).

Pyrolysis and gasification are two processes which denote the ideal pathways for energy converted from waste. This is due to their high efficiency, multifaceted feedstock and the potential scalability, and the consistency of the product, compared to other chemical or biochemical energy conversion techniques.

11.4.5 Production of Syngas (Synthetic Gas)

In contrast to bio-oil production, this is a more complex process as it has many stages in it, namely pre-treatment, gasification, gas cleaning, and storage. It begins with the decomposition of PSW into a mixture of gases such as ethane, ethylene, propane, and so on. The mixture is then oxidized with the help of gasification agents. Gasification agents help keep the temperatures at a raised level without the help of an external source (Al-Salem et al., 2009). Gasification takes place at a high temperature of 800°C–1000°C. Depending on the pollutants carbon and nitrogen, rich deposits and mixtures are formed. These waste products are filtered out carefully. It is concluded from a study of co-pyrolysis of PS with charcoal and materials containing lignocellulose that polystyrene does not have an effect on pyrolysis of lignocellulose whereas the biochar formed enhances styrene hydrogenation and other oligomers present in the material to increase the yield of hydrogenated aromatic compounds, such as ethylbenzene and toluene (Jakab et al., 2000). MSW (Municipal Solid Waste) does not contain only plastic waste. It contains a host of other wastes like paper waste, bio/food waste, sludge remnants, and so on. These wastes have low energy density values due to moisture and oxygen content. This poses a drawback during energy conversion, as the accumulation of other waste materials consumes most of the energy, and waste management becomes inefficient due to the usage of a high amount of resources. Co-pyrolysis offers a solution to this drawback, helping in better utilization of MSW and to get maximum energy/calorific value, reducing landfill, and the emission of greenhouse gases.

In MSW, PS is the largest and the most common non-recyclable plastic that can be found in landfills. The large energy density value is key to efficient energy production in the feedstock. The activation energy of polystyrene was calculated by

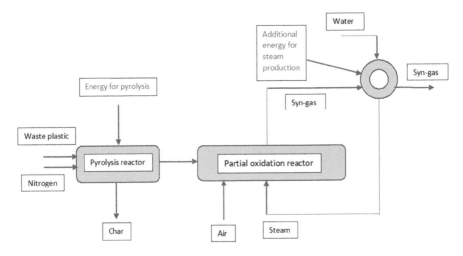

FIGURE 11.14 Schematic diagram of a syngas production plant (Martínez et al., 2014).

Thermogravimetric Analysis (TGA) and was found to be in the range of 168–286 kJ/mol. PS waste yielded high liquid output in the form of styrene, and polyethylene yielded high gaseous output as propenyl-benzene (Encinar and González, 2008). Figure 11.14 is a schematic representation of a typical syngas production plant.

Companies such as Lanza-Tech have made significant efforts to use similar techniques to produce liquid fuel and alcohol amongst other products. Gasification, pyrolysis, and co-pyrolysis offer efficient methodologies to obtain maximum syngas output.

11.5 ENERGY RECOVERY FROM PYROLYSIS OF PLASTIC WASTE

In a broad spectrum, pyrolysis is a key technique that aids in energy recovery. This was thought to be an alternative to decrease landfills and rising plastic levels.

From pyrolysis of non-recycled plastics (NRP), three main products are obtained as follows:

- Pyrolysis oil— bio-oil
  o Derived mainly from HDPE (high-density polyethylene)
- Tar— biochar
  o Derived majorly from LDPE (low-density polyethylene)
- Non-condensable gases termed as syngas (hydrocarbon gas, carbon dioxide, etc.)
  o Derived from PP (polypropylene).

PS is also part of the plastic feed in certain measurements, but when extracted does not yield any viable product that can be used or processed. PS is extruded as a whole to make commercially viable products (food packets). The energy obtained in plastic pyrolysis can be made into commercially viable products (Figure 11.15).

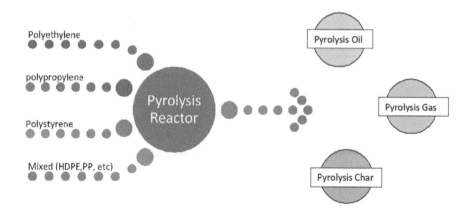

FIGURE 11.15 Pyrolysis of plastic waste to reuse of different energy sources (Singh et al., 2020).

Energy recovery in plastic pyrolysis is key to three factors:

- Sustainable source formation;
- Minimizing the effect of plastic pollution into a more efficient resource;
- Reusing and recycling the feedstock, efficiency is high and complete usage of feedstock is done to minimize further landfill of secondary products.

11.6 CONCLUSION

Due to an increase in consumer demand, the use of plastics has become unavoidable and inevitable. However, countries should devise their plan and policy to circumvent the gaps in plastic waste management, thereby reducing the burden on the environment. Creating wealth in the form of fuel, chemicals, and fertilizers from the waste has become a way of sustainable management of plastic wastes. Research intensification in these areas by finding appropriate green routes could further increase the diversity of product generation from the plastics discarded in landfills, thereby leveraging the attainment of sustainable development goals (SDG) as prescribed by the United Nations.

REFERENCES

Aguado, R., Olazar, M., Gaisán, B. (2002). Kinetic study of polyolefin pyrolysis in a conical spouted bed reactor, *Industrial & Engineering Chemistry Research*, 41: 4559–4566.

Ahmad, I., Ismail, K.M., Khan, H., Ishaq, M., Tariq, R., Kashif, G., Waqas, A. (2014). Pyrolysis study of polypropylene and polyethylene in to premium oil products, *International Journal of Green Energy*, 12: 663–671.

Aishwarya, K.N., Sindhu, N. (2016). Microwave assisted pyrolysis of plastic waste, *Procedia Technology*, 25: 990–997.

Al-Salem, S. M. (2019). Thermal Pyrolysis of High-Density Polyethylene (HDPE) in a novel fixed bed reactor system for the production of high value gasoline range hydrocarbons (HC). *Process Safety and Environmental Protection*, 127: 171–179.

Al-Salem, S. M., Lettieri, P., Baeyens, J. (2009). Recycling and recovery routes of plastic solid waste (PSW): A review. *Waste Management*, 29(10): 2625–2643.

Arabiourrutia, M., Elordi, G., Lopez, G., Borsella, E., Bilbao, J., Olazar, M. (2012) Characterization of the waxes obtained by the pyrolysis of polyolefin plastics in a conical spouted bed reactor. *Journal of Analytical and Applied Pyrolysis*, 94: 230–237, doi: 10.1016/j.jaap.2011.12.012.

Aracil, I., Font, R., Conesa, J. A. (2005). Semi volatile and volatile compounds from the pyrolysis and combustion of polyvinyl chloride. *Journal of Analytical and Applied Pyrolysis*, 74: 465–478.

Arthur O., Baraka K., John G., Karoli N., Mtui P. (2014). Potential of municipal solid waste, as renewable energy source - a case study of arusha, tanzania. *International Journal of Renewable Energy Technology Research*, 3: 1–9.

Arutselvy, B., Rajeswari, G., Jacob S. (2020). Sequential valorization strategies for dairy wastewater and water hyacinth to produce fuel and fertilizer. *Journal of Food Process Engineering*. doi: 10.1111/jfpe.13585.

Awasthi A.K., Shivashankar, M., Majumder, S. (2013). Department of Chemistry, School of Advanced Sciences, VIT University, Plastic solid waste utilization technologies: A Review. *IOP Conf. Series: Materials Science and Engineering*, 263:022024.

Bartoli, M., Giorcelli, M., Jagdale, P., Rovere, M. (2020). Towards traditional carbon fillers: biochar-based reinforced plastic, *IntechOpen*, 10.5772/intechopen.91962.

Begum, S.A., Rane, A.V., Kanny, K. (2020). Applications of compatibilized polymer blends in automobile industry. *Compatibilization of Polymer Blends*, 563–593.

Beyene H.D. (2014) Recycling of plastic waste into fuels, a Review. *International Journal of Science, Technology and Society*, 2:190–195. doi: 10.11648/j.ijsts.20140206.15.

Brems, A., Dewil, R., Baeyens, J., Zhang, R. (2013). Gasification of plastic waste as waste-to-energy or waste-to-syngas recovery route. *Natural Science*, 5: 695–704.

Bridgwater, A. (2003). Renewable fuels and chemicals by thermal processing of biomass. *Chemical Engineering Journal*, 91: 87–102.

Buekens A. G. (1978). Resource recovery and waste treatment in Japan, *Resource Recover Conservation*, 3 (3): 275–306.

Bujak, J. (2015). Thermal treatment of medical waste in a rotary kiln. *Journal of Environmental Management*, 162: 139–147.

Chen, D., Yin, L., Wang, H. (2014). Pyrolysis technologies for municipal solid waste: a review, *Waste Management*, 34: 2466–2486.

Chowdhury, R, Sarkar, A. (2012). Reaction kinetics and product distribution of slow pyrolysis of Indian textile wastes, *Industrial & Engineering Chemistry Research*, 10: 28–32.

Czajczyńska, D., Anguilano, L., Ghazal, H., Krzyżyńska, R., Reynolds, A. J., Spencer, N., Jouhara, H. (2017). Potential of pyrolysis processes in the waste management sector. *Thermal Science and Engineering Progress*, 3: 171–197.

Demirbas, A., and Arin, G. (2002). An overview of biomass pyrolysis. *Energy Sources*, 24: 471–482.

Ding, K., Zhong, Z., Zhong, D., Zhang, B., Qian, X. (2016). Pyrolysis of municipal solid waste in a fluidized bed for producing valuable pyrolytic oils. *Clean Technologies and Environmental Policy* 18: 1111–1121.

Elordi, G., Olazar, M., Lopez, G. (2009). Catalytic pyrolysis of HDPE in continuous mode over zeolite catalysts in a conical spouted bed reactor, *Journal Analytical Applied Pyrolysis*, 15: 345–351.

Encinar, J. M., González, J. F. (2008). Pyrolysis of synthetic polymers and plastic wastes. Kinetic study. *Fuel Processing Technology*, 89: 678–686.

EPIC, Environment & Plastics Industry Council (EPIC) a council of the Canadian Plastics Industry Association (2004): 1-5.

Fakhrhoseini, S.M., Dastanian, M. (2013). Predicting pyrolysis products of PE, PP, and PET using NRTL activity coefficient model. *Journal of Chemistry*, https://doi.org/10.1155/2013/487676.

Fantozzi, F., Colantoni, S., Bartocci, P., Desideri, U. (2007). Rotary Kiln slow pyrolysis for syngas and char production from biomass and waste—Part I: working envelope of the reactor, *Journal of Engineering for Gas Turbines and Power*, 129: 901.

Fivga, A., Ioanna, D. (2018). Pyrolysis of plastic waste for production of heavy fuel substitute: A techno-economic assessment. *Energy*, 149: 865–874.

Guddeti, R.R., Knight, R., Grossmann, E.D. (2000). Depolymerization of polypropylene in an induction-coupled plasma (ICP) reactor, *Industrial Engineering Chemical Research*, 39:1171–1176.

Hornung, A. (2013). Intermediate pyrolysis of biomass. In: *Biomass Combustion Science, Technology and Engineering*, Ed. Rosendahl L, 172–186, Elsevier.

Huo, E., Lei, H., Liu, C., Zhang, Y., Xin, L., Zhao Y, Ruan, R. (2020). Jet fuel and hydrogen produced from waste plastics catalytic pyrolysis with activated carbon and MgO. *Science of The Total Environment*. doi: https://doi.org/10.1016/j.scitotenv.2020.138411.

Ioannis, K., Guohong, T.S.G. (2007). The utilisation of oils produced from plastic waste at different pyrolysis temperatures in a DI diesel engine, *Energy*, **131:** 179–185.

Jakab, E., Várhegyi, G., Faix, O. (2000). Thermal decomposition of polypropylene in the presence of wood-derived materials. *Journal of Analytical and Applied Pyrolysis*, 56: 273–285.

Jamradloedluk, J., Lertsatitthanakorn, C. (2014). Characterization and utilization of char derived from fast pyrolysis of plastic wastes. *Procedia Engineering*, 69: 1437–1442.

Janusz, W. B. (2015). Thermal utilization (treatment) of plastic waste, *Energy*, 90: 1–10.

Jefferson H, Robert E (2009), Plastic recycling: Challenges and opportunities. *Philosophical Transaction Royal Society London B Biological Science*, 364: 2115–2126.

Kalargaris, I., Tian, G., Gu, S. (2017). Combustion, performance and emission analysis of a DI diesel engine using plastic pyrolysis oil, *Fuel Processing Technology*, 157:108–115. doi: 10.1016/j.fuproc.2016.11.016.

Khan, M. Z. H., Sultana, M., Al-Mamun, M. R., Hasan, M. R. (2016). Pyrolytic Waste Plastic Oil and Its Diesel Blend: Fuel Characterization. doi: https://doi.org/10.1155/2016/7869080.

Kiran, N., Ekinci, E., Snape, C. (2000). Recyling of plastic wastes via pyrolysis. *Resources, Conservation and Recycling*, 29: 273–283.

Kumar, G., Panda, A. K., Singh, R. K. (2010). Optimization of process for the production of bio-oil from eucalyptus wood. *Journal of Fuel Chemical Technology*, 38: 162–167.

Lahm, S. S., Chase H. A. (2012). A review on waste to energy processes using microwave pyrolysis, *Energies*, 5: 4209–4232.

Li, A. M., Li, S. Q., Li, X. D. (2011). Pyrolysis of solid waste in a rotary kiln: Influence of final pyrolysis temperature on the pyrolysis**,** *Journal of Analytical and Applied*, 2: 116–119.

Li, L., Rowbotham, J. S., Christopher Greenwell, H., Dyer, P. W. (2013). An Introduction to Pyrolysis and Catalytic Pyrolysis: Versatile Techniques for Biomass Conversion. New and Future Developments in Catalysis. doi: 10.1016/B978-0-444-53878-9.00009-6.

Martínez, J. D., Murillo, R., García, T., Arauzo, I. (2014). Thermodynamic analysis for syngas production from volatiles released in waste tire pyrolysis. *Energy Conversion and Management*, 81: 338–353.

Mašek, O. (2016). Biochar in thermal and thermochemical biorefineries-production of biochar as a co-product. In: *Handbook of Biofuels Production*, 2nd edition. Rafael L, Carol S.K.L., Wilson K., Clark J. (Eds.), pp. 655–671.

Mastellone, M.L. (1999). Thermal treatments of plastic wastes by means of fluidized-bed reactors. Ph.D. Thesis, *Department of Chemical Engineering*, Second University of Naples, Italy.

Maul, J., Frushour, B.G., Kontoff, J.R., Eichenauer, H., Ott, K., Schade, C. (2007). Polystyrene and Styrene Copolymers. In: *Organic Polymer Chemistry*, 2nd edition, Saunders, K. J. (Ed.), pp. 475–490

Miandad, R., Rehan, M., Nizami, A.-S., El-Fetouh Barakat, M. A., Ismail, I. M. (2016). The energy and value-added products from pyrolysis of waste plastics. In: *Recycling of Solid Waste for Biofuels and Bio-chemicals*, Karthikeyan O.P., Heimann K., Subramanian Senthilkannan Muthu S.S. (Eds.), pp. 333–355

Olazar, M., Lopez, G., Amutio, M., Elordi, G., Aguado, R., Bilbao, J., (2009). Influence of FCC catalyst steaming on HDPE pyrolysis product distribution. *Journal of Analytical and Applied Pyrolysis*, 85:359–365.

Paradela, F., Gulyurtlu, I., Cabrita, I. (2009). Study of the co-pyrolysis of biomass and plastic wastes, *Clean Technologies Environment Policy*, 11: 115–122.

Prathiba, R., Shruthi, M., Miranda, L. R. (2018). Pyrolysis of polystyrene waste in the presence of activated carbon in conventional and microwave heating using modified thermocouple. *Waste Management*, 76: 528–536.

Qu, Y., Zhu, O., Sarkis, J., Geng, Y., Zhong, Y. (2013). A review of developing an e-wastes collection system in Dalian, China. *Journal of Cleaner Production*, 17: 117.

Saebea, D., Chaiburi, C., Authayanun, S. (2019). Model based evaluation of alkaline anion exchange membrane fuel cells with water management. *Chemical Engineering Journal*, 374: 721–729.

Shao, Y., Ren, B., Jin, B., Zhong, W., Hu, H., Chen, X., Sha, C. (2013). Experimental flow behaviors of irregular particles with silica sand in solid waste fluidized bed. *Powder Technology*, 234: 67–75.

Sharuddin, S. D., Abnisa, F., Wan Daud, W. M. A., Aroua M. K. (2016). A review on pyrolysis of plastic wastes. *Energy Conversion and Management*, 115: 308–326.

Sharuddin, S. D. A., Abnisa, F., Wan Daud, W. M. A., Aroua, M. K. (2017). Energy recovery from pyrolysis of plastic waste: Study on non-recycled plastics (NRP) data as the real measure of plastic waste. *Energy Conversion and Management*, 148: 925–934.

Shukla, S. R., Harad, A. M. (2006). Aminolysis of polyethylene terephthalate waste. *Polymer Degradation and Stability*, 91: 1850–1854

Singh, R. K., Ruj, B., Sadhukhan, A. K., Gupta, P. (2020). Thermal degradation of waste plastics under non-sweeping atmosphere: Part 2: Effect of process temperature on product characteristics and their future applications. *Journal of Environmental Management*, 261: 110–112.

Smolders, K., Baeyens, J. (2004). Thermal degradation of PMMA in fluidised beds. *Waste Management*, 24: 849–857.

Tenenbaum, Laura (2020). "The Amount Of Plastic Waste Is Surging Because Of The Coronavirus Pandemic." *Forbes*. https://www.forbes.com/sites/lauratenenbaum/2020/04/25/plastic-waste-during-the-time-of-covid-19/#79e2dc537e48, accessed on 01/07/20.

TNN (2018). "Toxicity alert! Know the 7 types of plastic and which is the MOST dangerous!" *Times of India*. https://timesofindia.indiatimes.com/life-style/health-fitness/photo-stories/toxicity-alert-know-the-7-types-of-plastic-and-which-is-the-most-dangerous/photostory/64828668.cms, accessed on 03/03/20.

Tsiamis, D. A., Castaldi, M. J. (2018). The Effects of Non-Recycled Plastic (NRP) on Gasification: A Quantitative Assessment. https://plastics.americanchemistry.com/NRP-Gasification-Report.pdf

Undri, A., Rosi, L., Frediani, M., Frediani, P. (2011). Upgraded fuel from microwave assisted pyrolysis of waste tyre. *Fuel*, 115:600–608.

Undri, A., Rosi, L., Frediani, M., Frediani, P. (2014). Efficient disposal of waste polyolefins through microwave assisted pyrolysis. *Fuel*, 116: 662–671.

Wang, Z., Cao, J., Wang, J. (2009). Pyrolytic characteristics of pine wood in a slowly heating and gas sweeping fixed-bed reactor. *Journal of Analytical and Applied Pyrolysis*, 84: 179–184

Whiteley, K. S., Heggs, T. G., Koch, H., Mawer, R. L., Immel, W. (2000). Polyolefins. *Ullmann's Encyclopedia of Industrial Chemistry*. doi: https://doi.org/10.1002/14356007. a21_487

Wilk, V., Hofbauer, H. (2013). Conversion of mixed plastic wastes in a dual fluidized bed steam gasifier. *Fuel*, 107: 787–799

Williams, P. T., Williams, E. A. (1999). Fluidised bed pyrolysis of low density polyethylene to produce petrochemical feedstock. *Journal of Analytical and Applied Pyrolysis*, 51: 107–126.

Wu, C., Budarin, V. L., Gronnow, M. J., De Bruyn, M., Onwudili, J. A., Clark, J. H., Williams, P. T. (2014). Conventional and microwave-assisted pyrolysis of biomass under different heating rates. *Journal of Analytical and Applied Pyrolysis*, 107: 276–283.

Xiao, G., Ni, M., Huang, H. (2007). Fluidized-bed pyrolysis of waste bamboo, *Journal of Zhejiang University Science B*, 8: 1495–1499.

Zaccariello, L., Cremiato, R., Mastellone, M.L. (2015) Evaluation of municipal solid waste management performance by material flow analysis: Theoretical approach and case study. *Waste Management Research*, 33:871–885. doi: 10.1177/0734242X15595284.

Zeng, K., Minh, D. P., Gauthier D (2015). The effect of temperature and heating rate on char properties obtained from solar pyrolysis of beech wood, *Bioresource Technology*, 182: 114–119.

Zhang, G., Zhu, J. (2006). Prospect and current status of recycling waste plastics and technology for converting them into oil in China, *Resources, Conservation and Recycling*, 50: 231–239.

Zhang, Q., Cai, H., Yang, K., & Yi, W. (2017). Effect of biochar on mechanical and flame retardant properties of wood – Plastic composites. *Results in Physics*, 7: 2391–2395.

https://omnexus.specialchem.com/selection-guide/polyvinyl-chloride-pvc-plastic, accessed on 02/07/20.

https://polychem-usa.com/plastic-coding-system/, accessed on 04/07/20.

http://www.biogreen-energy.com/wp-content/uploads/2017/08/Biogreen-syngas-repartition. png, accessed on 06/07/20.

12 Biosurfactant
A Sustainable Replacement for Chemical Surfactants

Tamil Elakkiya Vadivel, Krishnan Ravi Shankar, Tholan Gajendran, Theresa Veeranan and Renganathan Sahadevan
Anna University, India

CONTENTS

12.1 SURFACTANT

Surfactants are the compounds that reduce the physical phenomenon, that is, interfacial surface tension between two liquids or between a liquid and solid. The surface-active agents must be partly hydrophilic, that is, water soluble and partly lipophilic, meaning soluble in lipids and oils. These are always known as amphiphilic or amphipathic molecules. The hydrophilic portion may contribute to the functional groups such as alcoholic (–OH), carboxylic acid (–COOH), sulfate (–SO_4) and quaternary ammonium (NH_4^+). The alkyl chains may contribute to the lipophilic portion of the molecules (Abdel-Mawgoud et al., 2010).

12.1.1 Disadvantages of Surfactants

- Lack of stability and safety;
- Easily decomposed by heat or light;
- Lower degradation, affects food products;
- Degradation during manufacture and storage;
- Persistent organic pollutant;
- Endocrine disruptor;
- Environmental risk of toxicity.

12.1.2 Need for an Alternative

Recently, preeminence has been given to the environmental impacts caused by chemical surfactants due to their toxicity and difficulty in being degraded within the environment. Increasing the environmental concerns, advances in biotechnology which results in the emanation of more stringent laws has led to biosurfactants being a possible alternative source for the chemical surfactants available on the market.

Biosurfactants offers various advantages over chemical surfactants like biodegradability due to their simple chemical structure, environmental compatibility, and low toxicity: this enables using them in cosmetic, pharmaceutical and food industries.

12.1.3 BIOSURFACTANTS

Biosurfactants are the compounds which are excreted extracellularly and are amphiphilic in nature. They have an excellent tendency to reduce surface and interfacial surface tension. They possess the characteristic property of reducing the surface and interfacial surface tension by using an equivalent mechanism because of the chemical surfactants (Plackett and Burman, 1946). Biosurfactants are produced by a spread of microorganisms including bacteria, filamentous fungi and yeast from different substrates including oils, sugars, and glycerol. Biosurfactants are a group of bio-based products with increasing scientific, environmental, and economic interest. These bio-based products are expected to partially replace conventional (oil based) surfactants and due to their unique structures, properties, low toxicity and high biodegradability may be suitable for novel applications in industrial and environmental biotechnology sectors. Microbial surface-active compounds are important for the physiology of the cells that produce them, as they are involved in cell motility (gliding and swarming motility and also as deadhesion from surfaces), cell-cell interaction (biofilm formation, maintenance and maturation, quorum sensing, amensalism and pathogenicity), cellular differentiation, substrate accession (via interfacial contact and pseudosolubilization of substrates), and also as avoidance of toxic elements and compounds (Rønning et al., 2015). Surface-active compounds produced by microorganisms are of two types:

- Those that reduce surface tension at the air-water interface (biosurfactants;)
- Those that reduce the interfacial tension between immiscible liquids or at the solid-liquid interface (bioemulsifiers).

Biosurfactants usually exhibit an emulsifying capacity but bioemulsifiers do not necessarily reduce physical phenomenon.

12.1.4 TYPES OF BIOSURFACTANT

The chemically synthesized surfactants are generally classified based on their polarity, whereas biosurfactants are usually categorized by their microbial origin and chemical composition. Biosurfactants can also be grouped into two categories, namely: low-moleucular-mass molecules (e.g. rhamnolipids, sophorolipids) with lower surface and interfacial tensions and high-moleucluar-mass polymers (e.g. Biodispersan), which bind tightly to surfaces (Figure 12.1)

12.1.5 PRODUCTION OF BIOSURFACTANTS FROM INEXPENSIVE RAW MATERIALS

Biosurfactant first discovered as extracellular, gains attention due to the emulsifying activity and solubility of insoluble hydrocarbons (Konishi et al., 2014). Natural surfactants, namely biosurfactants, have gained more attention in many industries like food and pharmaceuticals due to the use of renewable sources for their production in order to lower the cost. Certain carbon sources like molasses, glycerol, whey, frying oil, animal fat, soap stock and starch-rich wastes, for example potato wastes, play a main role in the production of biosurfactants to be applied in wide areas in research (Mulligan, 2005) (Table 12.1).

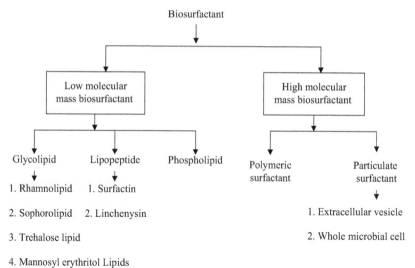

FIGURE 12.1   Types of biosurfactant.

TABLE 12.1

Maximum Yield of Biosurfactants from Inexpensive Raw Materials (Mulligan, 2005)

Low Cost or Waste of Raw Material	Biosurfactant Type	Producer Microbial Strain	Maximum Yield(g/L)
Rapeseed oil	Rhamnolipids	*Pseudomonas*sp. DSM2874	45
Babassu oil	Sophorolipids	*Candida lipolytica* IA1055	11.72
Turkish corn oil	Sophorolipids	*Candida bombicola* ATCC 22214	400
Sunflower &soyabean oil	Rhamnolipids	*Pseudomonas aeruginosa* DS10-129	4.31
Sunflower oil	Lipopeptide	*Serratia marcesecens*	2.98
Soyabean oil	Mannosylerythritol lipid	*Candida* sp. SY16	95
Oil refinery waste	Glycolipids	*Candida antarctica, Candida apicola*	10.5
Curd whey & distillate waste	Rhamnolipids	*Pseudomonas aeruginosa* strainBS2	0.92
Potato process effluents	Lipopeptide	*Bacillus subtilis*	2.7
Cassava flour wastewater	Lipopeptide	*Bacillus subtilis* ATCC21332, *Bacillus subtilis* LB5a	2.2

12.1.6   PRODUCTION

12.1.6.1   Screening Tests for Biosurfactant Production

Biosurfactants are structurally a very diverse group of biomolecules, for example glycolipids, lipopeptides, lipoproteins, lipopolysaccharides, or phospholipids. Therefore, most methods for a general screening of biosurfactant-producing strains are based on the physical effects of surfactants. Alternatively, the ability of strains to interfere with hydrophobic interfaces can be explored. On the other hand, specific screening methods like the colorimetric CTAB agar assay are suitable only to a limited group of biosurfactants. The screening methods can give qualitative and/or quantitative results. For a first screening of isolates, qualitative methods are generally sufficient (Singh, 2012) (Figure 12.2).

12.1.6.2   Hemolysis Method

This is often the primary test to spot the biosurfactant-producing organism. Each isolate was streaked on agar medium and incubated at 37°C for 24–48 h. Then the plates were visually inspected for the zone of inhibition round the colonies which indicates the assembly of biosurfactant.

12.1.6.3   Oil Spreading Test

In this test, oil was layered over the water in a Petri plate and a drop of cell-free extract was added to the surface. A diameter of clear zone on the oil surface was measured in three replications for each isolate and a water droplet was used as a negative control.

12.1.6.4   Emulsification Index Test

The emulsifying capacity of isolates was evaluated by an emulsifying index for hexane, xylene, and crude oil. In order to carry this out, 1.5 ml of hydrocarbon was added

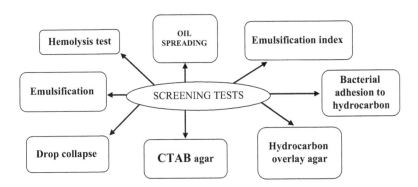

FIGURE 12.2   Different types of screening tests.

to 1.5 ml of cell-free broth in the test tube and vortexed at high speed for 2 min and allowed to stand for 24 h (undisturbed). Calculate the index by using the formula:

$$E24 = \frac{\text{Height of emulsion layer}}{\text{Total height}}$$

12.1.6.5 Bacterial Adhesion to Hydrocarbon Assay (BATH)

This assay is used to estimate the hydrophobicity of the cell surface 2ml of the cell suspension is vortexed with 100μL of hydrocarbon (hexane, xylene and crude oil) for 3 min and allowed to stand for 1h. Reduction in the phase was taken as the percentage of cells adhered to the hydrocarbons, calculated by using the formula:

$$\%\text{Hydrophobicity} = \frac{1 - \text{aqueous phase OD}}{\text{OD of initial cell suspension}}$$

12.1.6.6 Hydrocarbon Overlay Agar Method

In this method LB agar plate coated with crude oil was inoculated with O/N grown culture of isolates and incubated at 30°C for 48–72 h. A colony surrounded by emulsified halos was considered as positive for biosurfactant production.

12.1.6.7 CTAB Agar Plate Method

Agar plate containing cetyltrimethylammonium bromide (CTAB) and methylene blue was used to detect extracellular glycolipid production. Production of biosurfactant was identified by the dark blue halo formation around the colonies.

12.1.6.8 Drop Collapse Method

This method relies on the destabilization of a liquid drop on the hydrocarbon surface by a cell-free extract containing biosurfactants. In this method crude oil (hydrocarbon) was spread on the lid of the petriplate and drop of cell-free extract was placed over it. Water was used as a negative control.

12.1.6.9 Emulsification Assay

1 ml of cell-free culture broth was added to 5ml of 50mM tris buffer (pH 8.0) in a 30ml screw capped test tube and 5 ml of hydrocarbon (crude oil) was also added. The solution was vortexed for 1 min and allowed for phase separation. Then the aqueous phase was separated and OD was measured at 400nm. Emulsification activity per ml was calculated by using the formula:

$$\begin{aligned} &\text{Negative control} - \text{Buffer} + \text{Crude oil; Positive control} \\ &\quad - \text{Triton X} - 100 \left(*\text{DF} - \text{dilution factor} \right) \end{aligned}$$

12.1.7 Fermentation Process

There are two types of fermentation process which are mostly followed for the bio-surfactant production. The types are explained below.

12.1.7.1 Solid Fermentation Process

Solid state fermentation is the cultivation of microorganisms under controlled conditions in the absence of free water. Bran, expelier cake, agar and so on are used as a substrate for solid fermentation process.

- 5 g of substrate was added to 20ml of M9 medium in a 500ml Erlenmeyer flask.
- Mixed thoroughly and autoclaved at 121°C,15 lb pressure for 15 min (before autoclaving particular pH should be adjusted to the respective strain).
- Then the flask was cooled to room temperature and bacterial culture was inoculated to the medium under sterile condition and was maintained at various temperatures to analyze the biosurfactant production.

12.1.7.2 Submerged Fermentation Process

Submerged fermentation is the method of manufacturing the biomolecules in which enzymes and other reactive compounds submerge in a liquid such as alcohol, oil, and nutrient broth.

- Initially the biosurfactant production was carried out in a 250ml Erlenmeyer flask: this contains 50ml of M9 medium with 2.0% as a substrate;
- Batch fermentation was subjected with 3 l working volume in a 5:l laboratory scale fermenter which consists of foam/antifoam probe system;
- The agitation speed was 200 rpm provided by the centrifuge propeller where oxygen and pH were used for the control of the conditions;
- The optimum pH and temperature were adjusted in the fermenter which makes it suitable for the bacteria to produce biosurfactants;
- Then cells were collected at different time intervals (24h, 48h, 72h, 96h) and the cell-free supernatant was subjected for surface activity assessment and bio-surfactant separation.

(*Composition of M9 medium in g /L: Na_2HPO_4, 6.0; KH_2PO_4, 3.0; NH_4NO_3, 1.0; NaCl, 1.0; $CaCl_2$, 0.014; $MgSO_4.7H_2O$, 0.245; thiamine-HCl solution, 1.0 ml and 1 ml of micronutrients solution.)

12.1.8 Parameters Controlling the Production of Biosurfactant

The type and quantity of biosurfactants produced also depends on various factors like nitrogen and carbon source, and temperature, aeration and trace elements also affect their production. Factors like temperature, nitrogen and carbon source, aeration and

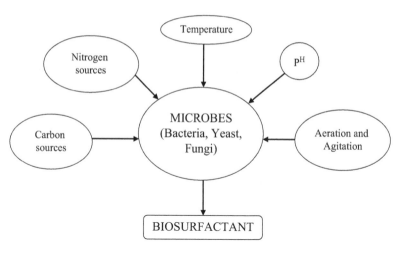

FIGURE 12.3 Responsible parameters for the production of biosurfactant.

trace elements contribute to the type and quantity of biosurfactant produced. The maximum production of biosurfactants can be achieved by optimizing the parameters under controlled conditions. In this connection, among the parameters including temperature, pH, medium composition, and salinity gain more attention for controlling and optimizing biosurfactant production (Figure 12.3).

The production of biosurfactants can be controlled by a combination of a variety of parameters like temperature, pH, aeration and agitation, and carbon and nitrogen sources. A variety of parameters are involved in the production, and optimization plays a vital role in enhancement of the desired product.

12.1.9 OPTIMIZATION FOR BIOSURFACTANT PRODUCTION

12.1.9.1 Response Surface Methodology

Response surface methodology is used to construct empirical model building which deals with the collection of mathematical and statistical techniques. The design is influenced by the output variable (response) and input variable (independent variables). The series of experiments is known as runs in which the modifications can be done in the input variable to obtain the desired output variable. The bottleneck of the RSM deals with the design optimization which is aimed at reducing the cost of expensive analysis methods and their numerical noise (errors) (Carley et al., 2004).

12.1.9.2 Plackett Burman Method

The plackett burman is an experimental design which is used to identify the most important factor early in the experimental phase when complete knowledge about the system is unavailable. The plackett burman designs are often used to screen important factors that influence output measures and the quality of the product. The bottleneck of the PB design deals with the screening of the factors which provides a higher yield of the product (Porob et al., 2013).

12.1.9.3   Extraction of Biosurfactant

Extraction is the process of isolating the desired product from a mixture of a product. The extraction may occur based upon the solubility of the desired product present in the mixture. Mostly the extraction of biosurfactant depends upon its solubility nature in chloroform. Hence, the methanol and chloroform extraction process is widely adopted for most types of biosurfactant (Pathaka and Nakhate, 2015) (Figure 12.4).

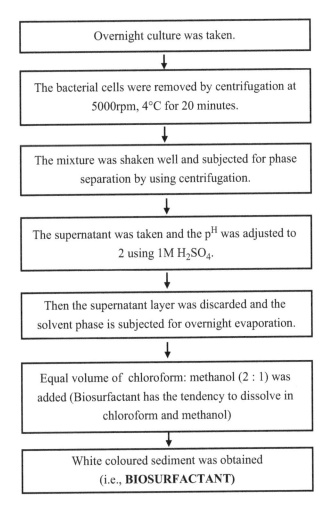

FIGURE 12.4   Extraction of biosurfactant.

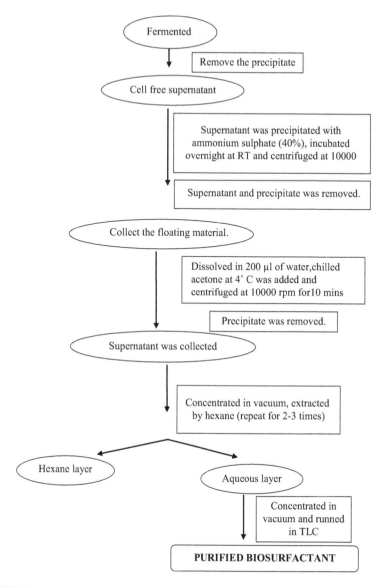

FIGURE 12.5 Purification of biosurfactant by ammonium sulphate method.

12.1.10 PURIFICATION OF BIOSURFACTANT

After the biosurfactant was extracted by a solvent extraction process the purification is done by various methods, though the biosurfactant possesses the tendency to dissolve in the chloroform, and some other products exist along with the biosurfactant which can dissolve in the chloroform. In order to extract the biosurfactant alone, the purification process is carried out by two methods (Pathaka and Nakhate, 2015).

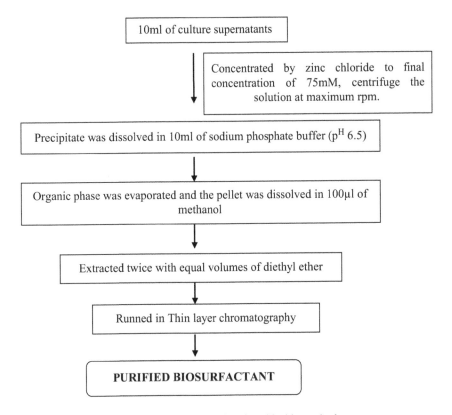

FIGURE 12.6   Purification of biosurfactant by zinc chloride method.

12.1.10.1   Ammonium Sulphate Precipitation Method

The supernatant containing BS was precipitated with 40% (w/v) ammonium sulphate and incubated overnight at 4°C. The precipitate formed was collected by centrifugation at 10000 rpm, 4°C for 30 min. The precipitate was dried until a dry, white-colored powder was obtained (Figure 12.5).

12.1.10.2   Zinc Chloride Precipitation Method

40% (w/v) zinc chloride was gradually added to the supernatant in order to precipitate the biosurfactant (BS). The mixture was then incubated for 24 h at 4°. The precipitate was collected by centrifugation at 4°C, 10000 rpm for 30 min, and then dried (Figure 12.6).

Apart from various other methods, chromatography technique plays a vital role in purification of the product produced from the microorganism. TLC, HPLC, GC are some of the categories that come under the chromatography technique.

(*TLC – thin layer chromatography, *HPLC – high performance liquid chromatography*GC – gas chromatography, etc.)

12.2   PRODUCTION AND PURIFICATION OF VARIOUS BIOSURFACTANT

12.2.1   SOURCES OF BIOSURFACTANT PRODUCTION (TABLE 12.2)

TABLE 12.2
Sources of Biosurfactant Production

S. No	Biosurfactants	Sources
1.	GLYCOLIPIDS	
	– Trehalolipids	*Rhodococcus erythropolis,Nocardiaerythropolis, Arthrobacter sp.,Mycobacterium sp.*
	– Trehalose dimycolates	*Mycobacterium sp., Nocardia sp.*
	– Trehalose dicorynomycolates	*Arthrobacter sp., Corynebacterium sp.*
	– Rhamnolipids	*Pseudomonas aeruginosa,Pseudomonas sp.*
	– Sophorolipids	*Torulopsis bombicola,Torulopsis apicola,Torulopsis petrophilum ,Torulopsis sp*
	– Cellobiolipids	*Ustilago zeae,Ustilago maydis*
2.	LIPOPEPTIDES AND LIPOPROTEINS	
	– Peptide-lipid	*Bacillus licheniformis*
	– Serawettin	*Serratia marcescens*
	– Viscosin	*Pseudomonas fluorescens*
	– Surfactin	*Bacillus subtilis*
	– Fengycin	*Bacillus sp.*
	– Arthrofactin	*Arthrobacter sp.*
	– Subtilisin	*Bacillus subtilis*
	– Gramicidin	*Bacillus brevis,Brevibacterium brevis*
	– Polymyxin	*Bacillus polymyxa,Brevibacteriumpolymyxa*
	– Lichenysin	*Bacillus licheniformis*
	– Ornithine lipidss	*Myroides sp. SM1,Pseudomonas sp.,Thiobacillus sp., Agrobacterium sp.,Gluconobacter sp.*
3.	FATTY ACIDS, PHOSPHOLIPIDS, AND NEUTRALLIPIDS	
	– Neutral lipids	*Nocardia erythropolis*
	– Phospholipids	*Thiobacillus thiooxidans*
	– Bile salts	*Myroides sp.*
	– Fatty acids	*Candida lepus,Acinetobacter sp.,Pseudomonas sp., Micrococcus sp.,Mycococcus sp., Candida sp.,Penicillium sp., Aspergillus sp.*
4.	POLYMERIC SURFACTANTS	
	– Emulsan	*Arethrobacter calcoaceticus*
	– Biodispersan	*Arethrobacter calcoaceticus*
	– Mannan lipid protein	*Candida tropicalis*
	– Liposan	*Candida lipolytica*
	– Carbohydrate protein lipid	*Pseudomonas fluorescens,Debaryomycespolymo rphus*
	– Protein PA	*Pseudomonas aeruginosa*
5.	PARTICULATE BIOSURFACTANTS	
	– Vesicles and fimbriae whole cells	*Arthrobacter calcoaceticus*

12.2.2 SCREENING, EXTRACTION AND PURIFICATION PROCESS OF VARIOUS BIOSURFACTANT

Most of the methods for a general screening of biosurfactant-producing strains are based on the physical effects of surfactants. Alternatively, the ability of strains to interfere with hydrophobic interfaces can be explored. On the other hand, specific screening methods like the colorimetric CTAB agar assay are suitable only to a limited group of biosurfactants. The screening methods can give qualitative and/or quantitative results. For a first screening of isolates, qualitative methods are generally sufficient (Table 12.3).

12.3 RECENT APPLICATIONS OF BIOSURFACTANT IN VARIOUS FIELDS (NWAGUMA ET AL., 2016)

- In addition to some of the classical applications, Biosurfactants can now be used effectively in various fields, including environmental protection, crude-oil recovery, food-processing industries, and in various fields of biomedicine (Pathaka and Nakhate, 2015).
- In cosmetics, mannosylerythritol lipid B and sodium surfactin play a promising role because of their high moisturizing function and high level of surface activating function, respectively.
- In agriculture, biosurfactants are used as a biopesticide and are used as for disease resistance and to develop immunity in plants. They are employed as a pest/insect control in agriculture and to enhance the soil remediation.
- In the pharma industry, biosurfactants are employed as they possess antimicrobial activity, anti-cancer activity, anti-adhesive agent, antiviral activity, and immunological adjuvants.

In modern genetic engineering they are used for gene delivery and to recover intracellular products, as well as acting as biological and immunological molecules.

In commercial laundry detergents, cyclic lipopeptide shows good emulsion formation capability and possesses excellent compatibility and stability for the formulation of laundry detergent.

In the petroleum industry, mostly rhamnolipids are used to extract crude oil from reservoirs and also as a cleaning agent in an oil storage tank.

In environmental cleaning, current bioremediation methods are In-situ soil flushing, Ex-situ washing, Heavy metal sequestration. Rhamnolipids are mostly used in these methods.

Biosurfactants act as an ecofriendly product for flocculation and dispersion of micro particles, which contain high solid content, and for the synthesis of silver nanoparticles.

In textile washing, biosurfactants are used to remove lipophilic preparations from fiber surfaces and is also used to enhance dye water solubility.

In mining and metallurgy, biosurfactants are used for bioleaching.

TABLE 12.3

Screening, Extraction and Purification Process of Various Biosurfactant

S.No.	Types of Biosurfactant	Screening Test	Extraction Method	Purification Method	Application
1	Glycolipid				
	Rhamnolipid	Oil displacement Hemolytic activity Blue agar plate method Surface tension (Bhat et al., 2015)	Solvent extraction (Plackett and Burman, 1946)	Silica gel column chromatography (De et al., 2015; DeOliveira et al., 2015)	Cosmetics Food processing Agriculture Pharmaceuticals Environmental bioremediation Enhanced oil recovery Cleaning of oil spills (Konishi et al., 2014)
	Trehalolipid	Oil spreading test Hemolytic activity	Solvent extraction (Franzetti et al., 2010)	Thin layer chromatography Column chromatoghapy	Bioremediation
	Sophorolipid	Drop collapse test Oil displacement test (Chandran and Das, 2011)	Solvent extraction (chloroform, methanol)	Column chromatography Dialysis	Adjuvent in formulation of herbicides Antifungal agents against plant pathogenic fungi Active ingredients in cosmetic product
	Mannosyl-erythritol lipid	Drop collapse method (Madslien et al., 2013)	Solvent extraction (Arutchevi et al., 2008)	Column chromatography	Topical moisturizer Repair damaged hair Antioxidants (Arutchelvi and Doble, 2010)
2	Lipopeptide				
	Surfactin	PCR screening (Siti Ratna Bint Mustafa, 2011)	Acid precipitation and solvent extraction (Jonathan Coronel et al., 2016)	Column chromatography (De et al., 2011)	Health care and biocontrol application Environmental application (Shoeb et al., 2015)
	Lichenysin	Oil displacement Haemolytic activity (Madslein, 2012)	Methanol extraction (Sen, 2010)	RP-HPLC (Grangemard et al., 1999)	Biomedical and environmental application (Kamal-Alahmad, 2015)

In the food industry, biosurfactants act as an excipient for food formulation, anti-adhesion, antioxidants, and emulsifiers.

In biofilms, rhamnolipid is used to maintain the biofilm structure which is based on a dynamic disruption of the continuous structure, allowing the diffusion of nutrients and gases in the cells within the biofilm. This suggests the biosurfactants might have a useful application in preventing the formation of biofilms on surfaces such as catheters or might be used to disrupt established biofilms on surfaces through their use in surface cleaning and formulations.

In wound healing, biosurfactants are used to aid wound healing. Once again this opens the possibility that biosurfactants could usefully be incorporated into a wide range of skin care products in place of chemical surfactants and that this would have the added benefit of aiding the healing of minor skin lesions.

12.4 CONCLUSION

Biosurfactants derived from renewable raw materials are progressively entering the market. The assembly costs and applicability will be the determining factors as to whether they can be used as renewable substrates for biosurfactant production. The assembly of biosurfactants with high added-value properties is the central part of future research. The commercial realization of biosurfactants which is restricted by higher production costs are often equalized by the optimized production conditions. This is essentially provided by utilization of the cheaper renewable substrates along with the application of novel and efficient multistep downstream processing method. It's expected that in the future, super-active microbial strains will be developed using gene-splicing for the production of biosurfactants at industrial level using renewable substrates as the staple.

Biosurfactants are finding importance in various fields. A promising approach seems to be the appliance of inoculants of biosurfactant-producing bacteria in the phytoremediation of hydrocarbon polluted soil. Application of the biosurfactants in phytoremediation on an outsized scale requires studies to spot their potential toxic effect on plants. Although the biosurfactants are thought to be ecofriendly, some experiments indicated that under certain circumstances they will be toxic to the environment. Nevertheless, careful and controlled use of those interesting surface-active molecules will surely help to remove toxic environmental pollutants.

Progress is currently being made in solving a number of the issues related to the tailoring of biosurfactant molecules to satisfy specific roles in product formulations. Products are already on the market containing biosurfactants and we can expect to ascertain further exploitation in the near future.

REFERENCES

Abdel-Mawgoud, A.M., Lepine, F. and Deziel, E. 2010. Rhamnolipids: Diversity of structures, microbial origins, and roles. *Applied Microbiology and Biotechnology* 86(5):1323–1336.

Arutchelvi, J. and Doble, M. 2010. Mannosylerythritol Lipids: Microbial Production and Their Applications. In *Biosurfactants*. Springer, Taylor and Francis Group, 145–177.

Arutchevi, J.I., Bhaduri, S., Uppara, P.V. and Doble, M. 2008. Mannosylerythritol lipids: A review. *Journal of Industrial Microbiology &Biotechnology* 35:1559–1570.

Bhat, R., Dayamani, K.J., Hathwar, S., Hegde, R. and Kush, A. 2015. Exploration on production of rhamnolipid biosurfactants using native *Pseudomonas aeruginosa* strains. *Journal of Bioscience and Biotechnology* 4(2):157–166.

Carley, K.M., Kamneva, N.Y. and Reminga, J. 2004. Response surface methodology (No. CMU-ISRI-04-136). 1–15.

Chandran, P. and Das, N. 2011. Characterization of sophorolipid biosurfactant produced by yeast species grown on diesel oil. *International Journal of Science and Nature* 2(1):63–71.

De Faria, A.F., Teodoro-Martinez, D.S., de Oliveira Barbosa, G.N., Vaz, B.G., Silva, Í.S., Garcia, J.S., Tótola, M.R., Eberlin, M.N., Grossman, M., Alves, O.L. and Durrant, L.R. 2011. Production and structural characterization of surfactin (C_{14}/Leu $_7$) produced by *Bacillus subtilis* isolate LSFM-05 grown on raw glycerol from the biodiesel industry. *Process Biochemistry* 46(10):1951–1957.

De Oliveira, M.R., Magri, A., Baldo, C., Camilios-Neto, D., Minucelli, T. and Celligoi, M.A.P.C. 2015. Review: Sophorolipids A Promising Biosurfactant and it's Applications. *International Journal of Advanced Biotechnology and Research (IJBR)* 6(2):161–174.

DeOliveira, M.R., Magri, A., Baldo, C., Camilios-Neto, D., Minucelli, T. and Celligoi, M.A.P.C. 2015. Review: Sophorolipids A Promising Biosurfactant and it's Applications. *International Journal of Advanced Biotechnology and Research (IJBR)* 6(2):161–174.

Franzetti, A., Gandolfi, I., Bestetti, G., Smyth, T.J. and Banat, I.M. 2010. Production and applications of trehalose lipid biosurfactants. *European Journal of Lipid Science and Technology* 112(6):617–627.

Grangemard, I., Bonmatin, J.M., Bernillon, J., Das, B.C. and Peypoux, F. 1999. Lichenysins G, a novel family of lipopeptide biosurfactants from *Bacillus licheniformis* IM 1307: Production, isolation and structural evaluation by NMR and mass spectrometry. *The Journal of Antibiotics* 52(4):363–373.

Jonathan Coronel, L., Angeles, M., and Marques, A. 2016. Lichenysin production and application in pharmaceutical field. *Recent Advances in Pharmaceutical Science* VI:147–163.

Kamal-Alahmad. 2015. The definition, Preparation and application of rhamnolipids as biosurfactants. *International Journal of Nutrition and Food Sciences* 4(6):613–623.

Konishi, M., Maruoka, N., Furuta, Y., Morita, T., Fukuoka, T., Imura, T. and Kitamoto, D. 2014. Biosurfactant-producing yeasts widely inhabit various vegetables and fruits. *Bioscience, Biotechnology, and Biochemistry* 78(3):516–523.

Madslein, F. 2012. Biosurfactant: Production and Application. *Journal of Petroleum & Environmental Biotechnology* 261: 106–113.

Madslien, E.H., Rønning, H.T., Lindbäck, T., Hassel, B., Andersson, M.A. and Granum, P.E. 2013. Lichenysin is produced by most *Bacillus licheniformis* strains. *Journal of Applied Microbiology* 115(4):1068–1080.

Mulligan, C.N. 2005. Environmental applications for biosurfactants. *Environmental Pollution* 133(2):183–198.

Nwaguma, I.V., Chikere, C.B. and Okpokwasili, G.C. 2016. Isolation, characterization, and application of biosurfactant by Klebsiella pneumoniae strain IVN51 isolated from hydrocarbon-polluted soil in Ogoniland, Nigeria. *Bioresources and Bioprocessing* 3(40):1–13.

Pathaka, A.N. and Nakhate, P.H. 2015. Optimisation of Rhamnolipid: A New Age Biosurfactant from *Pseudomonas aeruginosa* MTCC 1688 and its Application in Oil Recovery. *Heavy and Toxic Metals Recovery. Journal of Bioprocessing & Biotechniques* 5(5):14.

Plackett, R.L. and Burman, J.P. 1946. The design of optimum multifactorial experiments. *Biometrika* 33(4):305–325.

Porob, S., Nayak, S., Fernandes, A., Padmanabhan, P., Patil, B.A., Meena, R.M. and Ramaiah, N. 2013. PCR screening for the surfactin (sfp) gene in marine *Bacillus* strains and its molecular characterization from Bacillus tequilensis NIOS11. *Turkish Journal of Biology* 37(2):212–221.

Rønning, H.T., Madslien, E.H., Asp, T.N. and Granum, P.E. 2015. Identification and quantification of lichenysin-a possible source of food poisoning. *Food Additives & Contaminants: Part A* 32(12):2120–2130.

Sen, R. 2010. *Surfactin: Biosynthesis, Genetics and Potential Applications*. Springer, New York, 316–323.

Shoeb, E., Ahmed, N., Akhter, J., Badar, U., Siddiqui, K., Ansari, F., Waqar, M., Imtiaz, S., Akhtar, N., Shaikh, Q.U.A. and Baig, R. 2015. Screening and characterization of biosurfactant-producing bacteria isolated from the Arabian Sea coast of Karachi. *Turkish Journal of Biology* 39(2):210–216.

Singh, V. 2012. Biosurfactant–Isolation, production, purification & significance. *International Journal of Scientific Research Publications* 2(7):1–4.

Siti Ratna Bint Mustafa. 2011. *Solid substrate fermentation (SSF) and submerged fermentation (SmF) of Aspergillus versicolor A6 using pineapple waste as substrate for pectinase production*.

Xin, M. 2012. *Purification and structural characterization of Rhamnolipid surfactant produced by Pseudomonas aeruginosa CICC 10204*. *International Conference of Clinical Microbiology &Microbial Genomics*, Hilton San Antonio Airport, USA.

13 Nutraceutical Prospects of Green Algal Resources in Sustainable Development

*Prasanthkumar Santhakumaran and
Joseph George Ray*
Mahatma Gandhi University, India

CONTENTS

13.1 INTRODUCTION

In the current environmental scenario of global warming, traditional crops are failing (Mall et al. 2007). The sustainability of human development now depends on finding a successful solution to the challenge mentioned above, because more than half of the humans, most of whom are in developing countries, have no access to nutritionally balanced food (Bhattacharya, Currie, and Haider 2004). Moreover, environmental pollution and degradation of diverse kinds are disrupting all kinds of production systems and undermining the human ability to provide nutritionally rich food for fall, especially in the low-income and middle-income countries (Fore et al. 2020). Moreover, human health and quality of life throughout the world are challenged by the current global climate change, and people are becoming vulnerable to new diseases and food insecurity (Meybeck et al. 2018). The food insecurity issues are operating in different ways in different parts of the world. Thus, climate change is acting as a significant developmental impediment to humans. In the above context, human research must focus on nutritionally valuable alternative food as an imminent solution to the socio-economic and environmental crisis of development, globally. Recent studies emphasize the role of nutritionally and medicinally valuable food, known as nutraceuticals, in boosting human resilience to new diseases, especially in resisting even the attack of the recent global pandemic from the mutated coronaviruses (McCarty and DiNicolantonio 2020). Such a novel viral attack may be a natural outcome of climate change impact on humans.

The world population may reach 9.8 billion by 2050 (FAO 2017), and it will cause an increasing demand for agricultural products in the world. Excessive demand for natural resources is causing further depletion of non-renewable resources (Chel 2011). Excessive use of mineral nutrients for agricultural purposes can lead to further addition of nutrient load (Ray and Nidheesh 2019) in the soil and waters and toxic accumulation of heavy metals in soil (Sahodaran and Ray 2018). Therefore, the prospect of further intensification in conventional agriculture has become quite bleak and, hence, a sustainable approach toward development is the need of the hour.

Nutritional food sustainability is defined as "the ability of a food system to provide sufficient energy and essential nutrients in required quantities to maintain the good health of population without compromising the ability of future generations to meet their nutritional needs" (Swanson et al. 2013). Achieving sustainable development in the period of climate change, depleting resources, and increasing food demand is quite a Herculean task (Vermeulen, Campbell, and Ingram 2012). Besides, the current global socio-environmental scenario of the corona pandemic presents another serious challenge (Mori et al. 2020) to further human development and even survival on the earth. Therefore, new alternative species to overcome the limitations of conventional food production systems have become inevitable to achieve sustainable development globally. In this context, the fastest carbon-sequestering photoautotrophic organisms (Singh and Dhar 2019), microalgae, which are industrially mass-producible can be considered an important eco-technological tool (Santhakumaran, Ayyapan, and Ray 2020a) for attaining developmental sustainability from diverse environmental perspectives, especially as an alternative non-conventional means of providing a nutritionally valuable food resource to all people as per the growing demand of the same.

Microalgae are a definite alternative bioresource of food and medicinal applications that can ensure food security and environmental safety for the future (Wells et al. 2017). Microalgae are amenable to industrial production of biomass in a carbon-neutral and environmental remediating manner (Benedetti et al. 2018). The vast biodiversity of green microalgae (Guiry 2012; Singh and Saxena 2015; Ray, Santhakumaran, Kookal 2020) and their fast CO_2 sequestration potential (Singh and Dhar 2019), high biomass productivity (Mata, Martins, and Caetano 2010), presence of a diverse array of industrially valuable biochemicals in their biomass (Wells et al. 2017; Benedetti et al. 2018; Rizwan et al. 2018; Sun et al. 2018), availability of vast area of the open sea and non-arable lands for utilization in algal biomass production, the possibility of using eutrophic waters as a potential medium for biomass production (Kookal, Santhakumaran, and Ray 2017) are all factors favoring dependence on microalgae as a highly significant biomass resource for industrial progress.

Currently, a few species of microalgae are used as a sustainable bioresource for industrial production of nutritionally valuable products (Dixon and Wilken 2018), which are insufficient to meet the existing demand for the same. Therefore, the success of utilization of microalgae as an alternative bioresource depends on the bioprospecting of suitable microalgal species, which are highly productive and contain a high amount of nutraceutical compounds in the biomass, from local environments (Santhakumaran, Kookal, and Ray 2018; Santhakumaran et al. 2019). Thus, phycoprospecting (bioprospecting of microalgae) of indigenous microalgae and the experimental optimization of productivity in them (Santhakumaran et al. 2020b) have become inevitable for identification and scaling up of alternative algal bioresources for nutritional security for the future. Moreover, algal biomass is a multipurpose bioresource for fuel, food, feed, nutritional and medicinal products (Mathimani and Pugazhendhi 2019). However, phycoprospecting of nutraceutically valuable species has high relevance as the demand for the same is very high in the current global environmental scenario. The nutraceutical prospecting of microalgae also remains unique from the conservational, environmental, industrial, economic, social, and health perspectives. Therefore, the present chapter comprehensively reviews the relevant recent literature on the concept of nutraceuticals, algae as a source of nutraceuticals, roles of nutraceuticals, nutraceutical components in algal biomass, and industrial significance of algal nutraceuticals. The gaps and trends of different aspects of these areas of research are also discussed.

13.2 NUTRACEUTICALS

The concept of food with pharmaceutical value led to the formation of a new concept "Nutraceuticals" or "functional foods"; the term "nutraceutical" was coined from the words "nutrition" and "pharmaceutical" (DeFelice 1995). According to the same author, nutraceuticals are "food or a part of food that provides medical or health benefits, including the prevention or treatment of a disease". In other words, nutraceuticals are medicinally significant functional food or food supplements that have health benefits, including immunity build-up and prevention or control of diseases without any side effects. They are natural foods or eatables, valuable in terms of proteins or amino acids or fatty acids or vitamins or pigments or minerals or antioxidants or antimicrobials or other valuable components in them, consumption of which build up immunity or control diseases. Therefore, nutraceuticals can reduce the use of chemical pharmaceutics that have side effects. Nutraceuticals have high market value as they are expected to increase life expectancy, improve quality of life and overcome expensive health care treatments (Bigliardi and Galati 2013; Daliri and Lee 2015). However, research on nutraceuticals, especially publications on novel sources of such valuable biomaterials, are quite rare from many tropical countries.

13.3   ALGAE AS A FOOD RESOURCE

Microalgae have been in use as human food for thousands of years in different parts of the world (Borowitzka 1998). However, the commercial cultivation of algae as a source of nutritional food (Olaizola 2003) and other valuable products started recently. Recent findings of the significance of nutritional support for disease control or prevention of diseases have led to a different approach in medical care by recommending nutritionally valuable foods known as nutraceuticals to maintain good health (Hardy 2000). The freshwater green microalgae are considered as a promising renewable biomass resource, sustainably producible (Driver, Bajhaiya, and Pittman 2014) for a diverse array of nutraceutically significant biochemicals such as proteins, essential amino acids, omega fatty acids, pigments, minerals, vitamins, antioxidants, and antimicrobial compounds (Benedetti et al. 2018). Besides protein, lipids, pigments, and minerals, algae are a good source of antioxidants and antimicrobial agents (Natrah et al. 2007; Dixit et al. 2017). As an alternative food source with natural antioxidant, antimicrobial and nutritional components, algal biomass remains a valuable, sustainable bioresource.

Currently, the biomass of a few algae such as *Spirulina platensis, Chlorella* sp., *Dunaliella salina, Haematococcus pluvialis, Chlamydomonas* sp., and *Scenedesmus* sp. (Udayan et al. 2017; Wells et al. 2017), and single-cell organisms such as *Saccharomyces boulardii, Bifidobacterium* spp., and *Lactobacillus acidophilus* (Syngai et al. 2016) consisting of proteins, lipids, carbohydrates, pigments, and vitamins (Spolaore et al. 2006), are used for industrial production of various food products (Gangl et al. 2015). Similarly, the biomass of a few algae is identified with nutritionally valuable and medicinally significant components and are used as nutraceuticals (Udayan, Arumugam, and Pandey 2017). Moreover, microalgal biomass also contains a high concentration of minerals, and trace elements comparable to terrestrial bioresources (Khairy and El-Sheikh 2015). Industries of food and dietary supplements require non-traditional sources of nutrition and microalgae have naturally become an alternative bioresource (Clemens and Pressman 2020) for the same. In general, the concept of nutraceutically valuable food is conquering the world and is inspiring both science and industry to search for indigenous nutraceutically valuable microalgal species (Piwowar and Harasym 2020). Moreover, research on the nutraceutical value of biopeptides and polysaccharides, polyphenolic compounds such as flavonoids and isoflavonoids, phenolic compounds and antimicrobial compounds in microalgae (Falaise et al. 2016; Habibi, Imanpour Namin, and Ramezanpour 2018) represent the recent trends in microalgal research. However, publications dealing with the search for nutraceutically valuable algae from local environments are relatively few. Algae being a highly biodiverse group of nutritionally valuable organisms, phycoprospecting of freshwater green microalgae for nutritional purposes from local environments remains highly relevant.

13.4 ROLE OF NUTRACEUTICALS

13.4.1 IN HUMAN HEALTH

Changing lifestyles, along with rising chronic diseases, are causing considerable increase in demand for various dietary supplements across the world. Moreover, an increase in the consumption of junk food products and increasingly busy lifestyles are resulting in a low intake of required nutrients. Further, change in food habits leads to health hazards of diverse kinds such as obesity in the general population. Similarly, accumulation of free radicals in the human body from excessive consumption of junk food is associated with an increased outbreak of diseases like cancer, diabetes, atherosclerosis, and inflammatory disorders (Liguori et al. 2018). Naturally, nutraceutical foods with high nutritionally and medicinally valuable compounds are significant in overcoming such life-style and environmental diseases. The natural polyphenolic group of compounds consisting of natural-phenols and flavonoids present in many microalgae remain one of the best natural antioxidants to meet this challenge. Furthermore, the formation of free radicals in food leads to the degradation of lipids and proteins, which ultimately reduces the shelf-life and nutritional quality of human food (Zainol et al. 2003). Unlike the toxic synthetic-antioxidants such as BHT (butylated hydroxytoluene) and BHA (butylated hydroxyanisole) used in food preservation (Chen, Pearson, and Gray 1992), the natural antioxidants in nutraceuticals are relatively safe. The inclusion of polyphenolic compounds in human diets as nutraceuticals can contribute to a decrease in cardiovascular diseases and have other beneficial health effects (Gomez-Guzman et al. 2018). In general, the consumption of natural antioxidant-rich foods or nutraceuticals can increase the human lifespan and improve the quality of human life (Charles 2013). However, little information is currently available on microalgae rich in polyphenolic compounds (Goiris et al. 2012), which can be marketed for consumption as nutraceuticals.

The microalgal biomass is also a valuable mineral-rich bioresource. The biomass of many algae contains high amounts of significant elements such as phosphorus (P), sodium (Na), potassium (K), magnesium (Mg), and calcium (Ca), and trace elements such as zinc (Zn), iodine (I), manganese (Mn), copper (Cu), selenium (Se), cobalt (Co), and molybdenum (Mo) (Kovac et al. 2013; Udayan, Arumugam, and Pandey 2017). Microalgae thus serve as an exciting alternative to dietary sources of valuable minerals significant to human health. However, this aspect of microalgal biomass utilization remains entirely unexplored.

13.4.2 IN THE FOOD INDUSTRY

Food security is closely linked to the protection of valuable natural food from degradation, which is an essential aspect of food safety (Elkhishin, Gooneratne, and Hussain 2017) as well. The increase in global temperature due to climate change favors bacterial growth on food materials and thus reduces food safety (Hammond et al. 2015). Bacterial contamination of foods is a major cause of toxicity and food spoilage. The known toxicities of inorganic antibacterial agents currently in use such as sodium benzoate and sodium nitrate demand the search for alternative natural plant-based antimicrobial products. Moreover, the development of microbial resistance to antibiotics (Ventola 2015) also forces humans to search for new, unexplored antimicrobial compounds with fewer side effects. Such a search of the pharmaceutical and food industries for safe and novel antimicrobial compounds of natural origin (Hintz, Matthews, and Di 2015) also prompts investigating biochemicals in microalgae for novel compounds against microbial contamination and infection. Moreover, nutraceutically valuable algal biomass that contains consumable antibacterial compounds such as lipids can be a valuable alternative food-additive to improve the shelf-life of sensitive food.

13.5 THE RELEVANCE OF MICROALGAL BIOMASS AS ALTERNATIVE NUTRACEUTICAL RESOURCE

The mass-production of microalgae as an alternative nutritionally valuable food resource is considered to be a sustainable practice (Fernandez, Sevilla, and Grima 2017). In the conventional food system, certain vegetables and fruits are considered good sources of electrolytes, flavonoids, minerals, vitamins, carotenoid, and other bioactive compounds that have a significant benefit to human health (Slavin and Lloyd 2012) and fall into the category of functional foods. The functional molecules in such foods as mentioned above can help in the management of body weight, balancing blood sugar, cholesterol level, prevention of aging, reduction in the risk of cardiovascular diseases, inflammatory diseases, and diabetes. They can increase the immunity of the human body (Adefegha 2018). The discoveries of beneficial microorganisms in certain conventional natural foods such as honey, milk, and milk products, including yogurt, cheese, buttermilk, and ghee are now considered as prebiotics, which enables immunity build-up and increases disease resistance in diverse ways. Such foods are now called functional foods. Algae, as an alternative food resource, are significant from such a functional food perspective as well.

In general, consumption of functional foods helps in the reduction of cholesterol and triacylglycerols, protection against gastroenteritis, improvement against lactose intolerance, and stimulation of the immune system (Srinivasan 2010). Therefore, alternative food resources that have functional or prebiotic food value are of exceptionally high demand in the current food market. In general, the alternative food resources are expected to have high nutritional quality, functional food value, and prebiotic properties. Moreover, humans must be able to produce such foods in required quantities, in limited time, without causing any environmental contamination or not causing any imbalance in the natural ecosystem processes irrespective of where and how they are produced. It is in this context that the microalgal biomass appears to be the easiest, cheapest, and most eco-friendly producible bioresource because it can capture unlimited solar energy and convert the same into valuable products (Vonshak 1990).

Overall, microalgal biomass is a nutritionally valuable bioresource. Cultivation of nutraceutically valuable algae as a food resource can become a sustainable solution to the increasing nutritional requirements of a growing population (Caporgno and Mathys 2018). Thus, industrially cultivable microalgae are becoming a competent alternative food-grade biomass resource capable of solving the current food and health crisis of the world, because the biomass of certain algae can provide high amounts of protein, valuable lipids, minerals, antimicrobials and antioxidants in limited time. The presence of such nutritionally and medicinally valuable components makes the biomass of certain algal species a nutraceutically valuable food. However, for understanding the same potentials of microalgal biomass as a valuable alternative food resource, a detailed discussion of the literature on specific publications on nutraceutical potentials of the microalgal biomass becomes essential, which follows.

13.6 NUTRACEUTICALLY VALUABLE COMPOUNDS IN MICROALGAL BIOMASS

Microalgae were considered a source of food and a valuable biochemical in the early 1940s, and mass cultivation of algae for such purposes was suggested in the first "Algae Mass-Culture Symposium" (Davis et al. 1953). Unlike the conventional biomass, microalgal biomass is considered rich in nutritionally and medicinally valuable compounds. Algal biomass with such nutraceutical compounds grown locally can be consumed directly (Lee and Marino 2010). Algae typically have nutrient profiles superior to land plants, especially in the quality and quantity of proteins, fatty acids, pigments, antibacterials, antioxidants, minerals, vitamins (Wells et al. 2017; Koyande et al. 2019), and so on. Specific algal species are well known for such specific nutritional components in varying amounts. In this context, research publications on diverse specific nutritional or nutraceutical components of algae such as proteins, lipids, pigments, carbohydrates, minerals, vitamins, antioxidants, and antimicrobials are discussed one by one.

13.6.1 NUTRACEUTICALLY VALUABLE PROTEINS IN ALGAE

The utilization of specific microalgal biomass as nutraceuticals (Brower 1998) is mainly related to its high nutritional value, taking into account the high content of proteins (Soletto et al. 2005) with essential amino acids (EAAs) and easy digestibility. Proteins are essential macro-nutrients required for the overall growth of human beings. The currently used conventional plant or animal-based protein sources are insufficient to meet the existing and future demand for proteins (Wu et al. 2014). An increase in the production of proteins by conventional means is quite impossible for meeting the increasing demands (WHO 2015). Because the dependence on conventional animal-based proteins as the primary source of protein is not only a severe problem from a vegan point of view, but it is also a severe threat to the already delicate environmental stability of the world; in this context, microalgal proteins are gaining more prominence as an alternative plant-based protein source. Nutritionally valuable algal biomass is often considered as a wholesome protein source because it contains many essential amino acids which are relatively absent in conventional foods. Algae protein is used for promoting weight loss, reducing fatigue and anxiety, and in case of diabetes, attention deficit-hyperactivity disorder (ADHD), and premenstrual heart diseases, among other health issues. Besides, the algal resource is sustainably producible, and it has a high demand in the food market (Caporgno and Mathys 2018).

Certain species of algae viz. *Spirulina platensis* contains a high amount of protein in its biomass (46%–75%) (Ritala et al. 2017) with EAAs such as leucine, valine, isoleucine, phenylalanine, tyrosine, methionine, and cysteine. The microalgal proteins with essential amino acids, high protein efficiency ratio, and easy digestibility are equivalent to or better than the conventional plant-based protein source. Algal based proteins are often considered equivalent to conventional plant protein resources such as soya, and animal protein sources such as eggs, poultry meat, dairy products, and fish (Becker 2007). Moreover, algal proteins have already received the recognition of "Generally Recognized as Safe" (GRAS) grading by the Food and Drug Administration of the USA (FDA 2016). On comparing protein production from terrestrial crops such as wheat, pulses, and legumes, especially soya, known as vegetable meat, microalgae are higher (15-30 tons/ha/year) in protein yield (Krimpen et al. 2013). Therefore, the extraction of protein from microalgae is beneficial in terms of protein value and productivity. It is easy to control production environments of algae to make algal biomass resource-free of environmental pollutants.

Nutraceutical and toxicological evaluations of specific algal biomass have demonstrated microalgae as a valuable food-supplement, with high bioavailability, bioaccessibility, and bioactivity (Bleakley and Hayes 2017). Therefore, algal proteins can easily replace conventional protein supplements. Biomass of microalgae such as *Spirulina platensis* and *Chlorella vulgaris* is currently used as a nutritional supplement for humans and as feed for animals and fish (Habib et al. 2008). Other nutraceutically valuable algal components are bioactive peptides and amino acids that have therapeutic and enzymatic effects (Khan, Shin, and Kim 2018). The presence of a bioactive compound with a specific amino acid sequence of 2–20 amino acids in specific microalgal biomass provides antihypertensive, antioxidant, antithrombotic, hypocholesterolemic, appetite-suppressive, antimicrobial, immune-modulatory, and cyto-modulatory properties (Apone, Barbulova, and Colucci 2019). The green microalga *Chlorella pyrenoidosa* contains a water-soluble extract, "Chlorella Growth Factor" (CGF), a mixture of essential amino acids, peptides, proteins, vitamins, sugars, and nucleic acids (Merchant and Andre 2001), which are involved in wound healing, controlling serum lipid, and can stimulate the immune system (An et al. 2010). *Kirchneriella lunaris* was found to be a valuable source of proteins. The protein yield of microalga was observed to be 58.95% in the biomass (Santhakumaran, Kookal, and Ray 2018). In general, proteins from microalgae with well-balanced essential amino acids profiles, and higher digestibility than standard protein source (Niccolai et al. 2019), are mainly used as nutraceuticals or included in the formulation of a functional food (Khanraa et al. 2018). Therefore, nutraceutically valuable proteins from microalgal biomass have global significance. However, the gap of research on protein content and quality of freshwater green algae from local environments remains relatively high. Since the microalgal biomass has high significance in food, feed, and nutraceutical industries, research on microalgae should be global, being undertaken in all countries, before the biodiversity of sensitive species is eroded day by day from climate change impact.

13.6.2 Nutraceutically Valuable Oils in Algae

Lipids represent the most nutritionally valuable group of compounds present in microalgal biomass. Lipids serve as components of membranes, energy storage compounds, and as cell-signaling molecules. Certain algae are known to accumulate lipids up to 30%–75% of their dry cell weight (Mata, Martins, and Caetano 2010). Among the lipids, polyunsaturated fatty acids (PUFA) are considered as nutraceutically valuable (Lenihan-Geels, Bishop, and Ferguson 2013), whereas the saturated (SFA) and monounsaturated fatty acids (MUFA) are widely considered as antimicrobials (Desbois and Smith 2010), and also for the production of biofuels (Knothe 2005). Therefore, the proportions of saturated, monounsaturated, and polyunsaturated fatty acid in the biomass decide the food value or fuel value of algae.

The PUFAs are essential fatty acids and are introduced into our body through diet only. Based on the number of double bonds, PUFAs are categorized into four kinds such as omega-3 fatty acids, omega-6 fatty acids, omega-7 fatty acids, and omega-9 fatty acids. The major group of omega-3 fatty acids is α-linolenic acid (ALA; C18:3), which is the precursor of eicosapentaenoic acid (EPA; C 20:5), docosapentaenoic acid (C22:5), and docosahexaenoic acid (DHA; C22:6). The omega group of fatty acids have several medicinal applications such as (1) lowering triglyceride level and thereby reducing blood pressure and preventing cardiovascular diseases, (2) anti-inflammatory and anti-blood clotting actions, (3) reducing the risks of diabetes, and (4) beneficial visual functions (Ji, Ren, and Huang 2015; Benedetti et al. 2018).

The omega fatty acids from algae have great commercial significance, especially from the vegan point of view, because such medicinally valuable compounds are currently extracted from fish oils and animal fats. Moreover, the global diminishing of fish markets, over-dependence on specific seasons and locations, and environmental pollution risk factors from fish oils are adversely affecting the nutraceutical industry (Adarme-Vega et al. 2014; Ochsenreither et al. 2016). Several clinical studies have revealed the beneficial effect of omega-3 polyunsaturated fatty acids on human health. Many algae are already identified as rich sources of omega group of fatty acids. Such algae include *Nannochloropsis gaditana*, *Isochrysis galbana*, *Pavlova lutheri*, *Schizochytrium* spp., *Crypthecodinium cohnii* (commercially market as DHASCO™, have received "GRAS" status; rich in DHA and EPA), and *Tetraselmis suecica* (Liu, Sommerfeld, and Hu 2013; Ji, Ren, and Huang 2015; Ochsenreither et al. 2016). The omega group of fatty acids are involved in reducing the risk of type 2-diabetes (Brostow et al. 2011), preventing colorectal cancer (Cockbain, Toogood, and Hull 2012), cardiovascular diseases (Mohebi-Nejad and Bikdeli 2014), depression (Wani, Bhat, and Ara 2015), kidney disorders (Panahi et al. 2016), arthritis (Akbar et al. 2017), and asthma (Adams et al. 2019). Recent research shows that DHA is included as a supplement in infant formulas because DHA is helpful in the development of the retina and brain function in infants and children (Shahidi and Ambigaipalan 2018; Sahni et al. 2019). In general, microalgal species are found to be a rich source of nutraceutically valuable omega group of fatty acids. Recently, Kookal, Santhakumaran, and Ray (2017), Santhakumaran, Kookal, and Ray (2018), and Santhakumaran et al. (2019) identified several uninvestigated freshwater micro-algae from bloomed water bodies of Kerala, India as a rich source of omega group of fatty acids. The current trend in microalgal research is to identify desirable algal species for PUFAs as the best sustainable alternative source of the same in the food and pharmaceutical industry. Algal nutraceuticals are also significant from other nutra-ceutical components such as pigments. However, research on valuable lipids from freshwater green microalgae from the local environment remains relatively rare.

13.6.3 NUTRACEUTICALLY VALUABLE PIGMENTS IN ALGAE

Among the various natural bioresources for pigments, microalgae are considered as a promising bioresource for pigment production due to their easiness in cultivation, high pigment yield, and no competition with food production (Orosa et al. 2000). The three major classes of pigments in microalgae are chlorophyll, carotenoids, and phycobilins, of which phycobilins are absent in green algae. Microalgae are consid-ered as an alternative to conventional pigment-yielding plants, as they have high biomass productivity and pigment yield when compared to conventional plant sources. Algal pigments, such as chlorophyll and carotenoids, are mainly extracted from *Haematococcus pluvialis* -astaxanthin (Guerin, Huntley, and Olaizola 2003), *Dunaliella salina*-beta-carotene (Murthy et al. 2005), and *Muriellopsis* sp. and *Scenedesmus almeriensis* -lutein (Del Campo et al. 2007). These pigments are valu-able nutraceutical compounds with antioxidant properties. Moreover, microalgae being a renewable bioresource, pigment production and extraction of it from the same can be a sustainable process.

Chlorophylls and carotenoids are generally fat-soluble molecules, whereas phy-cobilins are water-soluble. Moreover, chlorophyll molecules on reaction with weak acids, light, temperature, and oxygen form chlorophyll derivatives (Cubas, Lobo, and Gonzalez 2008), having nutraceutical properties. The primary microalgal sources used for the production of Chlorophyll are *Chlorella vulgaris, Ankistrodesmus falca-tus,* and *Scenedesmus dimorphus* (Ferreira and Anna 2017). Recent research carried out by Santhakumaran et al. (2019) identified freshwater green microalgae such as *Sphaerocystis antoni-kadavilaii* and *Fasciculochloris boldii* as pigment-rich microal-gal species.

The two significant types of chlorophyll molecules found in green algae are chlo-rophyll-a and chlorophyll-b. Chlorophyll is used as a coloring agent. Chlorophyll-a and its derivatives such as pheophorbide-a and pheophytin-a are known for their health benefits, as antioxidants, in preventing oxidation of lipids and proteins and increasing the level of glutathione S-transferase in cells. Moreover, they prevent the inhibition of cytochrome P_{450} enzymes involved in the removal of carcinogenic compounds from the body (Mishra, Bacheti, and Husen 2011). Chlorophyll mole-cules are used to stimulate liver function and thereby increase bile secretion (Bishop and Zubeck 2012). Chlorophyll and its derivatives are currently used in food, cosmetics, pharmaceuticals, and nutraceutical industries as food additives, coloring agents, and as a potential source of antioxidants (Dufosse et al. 2005; Ferreira and Anna 2017).

Carotenoids are accessory pigments in algae mainly involved in light-harvesting processes. They are lipophilic compounds, which are widely used as food-colorants, feed-additives for aquaculture, components for cosmetics, skincare components, and antioxidants (Sathasivam et al. 2019). Most carotenoids share a typical C40 backbone structure of isoprene units and are divided into two groups: carotenes (hydro carbonated carotenoid) and xanthophylls (oxygenated derivatives of carotenes). Even though the chemical synthesis of carotenoids is well established, the use of synthetic carotenoid products in direct human consumption is limited, as synthetic carotenoids are dominantly trans-isomer compounds which causes inhibition in the absorption of the same and are not safe (Barreiro and Barredo 2012) for health. Therefore, carotenoids from natural origins are significant in the pigment industry. The powerful carotenoids of nutraceutical interest are astaxanthin, β-carotene, lutein, zeaxanthin, lycopene, fucoxanthin, and canthaxanthin (Koyande et al. 2019).

Astaxanthin (3, 3'- dihydroxy-β, β'-carotene-4, 4'-dione) is a red keto-carotenoid which belongs to the category of xanthophylls. Astaxanthin is widely used in the feed, cosmetic, aquaculture, nutraceutical, and pharmaceutical industries because of its high antioxidant potential. It is successfully commercialized by many industries worldwide through the cultivation of the green algae *Haematococcus pluvialis* (Lorenz and Cysewski 2000).

In the human body, carotenoids act as provitamin-A and are usually present in the range of 0.1%–0.2% of total dry matter of microalgae (Christaki, Florou-Paneri, and Bonos 2011). The β-carotene, a red-orange carotenoid is one of the vital carotenoids, used in industry as a coloring agent, antioxidant, and as a vitamin-A supplement (Barreiro and Barredo 2012). Moreover, such carotenoids are medicinal that prevent toxin build-up in the liver, improve the immune system, and have a preventative-role in eye diseases such as night-blindness and cataract (Bungau et al. 2019). The most commonly used microalgae for the production of β-carotene are *Dunaliella salina*, *Scenedesmus almeriensis*, and *Dunaliella bardawil* (Guedes et al. 2011).

Lutein is another class of vital carotenoid having antioxidant activity. The microalgae such as *Scenedesmus almeriensis*, *Chlorella protothecoides*, *Chlorella zofingiensis*, *Chlorococcum citriforme*, and *Neospongiococcum gelatinosum* are the familiar algal sources of lutein (Del Campo et al. 2000). Zeaxanthin is a yellow-colored carotenoid having application in the pharmaceutical, cosmetics, and food industries. The primary microalgal sources of Zeaxanthin are *Scenedesmus almeriensis* and *Nannochloropsis oculata* (Lubian et al. 2000; Granado-Lorencio et al. 2009). Moreover, the carotenoid fucoxanthin extracted from *Isochrysis galbana* is marketed for its anti-obesity, anticancer, and its potential anti-inflammatory activities (Kim et al. 2012). In general, carotenoids contribute to the overall antioxidative activity of microalgae (Banskota et al. 2019).

Although microalgal pigments are a rich source of nutraceutical compounds that have a global market value, intensive research in this line, especially publications dealing with pigment value of specific freshwater green algae from local environments, remain scarce in the literature.

13.6.4 Nutraceutically Valuable Carbohydrates in Algae

The valuable carbohydrates in algae include starch, cellulose, sugars, and other polysaccharides. The polysaccharides extracted from the microalgae, *Chlorella stigmatophora* and *Phaeodactylum tricornutum*, are known to have significant anti-inflammatory and immunomodulatory properties (Guzman et al. 2003). The ability of microalgal polysaccharides in promoting the growth of human gut-microflora and regulating the blood-glucose level makes microalgal biomass a source of prebiotic supplements (Ibanez and Cifuentes 2013). Moreover, the polysaccharides isolated from *Spirulina platensis* and *Porphyridium cruentum* are well known for their antioxidant activities (Sun et al. 2009; Zaid, Hammad, and Sharaf 2015). The algal species such as *Chlorella vulgaris*, *Nannochloropsis* sp., and *Arthrospira platensis* are used as prebiotics (Raposo, Morais, and Morais 2016). Besides, specific sulfated algal polysaccharides are reported to have therapeutically valuable immunomodulatory, anticoagulant, hypolipidemic and antithrombotic (Zaporozhets and Besednova 2016), anticancer, and anti-hypocholesterolemic activities (Lordan, Ross, and Stanton 2011; Raposo, Morais, and Morais 2016) agents. Moreover, the liquefiable carbohydrate-rich algal biomass of certain microalgal species such as *Chlamydomonas reinhardtii*, *Chlorella vulgaris*, and *Arthrospira platensis* are now considered as potential sources for the production of bioethanol (Markou et al. 2013; Rizza et al. 2017) and methane (Thangavel et al. 2018). A recent study by Santhakumaran et al. (2019) established biomass of *Chlorella zachariaii* as a good source of carbohydrate. However, specific research on the quality and quantity of valuable carbohydrates in specific freshwater green algae remain relatively scant.

13.6.5 Nutraceutically Valuable Vitamins and Minerals in Algae

The microalgal biomass represents a valuable source of nearly all the essential vitamins such as vitamin-A, B_1, B_2, B_6, B_{12}, C, E, niacin, biotin, folic acid, pantothenic acid, and also minerals such as Na, K, Ca, Mg, Fe, Zn and trace minerals (Koyande et al. 2019). The high levels of vitamin B_{12} and Iron in *Spirulina* spp. make it particularly suitable as nutritional supplements for vegetarians as vitamin B_{12} is not synthesized by higher plants (Watanabe and Bito 2018). In general, the vitamin content of an alga depends on the species, its growth stage, light intensity, harvesting, and the biomass drying methods (Brown et al. 1999). The natural food supplements rich in minerals such as Co, Ni, Cu, Zn, and Pb have broad-spectrum antibacterial properties (Yasuyuki et al. 2010). The Na/K ratio in algal biomass is generally below 1.5, and, therefore, algal biomass is a safe source of minerals for consumption because bioresources with high Na/K ratio causes hypertension (Astorga-Espana et al. 2015). Recently, it was observed that biomass of microalgae such as *Chlorococcum humicola*, *Bracteacoccus minor*, and *Radiococcus nimbatus* is a rich source of zinc, iron, and manganese, respectively (Santhakumaran, Ayyapan, and Ray 2020a). However, research on microalgae in this line, mainly algae rich in valuable minerals, is quite rare in the literature. Considering the significance of the high market value of mineral and vitamin-rich biomass resources, research on specific freshwater green algae from local environments remains quite relevant.

13.6.6 NUTRACEUTICALLY VALUABLE ANTIOXIDANTS AND ANTIMICROBIALS IN ALGAE

Antioxidants are valuable in preventing health disorders such as cancer, diabetes mellitus, neurodegenerative diseases, and inflammatory diseases (Wilson et al. 2017). The antibacterial activity of *Desmococcus olivaceus*, *Chlorella vulgaris*, and *Chlamydomonas* sp. against several human bacterial pathogens (Sanmukh et al. 2014; Ramar et al. 2017) reveals the nutraceutical potential of such species. The presence of the high volume of polyphenols, carotenoids, and flavonoids are responsible for the antioxidant and antibacterial potential of algae (Dixit et al. 2017). Moreover, the saturated and unsaturated fatty acids in algae, such as myristic, palmitic, oleic, and eicosapentaenoic acids (EPA), are associated with the antimicrobial potential of algal species (Desbois and Smith 2010; Lee, Woo, and Lee 2016). The studies carried out by Banskota et al. (2019) found a positive correlation between lipid content and antioxidant properties, whereas total phenolic content was found to be positively correlated to antioxidant activity (Bulut et al. 2019). A similar finding was observed by Santhakumaran, Ayyapan, and Ray (2020). The phenolic extracts of *Nannochloropsis* sp. and *Spirulina* sp. were found to be a potential source of antifungal and antimycotoxigenic agents, which are more efficient than the commercial fungicide- tebuconazole (Scaglioni, Garcia, and Badiale-Furlong 2019). Recent research carried out by Santhakumaran, Ayyapan, and Ray (2020) revealed the antimicrobial activities of green microalgae such as *Myrmecia bisecta*, *Pseudotetradesmus quaternaries*, *Scenedesmus bijuga*, and *Bracteacoccus minor* against diverse pathogenic bacterial species. The same authors observed a high antioxidant potential of the alga *Tetrastrum komarekki*. In general, the multipurpose features of algal biomass make it a suitable bioresource for nutraceuticals, pharmaceuticals and so on. However, specific research publications in this direction are relatively rare, and there is high relevance for the intensive search of specific freshwater green algal species rich in antioxidant compounds from local environments.

13.7 THE INDUSTRIAL SIGNIFICANCE OF THE SEARCH OF ALGAL NUTRACEUTICALS

The demand for nutraceuticals is now increasing among consumers because of its significance in providing the required resilience against the recent pandemic of COVID-19. In general, the decline recently seen in the consumption of poultry, meat, and seafood products in non-vegetarian countries across the globe (Ho 2020) can lead to an increase in the demand for plant-based and vegan products for nutritional supplements. Among them, seaweed and microalgae have emerged as potent, sustainable sources of nutritional compounds. However, compared to the limitations in the availability of seaweed, industrially producible freshwater microalgae are highly relevant in this regard. The significant advantage of microalgae is that the industrial cultivation of them can be combined with carbon-sequestration and purification of eutrophic wastewaters.

Moreover, the biomass produced may be utilized for biofuel extraction and the nutraceutically valuable proteins, minerals, and vitamins may be extracted from the residual biomass after lipid extraction or the residual mass rich in such valuable compounds may be utilized as animal, poultry, or fish feeds. Recent market research conducted by Transparency Market Research (www.transparencymarketresearch. com 2020) has estimated the market value of algal products to reach USD 1,365.8 million by 2027, at a CAGR (Calculated Compound Annual Growth Rate) of 7.42%. In terms of volume, the market is expected to expand at a 5.35% CAGR between 2019 and 2027. Naturally, the research on the identification of nutraceutically valuable algal species in a country has become essential to its industrial growth in the future.

Among the various algal products, algal oil, particularly the omega group of fatty acids, is witnessing a considerable demand (www.prnewswire.com 2020) due to its high nutritional value, expanding vegan population, and declining fish population. The market of algal omega-3 ingredients such as EPA and DHA may reach the estimate of USD 1.2 billion by 2024, as it is growing at a rate of 11.3%, potentially until 2024 (www.prnewswire.com 2020). Algal proteins are another group of nutraceutical compound of high market demand. Similar to traditional sources, algal-protein has been generating significant attraction in the nutraceuticals industry owing to its high % content in the biomass, quality, and digestibility. Therefore, companies are increasingly opting for algae as a nutraceutical ingredient. In 2020, the global algal-protein market value exceeded USD 734.4 million and is estimated to grow at over 6.6% CAGR until 2027 (www.gminsights.com 2020). Similarly, the global algal pigment (Beta-Carotene, Astaxanthin, Fucoxanthin, Phycocyanin and Phycoerythrin) market value is estimated to reach USD 452.4 million by 2025 (www.meticulousresearch.com) at a CAGR of 4.0% till 2025.

The geographical distribution of algal markets is spread over North America, Europe, Asia Pacific, the Middle East and Africa, and Latin America. Among them, North America comprises more than 135 algal companies distributed over the USA, Canada, and Mexico. Currently, more than 50% of the algae produced through various cultivation technologies are utilized in DHA production to manufacture chemical components for medicines, health foods, cosmetics, and food additives. Table 13.1 shows the major algal companies in the global algal market. It may be noted that India, being a biodiversity-rich tropical country, is relatively lagging in the promotion of algae-based nutraceutical industries. Lack of promotion of proper research for the identification and industrial standardization of production protocols may be the primary reason for the same. Table 13.1 below shows the general distribution of major algae-based nutraceutical industries in the world.

13.8 CONCLUSION

Although the health benefits of microalgae have been known for centuries, comprehensive research into the nutraceutical potential and industrial-scale production for the same started seriously, only recently, in the last decade. The feasibility of microalgae to grow under varying environmental conditions, industrial cultivability, high productivity with an array of nutritionally and medicinally valuable biochemicals in its biomass make algae a highly relevant, sustainable bioresource for food, feeds, and nutraceuticals. However, even today, relatively few microalgal species are

TABLE 13.1

List of Major Algal Industries across the Globe

Country	Company	Website	Algae	Product	Reference
Austria	Panmol/Madaus	www.panmol.com	*Spirulina* sp.	Vitamin B_{12}	Lee and Marino (2010)
Canada	OceanNutrition	www.oceannutrition.com	*Chlorella* sp.	Carbohydrate extract	Hallmann (2007)
Denmark	Danisco	www.food.dupont.com	Macroalga	Hexose oxidase	Hallmann (2007)
France	InnovalG	–	*Odontella aurita*	EPA	Enzing et al. (2014)
Germany	Nutrinova/Celanese	www.celanese.com	Ulkenia	DHA	Hallmann (2007)
Germany	Subitec GmbH	www.subitec.com/de	Microalgae (Undisclosed)	Fatty acids	Hallmann (2007)
India	Parry agro Industries	www.murugappa.com	*Spirulina*	Nutraceuticals	Shimamatsu (2004)
India	Parry nutraceuticals (E.I.D Parry)	www.parrynutraceuticals.com	*Spirulina, Chlorella vulgaris*	Lycopene, phycocyanin	Enzing et al. (2014)
Indonesia	Indonesian Seaweed Industry Association (APBIRI)	www.indonesiaseaweed.com	*Gracilaria*	Agar	Hallmann (2007)
Israel	Algatech	www.algatech.com	*Haematococcus pluvialis*	Astaxanthin	Lee and Marino (2010)
Israel	Nature Beta Technologies	www.nikken-miho.com	*Dunaliella bardawil*	Beta-carotene	Walker et al. (2005)
Monacco	Exsymol S.A.M.	www.exsymol.com/en	Microalgae (Undisclosed)	Cosmetics	Hallmann (2007)
Netherlands	Corbion	www.corbion.com/algae-portfolio	Microalgae (Undisclosed)	DHA, oleic acid oil	www.corbion.com (2019)
Spain	Algas de Asturias S.A.	www.cybercolloids.net	Macroalgae	Agar	Lee and Marino (2010)
Spain	PharmaMar	www.pharmamar.com/	Microalgae (Undisclosed)	Anticancer drugs	Walker et al. (2005)
USA	Algenol	www.algenol.com/	Microalgae (Undisclosed)	Nutraceuticals	www.algenol.com (2020)
USA	Cyanotech	www.cyanotech.com	*Haematococcus pluvialis*	Astaxanthin	Shimamatsu (2004)
USA	DSM Nutritional Products	www.dsm.com/markets/human-nutrition/en-ap/home.htm	Microalgae (Undisclosed)	Nutritional products	www.dsm.com (2019)
USA	Earthrise firms	www.earthrise.com	*Spirulina platensis*	Nutraceuticals	Shimamatsu (2004)
USA	MERA Pharmaceuticals INC.	www.nelha.hawaii.gov/our-clients	*Haematococcus pluvialis*	Astaxanthin	Walker et al. (2005)
USA	Triton Algae Innovative	www.tritonai.com	*Chlamydomonas reinhardtii*	Protein	www.tritonai.com (2019)
USA	algaVia	www.algavia.com	Microalgae	Protein, Lipid	www.algavia.com (2019)

industrially utilized for the same. Significantly, the freshwater green microalgae, many of which are fast-growing with the potential of blooming natural waters in response to a slight increase in nutrients, remain uninvestigated regarding its productivity and nutritionally and medicinally valuable compounds in its biomass. Therefore, the prospects of algal investigation from all local environments, especially to identify the fast-growing species and their biochemical profile, remain relatively high.

In general, the existing literature crystal proves that the biomass of many species of algae has broad nutritional and pharmaceutical properties, and such species of algae can be used in industrial applications suitable for sustainable development. Therefore, research on nutraceutically valuable algae, especially the search for quality and quantity of biochemical compounds in them, remains the primary task of phycologists the world over. Algae are already esteemed as sustainable bioresources, and are capable of solving the global socio-environmental challenges, including malnutrition and food security. In this regard, the nutraceutical prospecting of microalgal resources for solving both the global environmental and economic crisis remains the most urgent need today.

REFERENCES

Adams, Shahieda, Andreas L. Lopata, Cornelius M. Smuts, Roslynn Baatjies, and Mohamed F. Jeebhay. 2019. "Relationship between Serum Omega-3 Fatty Acid and Asthma Endpoints." *International Journal of Environmental Research and Public Health* 16: 1–14. doi:10.3390/ijerph16010043.

Adarme-Vega, T Catalina, Skye R Thoams-Hall, David KY Lim, and Peer M Schenk. 2014. "Effects of Long-Chain Fatty Acid Synthesis and Associated Gene Expression in Microalga Tetraselmis Sp." *Marine Drugs* 12: 3381–3398. doi:10.3390/md12063381.

Adefegha, Stephen Adeniyi. 2018. "Functional Foods and Nutraceuticals as Dietary Intervention in Chronic Diseases; Novel Perspectives for Health Promotion and Disease Prevention." *Journal of Dietary Supplements* 15 (6): 977–1009. 10.1080/19390211.2017.1401573.

Akbar, Umair, Melissa Yang, Divya Kurian, and Chandra Mohan. 2017. "Omega-3 Fatty Acids in Rheumatic Diseases: A Critical Review." *Journal of Clinical Rheumatology* 23 (6): 330–339. doi:10.1097/RHU.0000000000000563.

An, Hyo Jin, Hong Kun Rim, Hyun Ja Jeong, Seung Heon Hong, Jae Young Um, and Hyung Min Kim. 2010. "Hot Water Extracts of Chlorella Vulgaris Improve Immune Function in Protein-Deficient Weanling Mice and Immune Cells." *Immunopharmacology and Immunotoxicology* 32 (4): 585–592. doi:10.3109/08923971003604778.

Apone, Fabio, Ani Barbulova, and Maria Gabriella Colucci. 2019. "Plant and Microalgae Derived Peptides Are Advantageously Employed as Bioactive Compounds in Cosmetics." *Frontiers in Plant Science* 10: 756. doi:10.3389/fpls.2019.00756.

Astorga-Espana, M. S., B. Rodríguez Galdon, E. Rodriguez, M. Rodriguez, and C. Diaz Romero. 2015. "Mineral and Trace Element Concentrations in Seaweeds from the Sub-Antarctic Ecoregion of Magallanes (Chile)." *Journal of Food Composition and Analysis* 39: 69–76. doi:10.1016/j.jfca.2014.11.010.

Banskota, Arjun H., Sandra Sperker, Roumiana Stefanova, Patrick J. McGinn, and Stephen J.B. O'Leary. 2019. "Antioxidant Properties and Lipid Composition of Selected Microalgae." *Journal of Applied Phycology* 31(1): 309–318. doi:10.1007/s10811-018-1523-1.

Barreiro, Carlos, and Jose-Luis Barredo. 2012. "Carotenoids Production: A Healthy and Profitable Industry." In *Methods in Molecular Biology*, edited by Carlos Barreiro and Jose-Luis Barredo, 1852:41–59. New York: Humana Press. doi:10.1007/978-1-61779-918-1_2.

Becker, E. W. 2007. "Micro Algae as a Source of Protein." *Biotechnology Advances* 25: 207–210. doi:10.1016/j.biotechadv.2006.11.002.

Benedetti, Manuel, Valeria Vecchi, Simone Barera, and Luca Dall'Osto. 2018. "Biomass from Microalgae : The Potential of Domestication towards Sustainable Biofactories." *Microbial Cell Factories* 17 (173): 1–18. doi:10.1186/s12934-018-1019-3.

Bhattacharya, Jayanta, Janet Currie, and Steven Haider. 2004. "Poverty, Food Insecurity, and Nutritional Outcomes in Children and Adults." *Journal of Health Economics* 23 (4): 839–862. doi:10.1016/j.jhealeco.2003.12.008.

Bigliardi, Barbara, and Francesco Galati. 2013. "Innovation Trends in the Food Industry: The Case of Functional Foods." *Trends in Food Science and Technology* 31 (2): 118–129. doi:10.1016/j.tifs.2013.03.006.

Bishop, West M., and Heidi M. Zubeck. 2012. "Evaluation of Microalgae for Use as Nutraceuticals and Nutritional Supplements." *Journal of Nutrition & Food Sciences* 2 (5): 147. doi:10.4172/2155-9600.1000147.

Bleakley, Stephen, and Maria Hayes. 2017. "Algal Proteins: Extraction, Application, and Challenges Concerning Production." *Foods* 6 (33): 1–34. doi:10.3390/foods6050033.

Borowitzka, Michael A., 1998. "Algae as Food." In *Microbiology of Fermented Foods*, edited by Brian J.B Wood, 1:585–598. UK: Blackie Academic & Professional. doi:10.1017/CBO9781107415324.004.

Brostow, Diana P., Andrew O. Odegaard, Woon Puay Koh, Sue Duval, Myron D. Gross, Jian Min Yuan, and Mark A. Pereira. 2011. "Omega-3 Fatty Acids and Incident Type 2 Diabetes: The Singapore Chinese Health Study." *The American Journal of Clinical Nutrition* 94 (2): 520–526. doi:10.3945/ajcn.110.009357.

Brower, Vicki. 1998. "Nutraceuticals: Poised for a Healthy Slice of the Healthcare Market?" *Nature Biotechnology* 16 (8): 728–731. doi:10.1038/nbt0898-728.

Brown, M R, M Mular, I Miller, C Farmer, and C Trenerry. 1999. "The Vitamin Content of Microalgae Used in Aquaculture." *Journal of Applied Phycology* 11: 247–255.

Bulut, Onur, Dilan Akın, Çağla Sönmez, Ayşegül Öktem, Meral Yücel, and Hüseyin Avni Öktem. 2019. "Phenolic Compounds, Carotenoids, and Antioxidant Capacities of a Thermo-Tolerant Scenedesmus sp. (Chlorophyta) Extracted with Different Solvents." *Journal of Applied Phycology* 31 (3): 1675–1683. doi:10.1007/s10811-018-1726-5.

Bungau, Simona, Mohamed M. Abdel-Daim, Delia Mirela Tit, Esraa Ghanem, Shimpei Sato, Maiko Maruyama-Inoue, Shin Yamane, and Kazuaki Kadonosono. 2019. "Health Benefits of Polyphenols and Carotenoids in Age-Related Eye Diseases." *Oxidative Medicine and Cellular Longevity*, 9783429. doi:10.1155/2019/9783429.

Campo, Jose A. Del, Mercedes Garcia-Gonzalez, and Miguel G. Guerrero. 2007. "Outdoor Cultivation of Microalgae for Carotenoid Production: Current State and Perspectives." *Applied Microbial* 74: 1163–1174. doi:10.1007/s00253-007-0844-9.

Campo, Jose A. Del, Jose Moreno, Herminia Rodríguez, M. Angeles Vargas, Joaquín Rivas, and Miguel G. Guerrero. 2000. "Carotenoid Content of Chlorophycean Microalgae: Factors Determining Lutein Accumulation in Muriellopsis Sp. (Chlorophyta)." *Journal of Biotechnology* 76 (1): 51–59. doi:10.1016/S0168-1656(99)00178-9.

Caporgno, Martin P., and Alexander Mathys. 2018. "Trends in Microalgae Incorporation Into Innovative Food Products With Potential Health Benefits." *Frontiers in Nutrition* 5: 58. doi:10.3389/fnut.2018.00058.

Charles, DJ. 2013. *Antioxidant Properties of Spices, Herbs and Other Sources*, 1st ed. New York: Springer-Verlag. doi:10.1007/s13398-014-0173-7.2.

Chel, Kaushik. 2011. "Renewable Energy for Sustainable Agriculture." *Agronomy for Sustainable Development* 31 (1): 91–118. doi:10.1051/agro/2010029>.

Chen, Chihoung, AM Pearson, and JI Gray. 1992. "Effects of Synthetic Antioxidants (BHA, BHT and PG) on the Mutagenicity of IQ-like Compounds." *Food Chemistry* 43: 177–183.

Christaki, Efterpi, Panagiota Florou-Paneri, and Eleftherios Bonos. 2011. "Microalgae: A Novel Ingredient in Nutrition." *International Journal of Food Sciences and Nutrition* 62 (8): 794–799. doi:10.3109/09637486.2011.582460.

Clemens, Roger, and Peter Pressman. 2020. "Microalgae: A Sustainable Source of Nutrients?" *Food Technology Magazine*. https://www.ift.org/news-and-publications/food-technology-magazine/issues/2020/march/columns/microalgae-a-sustainable-source-of-nutrients.

Cockbain, A. J., G. J. Toogood, and Mark A. Hull. 2012. "Omega-3 Polyunsaturated Fatty Acids for the Treatment and Prevention of Colorectal Cancer." *Gut* 61 (1): 135–149. doi:10.1136/gut.2010.233718.

Cubas, Catalina, M. Gloria Lobo, and Monica Gonzalez. 2008. "Optimization of the Extraction of Chlorophylls in Green Beans (Phaseolus Vulgaris L.) by N, N-Dimethylformamide Using Response Surface Methodology." *Journal of Food Composition and Analysis* 21: 125–133. doi:10.1016/j.jfca.2007.07.007.

Daliri, Eric Banan Mwine, and Byong H. Lee. 2015. "Current Trends and Future Perspectives on Functional Foods and Nutraceuticals." In *Beneficial Microorganisms in Food and Nutraceuticals*, edited by Min-Tze Liong, 27:221–244. Switzerland: Springer. doi:10.1007/978-3-319-23177-8.

Davis, E. A., Jean Dedrick, C. S. French, H. W. Milner, Jack Myers, J. H. C. Smith, and H. A. Spoehr. 1953. "Laboratory Experiments on Chlorella Culture at the Carnegie Institution of Washington Department of Plant Biology." In *Algal Culture: From Laboratory to Pilot Plant*, 1st ed. ,edited by John S. Burlew, 105–53:600. Washington, DC: Carnegie Institution of Washington Publication

DeFelice, Stephen L., 1995. "The Nutraceutical Revolution: Its Impact on Food Industry R&D." *Trends in Food Science and Technology* 6 (2): 59–61. doi:10.1016/S0924-2244(00)88944-X.

Desbois, Andrew P., and Valerie J. Smith. 2010. "Antibacterial Free Fatty Acids: Activities, Mechanisms of Action and Biotechnological Potential." *Applied Microbiology and Biotechnology* 85 (6): 1629–1642. doi:10.1007/s00253-009-2355-3.

Dixit, Dhara C., C. R.K. Reddy, Nikunj Balar, Poornima Suthar, Tejal Gajaria, and Devesh K. Gadhavi. 2017. "Assessment of the Nutritive, Biochemical, Antioxidant and Antibacterial Potential of Eight Tropical Macro Algae Along Kachchh Coast, India as Human Food Supplements." *Journal of Aquatic Food Product Technology* 27 (1): 1–79. doi:10.1080/10498850.2017.1396274.

Dixon, Chelsea, and Lisa R. Wilken. 2018. "Green Microalgae Biomolecule Separations and Recovery." *Bioresources and Bioprocessing* 5: 1–24. doi:10.1186/s40643-018-0199-3.

Driver, Thomas, Amit Bajhaiya, and Jon K. Pittman. 2014. "Potential of Bioenergy Production from Microalgae." *Current Sustainable/Renewable Energy Reports* 1 (3): 94–103. doi:10.1007/s40518-014-0011-8.

Dufosse, Laurent, Patrick Galaup, Anina Yaron, Shoshana Malis, Philippe Blanc, N Chidambara Murthy, and Gokare A Ravishankar. 2005. "Microorganisms and Microalgae as Sources of Pigments for Food Use : A Scientific Oddity or an Industrial Reality ?" *Food Science & Technology* 16: 389–406. doi:10.1016/j.tifs.2005.02.006.

Elkhishin, Mohamed T., Ravi Gooneratne, and Malik A. Hussain. 2017. "Microbial Safety of Foods in the Supply Chain and Food Security." *Advances in Food Technology and Nutritional Sciences* 3 (1): 22–32. doi:10.17140/AFTNSOJ-3-141.

Enzing, Christien, Matthias Ploeg, Maria Barbosa, and Lolke Sijtsma. 2014. "Microalgae-Based Products for the Food and Feed Sector: An Outlook for Europe." Spain. doi:10.2791/3339.

Falaise, Charlotte, Cyrille François, Marie Agnès Travers, Benjamin Morga, Joël Haure, Réjean Tremblay, François Turcotte, et al. 2016. "Antimicrobial Compounds from Eukaryotic Microalgae against Human Pathogens and Diseases in Aquaculture." *Marine Drugs* 14: 1–27. doi:10.3390/md14090159.

FAO. 2017. *The Future of Food and Agriculture - Trends and Challenges*. Rome: Food and Agriculture Organization of the United Nations. doi:10.4161/chan.4.6.12871.

FDA. 2016. "Summary: Substances Generally Recognized as Safe (Final Rule)." US Food and Drug Administration (HHS). https://www.fda.gov/AboutFDA/ReportsManualsForms/Reports/EconomicAnalyses/ucm517103.htm.

Fernandez, Francisco Gabriel Acien, Jose Maria Fernandez Sevilla, and Emilio Molina Grima. 2017. "Microalgae: The Basis of Mankind Sustainability." In *Case Study of Innovative Projects - Successful Real Cases*, edited by Bernardo Llamas. Intech Open. doi:10.5772/67930.

Ferreira, Veronica da Silva, and Celso Sant Anna. 2017. "Impact of Culture Conditions on the Chlorophyll Content of Microalgae for Biotechnological Applications." *World Journal of Microbiology and Biotechnology 33* (20): 1–8. doi:10.1007/s11274-016-2181-6.

Fore, Henrietta H., Qu Dongyu, David M. Beasley, and Tedros A. Ghebreyesus. 2020. "Child Malnutrition and COVID-19: The Time to Act Is Now." *The Lancet* 396 (10250): 517–518. doi:10.1016/S0140-6736(20)31648-2.

Gangl, Doris, Julie A.Z. Zedler, Priscilla D. Rajakumar, Erick M.Ramos Martinez, Anthony Riseley, Artur Włodarczyk, Saul Purton, et al. 2015. "Biotechnological Exploitation of Microalgae." *Journal of Experimental Botany* 66 (22): 6975–6990. doi:10.1093/jxb/erv426.

Goiris, Koen, Koenraad Muylaert, Ilse Fraeye, Imogen Foubert, Jos De Brabanter, and Luc De Cooman. 2012. "Antioxidant Potential of Microalgae in Relation to Their Phenolic and Carotenoid Content." *Journal of Applied Phycology* 24 (6): 1477–1486. doi:10.1007/s10811-012-9804-6.

Gomez-Guzman, Manuel, Alba Rodriguez-Nogales, Francesca Algieri, and Julio Galvez. 2018. "Potential Role of Seaweed Polyphenols in Cardiovascular-Associated Disorders." *Marine Drugs* 16 (8): 250. doi:10.3390/md16080250.

Granado-Lorencio, F., C. Herrero-Barbudo, G. Acien-Fernández, E. Molina-Grima, J. M. Fernandez-Sevilla, B. Perez-Sacristán, and I. Blanco-Navarro. 2009. "In Vitro Bioaccesibility of Lutein and Zeaxanthin from the Microalgae Scenedesmus Almeriensis." *Food Chemistry* 114 (2): 747–752. doi:10.1016/j.foodchem.2008.10.058.

Guedes, Ana Catarina, Helena M Amaro, and Francisco Xavier Malcata. 2011. "Microalgae as Sources of Carotenoids." *Marine Drugs* 9: 625–644. doi:10.3390/md9040625.

Guerin, Martin, Mark E. Huntley, and Miguel Olaizola. 2003. "Haematococcus Astaxanthin: Applications for Human Health and Nutrition." *Trends in Biotechnology* 21 (5): 210–216. doi:10.1016/S0167-7799(03)00078-7.

Guiry, Michael D. 2012. "How Many Species of Algae Are There ?" *Journal of Phycology* 48: 1057–1063. doi:10.1111/j.1529-8817.2012.01222.x.

Guzman, S., A. Gato, M. Lamela, M. Freire-Garabal, and J. M. Calleja. 2003. "Anti-Inflammatory and Immunomodulatory Activities of Polysaccharide from Chlorella Stigmatophora and Phaeodactylum Tricornutum." *Phytotherapy Research* 17 (6): 665–670. doi:10.1002/ptr.1227.

Habib, M.A.B., M. Parvin, T.C. Huntington, and M.R. Hasan. 2008. "A Review on Culture, Production and Use of Spirulina as Food for Humans and Feeds for Domestic Animals and Fish." Vol. 1034.

Habibi, Z, J Imanpour Namin, and Z Ramezanpour. 2018. "Evaluation of Antimicrobial Activities of Microalgae Scenedesmus Dimorphus Extracts against Bacterial Strains." *Caspian Journal of Environmental Sciences* 16 (1): 25–36. doi:10.22124/cjes.2018.2779.

Hallmann, Armin. 2007. "Algal Transgenics and Biotechnology." *Transgenic Plant Journal* 1 (1): 81–98.

Hammond, Sean T., James H. Brown, Joseph R. Burger, Tatiana P. Flanagan, Trevor S. Fristoe, Norman Mercado-Silva, Jeffrey C. Nekola, and Jordan G. Okie. 2015. "Food Spoilage, Storage, and Transport: Implications for a Sustainable Future." *BioScience* 65 (8): 758–768. doi:10.1093/biosci/biv081.

Hardy, Gil. 2000. "Nutraceuticals and Functional Foods: Introduction and Meaning." *Nutrition* 16: 688–689.

Hintz, Tana, Karl K. Matthews, and Rong Di. 2015. *"The Use of Plant Antimicrobial Compounds for Food Preservation."* BioMed Research International 2015, Hindawi Publishing Corporation, 1–12. doi:10.1155/2015/246264.

Ho, Sallly. 2020. "Coronavirus Triggers Biggest Decline In Global Meat Consumption In Decades." *Green Queen.* https://www.greenqueen.com.hk/coronavirus-triggers-biggest-decline-in-global-meat-consumption-in-decades.

Ibanez, Elena, and Alejandro Cifuentes. 2013. "Benefits of Using Algae as Natural Sources of Functional Ingredients." *Journal of the Science of Food and Agriculture* 93: 703–709. doi:10.1002/jsfa.6023.

Ji, Xiao-Jun, Lu-Jing Ren, and He Huang. 2015. "Omega-3 Biotechnology: A Green and Sustainable Process for Omega-3 Fatty Acids Production." *Frontiers in Bioengineering and Biotechnology* 3: 158. doi:10.3390/nu5041301.

Khairy, Hanan M., and Mohamed A. El-Sheikh. 2015. "Antioxidant Activity and Mineral Composition of Three Mediterranean Common Seaweeds from Abu-Qir Bay, Egypt." *Saudi Journal of Biological Sciences* 22 (5): 623–630. doi:10.1016/j.sjbs.2015.01.010.

Khan, Muhammad Imran, Jin Hyuk Shin, and Jong Deog Kim. 2018. "The Promising Future of Microalgae: Current Status, Challenges, and Optimization of a Sustainable and Renewable Industry for Biofuels, Feed, and Other Products." *Microbial Cell Factories* 17 (36): 1–21. doi:10.1186/s12934-018-0879-x.

Khanraa, Saumyakanti, Madhumanti Mondal, Gopinath Halder, O. N. Tiwari, Kalyan Gayen, and Tridib Kumar Bhowmick. 2018. "Proteins from Microalgae Are Mainly Used as Nutraceuticals or Included in the Formulation of Functional Food." *Food and Bioproducts Processing* 110: 60–84.

Kim, Sang Min, Suk Woo Kang, O. Nam Kwon, Donghwa Chung, and Cheol Ho Pan. 2012. "Fucoxanthin as a Major Carotenoid in Isochrysis Aff. Galbana: Characterization of Extraction for Commercial Application." *Journal of the Korean Society for Applied Biological Chemistry* 55 (4): 477–483. doi:10.1007/s13765-012-2108-3.

Knothe, Gerhard. 2005. "Dependence of Biodiesel Fuel Properties on the Structure of Fatty Acid Alkyl Esters." *Fuel Processing Technology* 86: 1059–1070. doi:10.1016/j.fuproc.2004.11.002.

Kookal, Santhosh Kumar, Prasanthkumar Santhakumaran, and Joseph George Ray. 2017. "Biomass Yield, Oil Productivity and Fatty Acid Profile of Chlorella Lobophora Cultivated in Diverse Eutrophic Wastewaters." *Biocatalysis and Agricultural Biotechnology* 11: 338–344. doi:10.1016/j.bcab.2017.08.006.

Kovac, Dajana J, Jelica B Simeunovic, Olivera B Babic, Aleksandra C Misan, and Ivan Lj Milovanovic. 2013. "Algae in Food and Feed." *Food and Feed Research* 40 (1): 21–31.

Koyande, Apurav Krishna, Kit Wayne Chew, Krishnamoorthy Rambabu, Yang Tao, Dinh-Toi Chu, and Pau-Loke Show. 2019. "Microalgae: A Potential Alternative to Health Supplementation for Humans." *Food Science and Human Wellness* 8: 16–24. doi:10.1016/j.fshw.2019.03.001.

Krimpen, van MM., P. Bikker, I. M. van der Meer, C.M.C. van der Peet- Schwering, and J.M. Vereijken. 2013. "Cultivation, Processing and Nutritional Aspects for Pigs and Poultry of European Protein Sources as Alternatives for Imported Soybean Products." Wageningen. https://library.wur.nl/WebQuery/wurpubs/437524.

Lee, Chacon T. L., and Gonzalez G. E. Marino. 2010. "Microalgae for 'Healthy' Foods-Possibilities and Challenges." *Comprehensive Reviews in Food Science and Food Safety* 9: 655–675. doi:10.1111/j.1541-4337.2010.00132.x.

Lee, Wonjong, Eun-Rhan Woo, and Dong Gun Lee. 2016. "Phytol Has Antibacterial Property by Inducing Oxidative Stress Response in Pseudomonas Aeruginosa." *Free Radical Research* 50 (12) : 1309–1318. doi:10.1080/10715762.2016.1241395.

Lenihan-Geels, Georgia, Karen S. Bishop, and Lynnette R. Ferguson. 2013. "Alternative Sources of Omega-3 Fats: Can We Find a Sustainable Substitute for Fish?" *Nutrients* 5 (4): 1301–1315. doi:10.3390/nu5041301.

Liguori, Ilaria, Gennaro Russo, Francesco Curcio, Giulia Bulli, Luisa Aran, David Della-Morte, Gaetano Gargiulo, et al. 2018. "Oxidative Stress, Aging, and Diseases." *Clinical Interventions in Aging* 13: 757–772. doi:10.2147/CIA.S158513.

Liu, Jin, Milton Sommerfeld, and Qiang Hu. 2013. "Screening and Characterization of Isochrysis Strains and Optimization of Culture Conditions for Docosahexaenoic Acid Production." *Applied Microbiology and Biotechnology* 97: 4785–4798. doi:10.1007/s00253-013-4749-5.

Lordan, Sinead, R. Paul Ross, and Catherine Stanton. 2011. "Marine Bioactives as Functional Food Ingredients: Potential to Reduce the Incidence of Chronic Diseases." *Marine Drugs* 9 (6): 1056–1100. doi:10.3390/md9061056.

Lorenz, R Todd, and Gerald R Cysewski. 2000. "Commercial Potential for Haematococcus Microalgae as a Natural Source of Astaxanthin." *Trends in Biotechnology in Biotechnology* 18 (4): 160–167.

Lubian, Luis M., Olimpio Montero, Ignacio Moreno-Garrido, I. Emma Huertas, Cristina Sobrino, Manuel Gonzalez-Del Valle, and Griselda Pares. 2000. "Nannochloropsis (Eustigmatophyceae) as Source of Commercially Valuable Pigments." *Journal of Applied Phycology* 12: 249–255. doi:10.1023/A:1008170915932.

Mall, R. K., Ranjeet Singh, Akhilesh Gupta, G. Srinivasan, and L. S. Rathore. 2007. "Impact of Climate Change on Indian Agriculture: A Review." *Climatic Change* 78 (1–2): 445–478. doi:10.1007/s10584-006-9236-x.

Markou, Giorgos, Irini Angelidaki, Elias Nerantzis, and Dimitris Georgakakis. 2013. "Bioethanol Production by Carbohydrate-Enriched Biomass of Arthrospira (Spirulina) Platensis." *Energies* 6 (8): 3937–3950. doi:10.3390/en6083937.

Mata, Teresa M., Antonio A. Martins, and Nidia. S. Caetano. 2010. "Microalgae for Biodiesel Production and Other Applications: A Review." *Renewable and Sustainable Energy Reviews* 14 (1): 217–232. doi:10.1016/j.rser.2009.07.020.

Mathimani, Thangavel, and Arivalagan Pugazhendhi. 2019. "Utilization of Algae for Biofuel, Bio-Products and Bio-Remediation." *Biocatalysis and Agricultural Biotechnology* 17: 326–330. doi:10.1016/j.bcab.2018.12.007.

McCarty, Mark F., and James J. DiNicolantonio. 2020. "Nutraceuticals Have Potential for Boosting the Type 1 Interferon Response to RNA Viruses Including Influenza and Coronavirus." *Progress in Cardiovascular Diseases* 63 (3): 383–385. doi:10.1016/j.pcad.2020.02.007.

Merchant, Randall E., and Cynthia A. Andre. 2001. "A Review of Recent Clinical Trials of the Nutritional Supplement Chlorella Pyrenoidosa in the Treatment of Fibromyalgia, Hypertension, and Ulcerative Colitis." *Alternative Therapies* 7 (3): 79–90.

Meybeck, Alexandre, Elizabeth Laval, Rachel Levesque, and Genevieve Parent. 2018. *"Food Security and Nutrition in the Age of Climate Change."* In *Proceedings of the International Symposium Organized by the Government of Québec in Collaboration with FAO*, 132. Quebec City: FAO.

Mishra, V.K., R.K. Bacheti, and A. Husen. 2011. "Medicinal Uses of Chlorophyll: A Critical Overview." In *Chlorophyll: Structure, Production and Medicinal Uses*, edited by Hua Le and Elisa Salcedo, 177–196. Hauppauge, NY: Nova Science Publishers, Inc.

Mohebi-Nejad, Azin, and Behnood Bikdeli. 2014. "Omega-3 Supplements and Cardiovascular Diseases." *Tanaffos* 13 (1): 6–14.

Mori, Hideyuki, Yasuo Takahashi, Eric Zusman, André Mader, Erin Kawazu, Takashi Otsuka, Mustafa Moinuddin, et al. 2020. "Implications of COVID-19 for the Environment and Sustainability." *Institute for Global Environmental Strategies*, no. May: 1–12. http://repositorio.unan.edu.ni/2986/1/5624.pdf.

Murthy, K. N. Chidambara, A. Vanitha, J. Rajesha, M. Mahadeva Swamy, P. R. Sowmya, and
Gokare A. Ravishankar. 2005. "In Vivo Antioxidant Activity of Carotenoids from
Dunaliella Salina — a Green Microalga." *Life Sciences* 76: 1381–1390. doi:10.1016/j.
lfs.2004.10.015.

Natrah, F. M., F. M. Yusoff, M. Shariff, F. Abas, and N. S. Mariana. 2007. "Screening of
Malaysian Indigenous Microalgae for Antioxidant Properties and Nutritional Value."
Journal of Applied Phycology 19 (6): 711–718. doi:10.1007/s10811-007-9192-5.

Niccolai, Alberto, Graziella Chini Zittelli, Liliana Rodolfi, Natascia Biondi, and Mario R.
Tredici. 2019. "Microalgae of Interest as Food Source: Biochemical Composition and
Digestibility." *Algal Research* 42: 101617. doi:10.1016/j.algal.2019.101617.

Ochsenreither, Katrin, Claudia Glück, Timo Stressler, Lutz Fischer, and Christoph Syldatk.
2016. "Production Strategies and Applications of Microbial Single Cell Oils." *Frontiers
in Microbiology* 7: 1539. doi:10.3389/fmicb.2016.01539.

Olaizola, Miguel. 2003. "Commercial Development of Microalgal Biotechnology: From the
Test Tube to the Marketplace." *Biomolecular Engineering* 20: 459–466. doi:10.1016/
S1389-0344(03)00076-5.

Orosa, M., E. Torres, P. Fidalgo, and J. Abalde. 2000. "Production and Analysis of Secondary
Carotenoids in Green Algae." *Journal of Applied Phycology* 12: 553–556.

Panahi, Yunes, Simin Dashti-Khavidaki, Farahnoosh Farnood, Hamid Noshad, Mahsa Lotfi,
and Afshin Gharekhani. 2016. "Therapeutic Effects of Omega-3 Fatty Acids on Chronic
Kidney Disease-Associated Pruritus: A Literature Review." *Advanced Pharmaceutical
Bulletin* 6 (4): 509–514. doi:10.15171/apb.2016.064.

Piwowar, Arkadiusz, and Joanna Harasym. 2020. "The Importance and Prospects of the Use
of Algae in Agribusiness." *Sustainability* 12 (14): 1–13. doi:10.3390/su12145669.

Ramar, Dineshkumar, Narendran Rajendran, P. Jayasingam, and Sampathkumar Pichai. 2017.
"Cultivation and Chemical Composition of Microalgae Chlorella Vulgaris and Its
Antibacterial Activity against Human Pathogens." *Journal of Aquaculture & Marine
Biology* 5 (3): 1–8. doi:10.15406/jamb.2017.05.00119.

Raposo, Maria Filomena de Jesus, Alcina Maria Miranda Bernardo de Morais, and Rui Manuel
Santos Costa de Morais. 2016. "Emergent Sources of Prebiotics: Seaweeds and
Microalgae." *Marine Drugs* 14 (2): 1–27. doi:10.3390/md14020027.

Ray, J.G., P. Santhakumaran, and S. K. Kookal. 2020. "Phytoplankton Communities of
Eutrophic Freshwater Bodies (Kerala, India) in Relation to the Physicochemical Water
Quality Parameters." *Environment, Development and Sustainability*. doi:https://doi.
org/10.1007/s10668-019-00579-y.

Ray, J G, and KS Nidheesh. 2019. "Assessment of Soil Fertility Characteristics of Chemical-
Fertilized Banana Fields of South India." *Communications in Soil Science and Plant
Analysis* 50 (3): 275–286. doi:10.1080/00103624.2018.1559331.

Ritala, Anneli, Suvi T. Hakkinen, Mervi Toivari, and Marilyn G. Wiebe. 2017. "Single Cell
Protein-State-of-the-Art, Industrial Landscape and Patents 2001-2016." *Frontiers in
Microbiology* 8: 2009. doi:10.3389/fmicb.2017.02009.

Rizwan, Muhammad, Ghulam Mujtaba, Sheraz Ahmed Memon, Kisay Lee, and Naim Rashid.
2018. "Exploring the Potential of Microalgae for New Biotechnology Applications and
beyond: A Review." *Renewable and Sustainable Energy Reviews* 92: 394–404.
doi:10.1016/j.rser.2018.04.034.

Rizza, Lara Sanchez, Maria Eugenia Sanz Smachetti, Mauro Do Nascimento, Graciela Lidia
Salerno, and Leonardo Curatti. 2017. "Bioprospecting for Native Microalgae as an
Alternative Source of Sugars for the Production of Bioethanol." *Algal Research* 22:
140–147. doi:10.1016/j.algal.2016.12.021.

Sahni, Prashant, Poonam Aggarwal, Savita Sharma, and Baljit Singh. 2019. "Nuances of
Microalgal Technology in Food and Nutraceuticals: A Review." *Nutrition & Food
Science*. doi:10.1108/nfs-01-2019-0008.

Sahodaran, Nidheesh Kammadavil, and Joseph George Ray. 2018. "Heavy Metal Contamination in 'chemicalized' Green Revolution Banana Fields in Southern India." *Environmental Science and Pollution Research* 25 (27): 26874–26886. doi:10.1007/s11356-018-2729-0.

Sanmukh, Swapnil, Benedict Bruno, Udhaya Ramakrishnan, Krishna Khairnar, Sandhya Swaminathan, and Waman Paunikar. 2014. "Bioactive Compounds Derived from Microalgae Showing Antimicrobial Activities." *Journal of Aquaculture Research and Development* 5 (3): 3–6. doi:10.4172/2155-9546.1000224.

Santhakumaran, Prasanthkumar, Sunil Meppath Ayyapan, and Joseph George Ray. 2020a. "Nutraceutical Applications of Twenty-Five Species of Rapid-Growing Green-Microalgae as Indicated by Their Antibacterial, Antioxidant and Mineral Content." *Algal Research* 47: 101878. doi:https://doi.org/10.1016/j.algal.2020.101878.

Santhakumaran, Prasanthkumar, Santhosh Kumar Kookal, Linu Mathew, and Joseph George Ray. 2019. "Bioprospecting of Three Rapid-Growing Freshwater Green Algae, Promising Biomass for Biodiesel Production." *BioEnergy Research* 12 (3): 680–693. doi:10.1007/s12155-019-09990-9.

Santhakumaran, Prasanthkumar, Santhosh Kumar Kookal, Linu Mathew, and Joseph George Ray. 2020b. "Experimental Evaluation of the Culture Parameters for Optimum Yield of Lipids and Other Nutraceutically Valuable Compounds in Chloroidium Saccharophilum (Kruger) Comb. Nov." *Renewable Energy* 147 (P1): 1082–1097. doi:10.1016/j.renene.2019.09.07.

Santhakumaran, Prasanthkumar, Santhosh Kumar Kookal, and Joseph George Ray. 2018. "Biomass Yield and Biochemical Profile of Fourteen Species of Fast-Growing Green Algae from Eutrophic Bloomed Freshwaters of Kerala, South India." *Biomass and Bioenergy* 119: 155–165. doi:10.1016/j.biombioe.2018.09.021.

Sathasivam, Ramaraj, Ramalingam Radhakrishnan, Abeer Hashem, and Elsayed F. Abd Allah. 2019. "Microalgae Metabolites: A Rich Source for Food and Medicine." *Saudi Journal of Biological Sciences* 26 (4): 709–722. doi:10.1016/j.sjbs.2017.11.003.

Scaglioni, Priscila Tessmer, Sabrinade Oliveira Garcia, and Eliana Badiale-Furlong. 2019. "Inhibition of in Vitro Trichothecenes Production by Microalgae Phenolic Extracts." *Food Research International* 124: 175–180.

Shahidi, Fereidoon, and Priyatharini Ambigaipalan. 2018. "Omega-3 Polyunsaturated Fatty Acids and Their Health Benefits." *Annual Review of Food Science and Technology* 9 (1): 345–381. doi:10.1146/annurev-food-111317-095850.

Shimamatsu, Hidenori. 2004. "Mass Production of Spirulina, an Edible Microalga." *Hydrobiologia* 512: 39–44.

Singh, Jasvinder, and Rakesh Chandra Saxena. 2015. "An Introduction to Microalgae : Diversity and Significance." In *Handbook of Marine Microalgae-Biotechnology Advances*, edited by Se-Kwon Kim, 11–24. USA: Academic Press. doi:10.1016/B978-0-12-800776-1.00002-9.

Singh, Jyoti, and Dolly Wattal Dhar. 2019. "Overview of Carbon Capture Technology: Microalgal Biorefinery Concept and State-of-the-Art." *Frontiers in Marine Science* 6 (29): 1–9. doi:10.3389/fmars.2019.00029.

Slavin, Joanne L, and Beate Lloyd. 2012. "Health Benefits of Fruits and Vegetables." *Advances in Nutrition* 3: 506–516. doi:10.3945/an.112.002154.506.

Soletto, D., L. Binaghi, A. Lodi, J. C.M. Carvalho, and A. Converti. 2005. "Batch and Fed-Batch Cultivations of Spirulina Platensis Using Ammonium Sulphate and Urea as Nitrogen Sources." *Aquaculture* 243: 217–224. doi:10.1016/j.aquaculture.2004.10.005.

Spolaore, Pauline, Claire Joannis-Cassan, Elie Duran, Arsène Isambert, Laboratoire De Génie, and Ecole Centrale Paris. 2006. "Commercial Applications of Microalgae." *Journal of Bioscience and Bioengineering* 101 (2): 87–96. doi:10.1263/jbb.101.87.

Srinivasan, Krishnapura. 2010. "Traditional Indian Functional Foods." In *Functional Foods of the East*, edited by John Shi, Chi-Tang Ho, and Fereidoon Shahidi, 1st ed., 51–84. Florida: CRC press. doi:10.1201/b10264-4.

Sun, Han, Weiyang Zhao, Xuemei Mao, Yuelian Li, Tao Wu, and Feng Chen. 2018. "High-Value Biomass from Microalgae Production Platforms: Strategies and Progress Based on Carbon Metabolism and Energy Conversion." *Biotechnology for Biofuels* 11 (227): 1–23. doi:10.1186/s13068-018-1225-6.

Sun, Liqin, Changhai Wang, Quanjian Shi, and Cuihua Ma. 2009. "Preparation of Different Molecular Weight Polysaccharides from Porphyridium Cruentum and Their Antioxidant Activities." *International Journal of Biological Macromolecules* 45 (1): 42–47. doi:10.1016/j.ijbiomac.2009.03.013.

Swanson, Kelly S., Rebecca A. Carter, Tracy P. Yount, Jan Aretz, and Preston R. Buff. 2013. "Nutritional Sustainability of Pet Foods." *Advances in Nutrition* 4 (2): 141–150. doi:10.3945/an.112.003335.

Syngai, Gareth Gordon, Ragupathi Gopi, Rupjyoti Bharali, Sudip Dey, G. M.Alagu Lakshmanan, and Giasuddin Ahmed. 2016. "Probiotics - the Versatile Functional Food Ingredients." *Journal of Food Science and Technology* 53 (2): 921–933. doi:10.1007/s13197-015-2011-0.

Thangavel, Kalaiselvi, Preethi Radha Krishnan, Srimeena Nagaiah, Senthil Kuppusamy, Senthil Chinnasamy, Jude Sudhagar Rajadorai, Gopal Nellaiappan Olaganathan, and Balachandar Dananjeyan. 2018. "Growth and Metabolic Characteristics of Oleaginous Microalgal Isolates from Nilgiri Biosphere Reserve of India." *BMC Microbiology* 18 (1): 1–17. doi:10.1186/s12866-017-1144-x.

Udayan, A., M. Arumugam, and A. Pandey. 2017. "Nutraceuticals From Algae and Cyanobacteria." In *Algal Green Chemistry: Recent Progress in Biotechnology*, 1st ed., edited by Rajesh Rastogi, Datta Madamwar, and Ashok Pandey, 65–89. doi:10.1016/B978-0-444-63784-0.00004-7.

Ventola, C Lee. 2015. "The Antibiotic Resistance Crisis Part 1: Causes and Threats." *Pharmacy & Therapeutics* 40 (4): 277–283.

Vermeulen, Sonja J., Bruce M. Campbell, and John S.I Ingram. 2012. "Climate Change and Food Systems." *Annual Review of Environment and Resources* 37: 195–222. doi:10.1146/annurev-environ-020411-130608.

Vonshak, Avigad. 1990. "Recent Advances in Microalgal Biotechnology." *Biotechnology Advances* 8: 709–727.

Walker, Tara L., Saul Purton, Douglas K. Becker, and Chris Collet. 2005. "Microalgae as Bioreactors." *Plant Cell Reports* 24: 629–641. doi:10.1007/s00299-005-0004-6.

Wani, Ab Latif, Sajad Ahmad Bhat, and Anjum Ara. 2015. "Omega-3 Fatty Acids and the Treatment of Depression: A Review of Scientific Evidence." *Integrative Medicine Research* 4 (3): 132–141. doi:10.1016/j.imr.2015.07.003.

Watanabe, Fumio, and Tomohiro Bito. 2018. "Vitamin B12 Sources and Microbial Interaction." *Experimental Biology and Medicine* 243 (2): 148–158. doi:10.1177/1535370217746612.

Wells, Mark L, James S Philippe Potin, John A Craigie, Sabeeha S Raven, Katherine E Merchant, Alison G Smith Helliwell, Camire Mary Ellen, and Brawley Susan H. 2017. "Algae as Nutritional and Functional Food Sources : Revisiting Our Understanding." *Journal of Applied Phycology* 29: 949–982. doi:10.1007/s10811-016-0974-5.

WHO. 2015. "Global and Regional Food Consumption Patterns and Trends." https://www.who.int/nutrition/topics/3_foodconsumption/en/index4.html.

Wilson, Douglas W, Paul Nash, Harpal Singh Buttar, Keith Griffiths, Ram Singh, Fabien De Meester, Rie Horiuchi, and Toru Takahashi. 2017. "The Role of Food Antioxidants, Benefits of Functional Foods, and Influence of Feeding Habits on the Health of the Older Person: An Overview." *Antioxidants* 6 (81): 1–20. doi:10.3390/antiox6040081.

Wu, Guoyao, Jessica Fanzo, Dennis D. Miller, Prabhu Pingali, Mark Post, Jean L. Steiner, and Anna E. Thalacker-Mercer. 2014. "Production and Supply of High-Quality Food Protein for Human Consumption: Sustainability, Challenges, and Innovations." *Annals of the New York Academy of Sciences* 1321 (1): 1–19. doi:10.1111/nyas.12500.

www.algavia.com. (2019). algaVia. www.algavia.com. Accessed 6 July 2019

www.corbion.com. (2019). Corbion. www.corbion.com/algae-portfolio. Accessed 10 January 2019

www.dsm.com. (2019). DSM. www.dsm.com/markets/human-nutrition/en-ap/home.html. Accessed 25 June 2019

www.gminsights.com. 2020. www.gminsights.com/industry-analysis/algae-protein-market. Accessed 2 August 2020

www.prnewswire.com. 2020. https://www.prnewswire.com/news-releases/algae-omega-3-in-gredients-market---growth-trends-and-forecasts-2019---2024. Accessed 2 August 2020

www.transparencymarketresearch.com. 2020. grandviewresearch.com/industry-analysis/algae-protein-market. Accessed 2 August 2020

www.tritonai.com. (2019). TRITON. www.tritonai.com. Accessed 2 June 2019

Yasuyuki, Miyano, Koyama Kunihiro, Sreekumari Kurissery, Nandakumar Kanavillil, Yoshiro Sato, and Yasushi Kikuchi. 2010. "Antibacterial Properties of Nine Pure Metals: A Laboratory Study Using Staphylococcus Aureus and Escherichia Coli." *Biofouling: The Journal of Bioadhesion and Biofilm Research* 26 (7): 851–858. doi:10.1080/08927014.2010.527000.

Zaid, Abeer A. Abu, Doaa M. Hammad, and Eman M. Sharaf. 2015. "Antioxidant and Anticancer Activity of Spirulina Platensis Water Extracts." *International Journal of Pharmacology* 11 (7): 846–851. doi:10.3923/ijp.2015.846.851.

Zainol, M. K., A. Abd-Hamid, S. Yusof, and R. Muse. 2003. "Antioxidative Activity and Total Phenolic Compounds of Leaf, Root and Petiole of Four Accessions of Centella Asiatica (L.) Urban." *Food Chemistry* 81 (4): 575–581. doi:10.1016/S0308-8146(02)00498-3.

Zaporozhets, Tatyana, and Natalia Besednova. 2016. "Prospects for the Therapeutic Application of Sulfated Polysaccharides of Brown Algae in Diseases of the Cardiovascular System: Review." *Pharmaceutical Biology* 54 (12): 3126–3135. doi:10.1080/13880209.2016.1185444.

14 Green Algae as a Bioenergy Resource with the Eco-technological Potential for Sustainable Development

Santhosh Kumar Kookal
International Centre for Genetic Engineering and
Biotechnology, India

Joseph George Ray
Mahatma Gandhi University, India

CONTENTS

14.1 INTRODUCTION

Human development is an energy-intensive process of natural transformation that intends socioeconomic progress. In the past, development through fast transformations of nature became possible by technological inventions for energy procurement and production process of diverse kinds. However, now the conventional means of developmental prospects remain bleak for humans, because the finite energy sources are being depleted and the unexpected environmental damages which emanate from conventional technological means of development are becoming detrimental to the sustainability of the already achieved progress globally. Global warming as an outcome of our ill-ecological technological developmental process is threatening human survival on the earth. In the above context, ecologically safe novel technologies have become inevitable for ensuring the sustainability of development. Technology for ensuring renewable energy availability, energy-efficient production processes, and environmental remediation have become more significant in this regard. All such eco-friendly technologies come under the standard label of ecotechnologies. Ecotechnologies are inevitable for achieving sustainable development.

The environmental impacts of excessive consumption of fossil fuels and radioactive minerals for energy cause environmental diseases, ill-health in humans, decrease the productivity of crops, and are detrimental to global environment stability (Sahoo et al., 2012). Moreover, the depletion of finite fuel reserves is reducing availability to critical levels and escalating the price of fuel continuously. As a result, the economic growth of nations, especially the poor and developing nations, remains severely impacted. Although less energy-intensive alternate pursuits of developmental models are emerging, a steady increase in the demand for sustainable energy forms needs immediate scientific attention. Thus, the current environment and energy scenario are inspiring researchers to discover and harness environmentally safe alternate sources of fuels and technology to ensure its continuous availability for supporting sustainable development.

Algal technology is one of the emerging fields of ecotechnology that explores the application and utilization of algal biomass for sustainable industrially oriented products (Fabris et al., 2020). The algae-based technologies are more adapted to and efficient for local environmental conditions (Bhola et al., 2014; Kamani et al., 2019), and are gaining more attention worldwide as a sustainable, eco-friendly green technology. Technologies which do not involve excessive carbon emission or other pollutants and do not create environmental mishappenings in the future are also called green technologies. The significant advancement and research area in the field of utilization of algae for sustainable development include the sustainable production of biomass for bioenergy or biofuels from potential microalgal resources (Chia et al., 2018; Rakesh et al., 2020; Jo et al., 2020). It focuses on the eco-friendly conversion of algal biomass to renewable fuel. Algal biomass can be converted both physically and biochemically into diverse kinds of fuel products such as biodiesel (Pugliese et al., 2020; Patel et al., 2020), ethanol (Khan et al., 2018a), syngas (Saad et al., 2019), biogas (Zabed et al., 2020), bio hydrogen (Mona et al., 2020; Chen et al., 2020), and so on. As a green technological procedure, algal fuel technology mainly focuses on environmental health, minimizing fossil fuel consumption, control of water pollution, and reducing carbon emission along with fuel production and consumption.

Biofuel is an alternative energy resource which is quite amenable to modern uses, including aviation fuel purposes (Hassan and Kalam, 2013). The different kinds of biofuels include bio-diesels (which are trans-esterified bio-oils), bioethanol (produced by fermentation of economically cheap carbohydrate resources), biocrude (liquified wet biomass under high pressure and temperature), biogas (methane from anaerobic fermentation of biomass) and so on. The use of such biofuel for energy purposes can reduce the emission of unburned hydrocarbons, carbon monoxide, sulfates, nitrates, and polycyclic aromatic hydrocarbons into the environment (Shalaby, 2011).

Based on sources, biofuels are categorized into first-generation, second-generation, and third- and fourth-generation biofuels. Biodiesel generated using consumable vegetable oils, and bioethanol produced by fermentation of edible carbohydrate resources belong to the first-generation biofuels. Although biodiesel and bioethanol produced from such resources have good fuel quality, overdependence on edible resources for fuel purposes has certain limitations. The competition of such resources for food and fuel purposes will lead to over utilization of land resources for agricultural production and consequent environmental issues and high production costs (Chen et al., 2018). Moreover, such a means of meeting fuel from edible resources cannot meet the increased demands for food or fuel resources. In order to overcome the situation, the search for non-food based and season independent alternate feedstock for biofuel such as Jatropha seed-oil for biodiesel was initiated, which became known as second-generation biofuels. A considerable volume of non-edible oil resources from diverse plants could be produced using barren land for biofuel purposes. But, there is a limit to the availability of even wastelands and water resources to produce sufficient amount of oil seeds or other kinds of energy-rich crop biomass to meet the tremendously growing demands of fuel resources. Thus, it became clear that the cost of first- and second-generation biofuel production cannot keep them really 'green' in nature. It was in this context that research for a third-generation fuel became highly necessary, and microalgae are emerging as the third- and fourth-generation biomass resources that have the potential to meet all the criteria of a truly alternative, cost-effective biofuel (Lackner, 2016). If the third-generation biofuel was from species-specific algae, the fourth-generation biofuels are obtained from genetically modified species of algae (Lü et al., 2011). Thus, algal biotechnology and algal genetic engineering remain a crucial means of algal biomass and biofuel production systems. These have become the thrust area of fuel research, which are currently going on worldwide.

The significance of microalgal biofuel as an alternative source of bioenergy is great. First, as the fastest growing photosynthetic microorganism, microalgae can provide renewable oil-rich biomass within a limited time and in large quantities (Khan et al., 2018b). Hence microalgae can ensure increasing demand for oil resources for the future. Another advantage is that the production process is quite environmentally feasible and will not create any environmental contradictions, but can even lead to the simultaneous cleaning of water (Chou et al., 2019; Tayari et al., 2020). Microalgal production can be sustainable, and the oil produced is entirely biodegradable. Many species of algae can produce a high amount of biomass with a high percentage of lipids and other hydrocarbons, which makes possible a comparison of algal fuel with that of petroleum fuel (Patel et al., 2017; Shi et al., 2018).Thus, dependence on algae for energy can lead to a reduction of harmful emissions of petroleum fuels and ultimately reduce the greenhouse effects and global warming.

Recent research on the algal technology and bioconversion of algal biomass focuses on the utilization of algal biomass as a multipurpose bioresource for industrial production of energy to fine chemicals (Aratboni et al., 2019). Experimental and pilot-scale studies in algal genetic engineering have shown that algal biomass production can meet current energy demands by using specific single species and has suitability toward fast carbon sequestration, which is quite essential in controlling climate change (Aratboni et al., 2019; Radakovits et al., 2010; Sharma et al., 2018). Thus, the third- and fourth-generation biofuel research programs globally focus on the downstream aspects of algae biomass such as production, harvesting techniques, conversion, and the chemistry of biofuel productions.

Algal bioconversion and technology perspectives encompass the search and collection of unique microalgal strains and exploitation of their biofuel potential in addition to their sustainable value-added products (Khan et al., 2018a). Therefore, the main objective of this chapter is to explore the current strategies focusing on the identification of lipid-rich species of algae, the simultaneous use of algae for biomass production and freshwater phyco-remediation, and the bioconversion technologies for the generation of biofuel from microalgae, which are discussed in detail. The research gap in the area of bioconversion of algal biomass and up-gradation of algal biofuel quality through technological means is also discussed in detail. The different themes under these headings are reviewed systematically below.

14.2 RESEARCH ON LIPID-YIELDING MICROALGAE

Microalgae are the fastest growing and most productive aquatic photosynthetic microorganisms, which are cosmopolitan in distribution. Compared to land plants, microalgae have a very high potential for multiplication, rate of production, and doubling of their biomass (Chisti, 2007). Because of its extraordinary productivity, microalgal cells may be called 'green factories'. Microalgae are quite amenable to mass production technologies (Shen et al., 2009). They can be easily streamlined genetically to industrially productive suitable strains for specific purposes through biotechnological means (Fu et al., 2019). They can be grown on an industrial-scale throughout the year (Abhishek et al., 2014) and can yield the highest amount of biomass per unit time, space, and nutrient supply (Saad et al., 2019).

More than 7,000 species of green algae are currently known (Hoek et al., 1995). The bioenergy potential of the majority of them is not well explored (Yoshida et al., 2012). The biochemical, as well as the biotechnological, application of many species and strains of indigenous microalgae needs further investigations (Odjadjare et al., 2017). Microalgae, being photoautotrophic, remain the best candidates for designing and developing cell factories (Radakovits et al., 2010). Among the thousands of species that are already identified in the world, the biotechnological applications of quite a few have been investigated, and only a few species are currently cultivated for industrial purposes (Olaizola, 2003). Among the several industrial applications, the most widely accepted application is that of industrial oil extraction from algal biomass (Anders et al., 2007).

According to the US Department of Energy Report (Khan et al., 2009), microalgae are capable of producing 30 times more oil per unit area of land compared to common terrestrial oilseed crops, such as canola or sunflowers. Many microalgal species are proven to be highly industrially valuable for the production of lipids and biomass (Adarme-Vega et al., 2012; Josephine et al., 2015). Oil from algae has the potential to yield better quality biofuel than that from conventional biomass. The productivity and quality of algal oils vary from species to species (Nascimento et al., 2014). Growth conditions such as nutrient availability and their limitations, pH, light, temperature, and growth phases are the critical factors that decide the quality and quantity of fatty acid compositions in algal oils (Su et al., 2011). Therefore, the selection of suitable green algae from naturally biodiverse-rich zones in terms of the nature and standardization of their growth conditions remains the most relevant research for the development of algal technology. It is essential to determine the energy efficiency (lipid content) of high biomass-yielding species and to understand the most cost-efficient role of them in biomass and bio-oil productions.

Algal oils are mainly stored in the algal cells as triacylglycerols, which is grouped into two categories: non-polar storage lipids and polar structural lipids (Sharma et al., 2012). Storage lipids are mainly stored cytosol in the form of tri-acyl glycerol (TAG) often 30%–60% of their dry weight, which is made of predominantly saturated fatty acids and unsaturated fatty acids (Figure 14.1). Depending upon the algae and their growing conditions, the concentration of such structural and storage lipids,

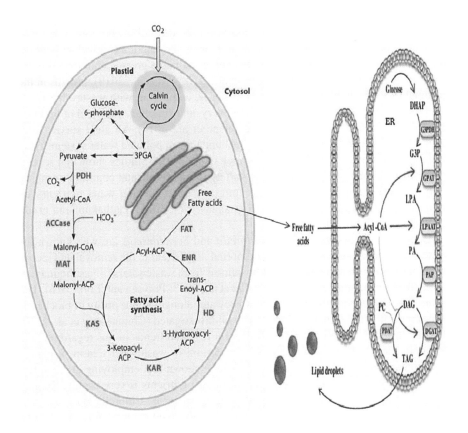

FIGURE 14.1   Lipid biosynthesis in microalgae (Radakovits et al., 2010).

which include saturated, monounsaturated, and polyunsaturated fatty acids, can be seen as different in different species (Chen et al., 2018; Goecke et al., 2010; Ikawa, 2004).The percentage level of these fatty acids determines the food and fuel value application of algal species (Bounnit et al., 2020). Table 14.1 shows the lipid content of industrially valuable species of algae.

However, the significant challenges in algal oil production are not only the identi-fication of oil-rich local species, but includes also the comparatively high investment costs for preparing the mass production facilities, biomass harvesting, and the requirement of high biomass producing strains in specific environments for a particu-lar oil quality (Chiaramonti et al., 2015; Gendy and El-Temtamy, 2013). Further, the success of this endeavor depends on the identification of industrially valuable spe-cific strains, and the development of biotechnological means to improve production levels, as well as establishing a pure culture of specific algae for desirable biomass quality.

14.3   RESEARCH ON ALGAL TECHNOLOGY AS A MEANS OF POLLUTION CONTROL

Pollution from organic or inorganic sources causes severe environmental issues faced by industrial production zones in the world and is difficult to be controlled. The application of algal technology such as algal biomass production coupled with indus-trial wastewater is the most cost-effective and sustainable green technology used to deal with wastewater problems (Higgins et al., 2018; Wu et al., 2019). It helps in the sustainable production of biomass for bioenergy, combining with wastewater treat-ment plants, nutrient removal, resource recovery, and simultaneous yield of biomass and bio-oils (Park et al., 2011; Makareviciene et al., 2014). An algal survey from natural habitats reported that they grow in abundance in polluted water systems (Ray et al., 2020; Pittman et al., 2011; Molazadeh et al., 2019). Studies show that several species have already adapted to a wide variety of environmental conditions ranging from high pH, salinity, nitrogen, potassium, and other running water nutrient-rich environments (Deviram et al., 2011; Ramanan et al., 2016; Santhakumaran et al., 2018; Chou et al., 2019).

Among the green algae species, *Chlorella* and *Scenedesmus* are well known for wastewater treatments and are already reported as efficient in the removal and recov-ery of nutrients such as nitrogen and phosphorus from wastewaters (Sivasubramanian et al., 2009; Venkataraman et al., 1982; Cai et al., 2013; Praveen and Loh, 2016). The biomass and lipid from algae grown in wastewater along with the utility of biochemi-cal components for biofuel production is well investigated (Chinnasamy et al., 2010; Hena et al., 2015; Tayari et al., 2020). Recent research trends in this regard include the application of microalgae-bacterial consortium in wastewater treatment plants. Experimental and pilot studies in this vein have shown that employing algae in ter-tiary wastewater treatment for recovery of the lost nutrients is very useful (Higgins et al., 2018). The algal-bacterial interaction such as Chlorella-Bacillus, and Microcystis-Bacillus and their co-cultivation are also well explored (Ji et al., 2018). The algal-bacteria symbiotic relationship helps the adaption of wastewater and stress tolerance in wastewater treatment plants (Yao et al., 2018).

TABLE 14.1
Lipid Contents of Industrially Oriented Microalgal Species

Microalgae species	Lipid % (DW)	SFA %	MUSFA %	PUFA %	References
Ankistrodesmus sp.	24–46	20–25.3	39–41.9	26–29.2	Cobos et al. (2017)
Botryococcus braunii	25–75	9.8–38.7	16.7–79.6	10.5–44.9	Abba et al. (2017); Nascimento et al. (2013)
Chaetoceros muelleri	33–43	38.4–49.3	39.8–45.3	16.7–21.6	Jesús-Campos et al. (2020); Wang et al. (2014)
Chlamydomonas reinhardtii	12–64	21.7–75.9	16.5–17.8	5.4–43.3	James et al. (2011); Thao et al. (2017)
Chlorella protothecoides	14–57	7.5–22.8	19.2–64.9	27.4–36.2	Makareviciene et al. (2014)
Chlorella sorokiniana	19–22	30.7–61.7	17.7–34.9	9.5–60.2	Ngangkham et al. (2012); Thao et al. (2017)
Chlorella pyrenoidosa	45–56	30.8–39.4	9.9–14.3	41.9–53.9	Dong et al. (2020)
Chlorella vulgaris	05–58	25.1–52.1	24.8–37.5	10.3–45.9	Nascimento et al. (2013)
Chloroidium saccharophillum	33–40	27.0–46.4	3.6–13.0	42.0–68.0	Santhakumaran et al. (2020)
Chlorococcum infusionum	19–31	25.3–32.5	23.5–51.8	16.4–48.2	Satpati et al. (2016); Sun et al. (2014)
Dunaliella sp.	16–71	13.4–49.4	19.5–24.7	31.0–48.7	Cakmak et al. (2014)
Chlorella ellipsoida	27–48	26.5–37.8	23.8–46.2	27.1–38.3	Satpati et al. (2016)
Haematococcus pluvialis	15–34	27.8–30.3	18.9–20.1	43.1–47.2	Damiani et al. (2010)
Isochrysis galbana	07–40	52.3–58.0	23.1–28.8	12.5–24.6	Gnouma et al. (2018); Ohse et al. (2015)
Kirchneriella lunaris	15–17	31.2–32.06	10.6–23.11	44.8–58.1	Nascimento et al. (2013); Santhakumaran et al. (2018)
Nannochloris sp.	20–68	29.1–40.5	8.7–50.0	9.5–30.5	Islam et al. (2013); Thao et al. (2017)
Neochloris oleoabundans	29–65	24.7–31.0	28.4–46.7	21.4–33.6	Singh et al. (2017); Sun et al. (2014)
Parachlorella Kessleri	40–60	20.2–35.0	18.8–41.0	24.1–42.8	Piasecka et al. (2017); Shaikh et al. (2019)
Pavlova lutheri	35–40	33.1–35.9	18.2–23.4	40.5–47.6	Guihéneuf et al. (2010)
Phaeodactylum tricornutum	18–57	28.1–46.8	4.2–54.1	17.8–18.3	Islam et al. (2013); Ohse et al. (2015)
Scenedesmus sp.	19–55	18.9–70.8	17.4–22.8	38–63.7	Gour et al. (2016); Islam et al. (2013); Nascimento et al. (2013)
Thalassiosira sp.	15–21	36.7–42.5	14.9–44.7	12.9–42.44	Cong et al., (2018); Ohse et al. 2015)
Tetraselmis sp.	10–49	28.6–44.6	30.8–48.5	16.1–46.5	Dammak et al. (2016); Guzmán et al. (2010)

The most common algae-based treatment-cum-cultivation systems include open race-way-pond, closed tubular photobioreactors, flat-flow-systems, high-rate-algal-ponds (HRAP), airlift-tubular-photobioreactors (AT-PBR), hybrid-anaerobic-baffled-reactor (HABR), and waste-stabilization-ponds (WSPs) (Craggs et al., 2014). The algae and bacterial consortium methods are most widely used in these types of wastewater treatment systems. Here algae supply oxygen for aerobic bacteria, and the bacteria convert complex mineral compounds to a freely available form of nutrients to algae (Craggs et al., 2014). Among the treatment plant cultivation systems, HABR photobioreactors with mixed algae show that they are capable of removing most of the organic and solids from wastewater and able to produce a healthy feedstock for algal biofuel production (Khalekuzzaman et al., 2019). The alga *Chlorella minutissima* under saline wastewater can produce biomass of 4.77 g/L and 0.55 g/L/day and can effectively utilize nitrogen and phosphorus from wastewater (Hawrot-Paw et al., 2020).

In the current global scenario, the algal technology is focusing on the exploitation of algae simultaneously for many purposes such as fuel and industrial products, which at the same time sequester carbon and manage wastewater treatment (Molazadeh et al., 2019). It is a green and sustainable technology for reducing the production costs of biodiesel, fertilizers, wastewater treatment, and flue gas treatments. It will also lead to a reduction in greenhouse gases in an efficient way. In the future, large-scale research on algae-based treatment programs could minimize the current technical or economic barriers of microalgae-based biofuel production. The application of algal-bacterial symbiosis on wastewater treatment can be commercialized and can become an eco-friendly green technology in the future. However, industries are looking forward to algal technology that can ensure continuous waste-recycling and biomass production under natural environmental conditions.

14.4   RESEARCH ON BIOCONVERSION TECHNOLOGIES

Algal biofuel is believed to be a clean, sustainable, and eco-friendly fuel, which is a better alternative to fossil fuels. Among the other sources of biomass with biofuel, algal biomass is considered as one of the best energy resources for sustainable development (Nascimento et al., 2013). Algal biomass and lipids of microalgae are employed in different types of fuel productions such as aviation fuel, biodiesel, gasoline, bioethanol and bio hydrogen (Benemann, 2000; Darzins et al., 2010; Laurens et al., 2015; Pulz and Gross, 2004). The energy efficiency and physicochemical properties of algal biofuels are found to be similar to that of petroleum fuels (Kim et al., 2013). However, the biofuel standards and user guidelines for testing and application of algal biofuel are different in different countries (Knothe, 2006; Stansell et al., 2012). Experimental studies based on the physicochemical characterization of algal biofuels and oil have shown that algal fatty acids and biomass components play an essential role in the production of quality biofuels (Lapuerta et al., 2009; Lopes et al., 2008; Stansell et al., 2012; Tong et al., 2011). However, the nature and quality of biofuel that may be produced from algal biomass depend on the kind of bioconversion technology employed for the purpose.

Bioconversion technologies are different types of physical and chemical methods used to convert biomass into fuels. Bioconversion methods include pyrolysis, gasification, hydrothermal liquefaction (HTL), transesterification, biogas generation or biomethanation, biohydrogen production, and fermentation processes. Currently, for industrial applications, four types of conversion technologies are widely used in the bioconversion of algal biomass. They are thermochemical, biochemical, transesterification, and photosynthetic microbial fuel cell conversions.

14.4.1 THERMOCHEMICAL CONVERSION OF ALGAL BIOMASS

Thermochemical conversion is the thermal breakdown of biomass to biofuels using techniques such as pyrolysis, gasification, and HTL. Pyrolysis helps to convert biomass to pyrolytic gas, bio-oil, and biochar. Based on heating temperature, pyrolysis is of three types: slow pyrolysis (~01°C–1°C for long duration), fast pyrolysis (~10°C–200°C for a short duration), and the flash pyrolysis (~1000°C/s for a brief duration). Among the pyrolytic techniques, enhanced microwave pyrolysis is reported as a very efficient method for bio-oil productions (Zhang et al., 2017). In the gasification process in the presence of air, oxygen, and water vapor, the biomass is heated to a high temperature (~800°C–1200°C), and in the process, the algal biomass is converted to bio-syngas. It consists of methane, hydrogen, carbon monoxide, water, and ashes (Saad et al., 2019).

In the hydrothermal liquefaction (HTL) process, the wet algal biomass is converted to biocrude under high temperature (~300°C–400°C) and pressure (40–200 bars) (Gollakota et al., 2018). The oily biocrude thus produced is a rich source of hydrocarbons, which can be converted to kerosene, diesel, and gasoline (Saad et al., 2019). In a study of the bioconversion of *Nannochloropsis* species using HTL under different temperatures and pressure, an 80% yield of biocrude is noticed (Valdez et al., 2012). The hydrothermal liquefaction of microalgae *Dunaliella tertiolecta* provides a yield of a maximum 25.8% bio-oil at a reaction temperature of 360 C° and 5% Na_2CO_3 as a catalyst (Shuping et al., 2010). The oil yield from HTL processing of algal biomass is reported as higher than that of a pyrolytic way of bio-oil production (Hu et al., 2017). In the HTL conversion process, the whole algae biomass is used, and therefore it is not necessary to promote lipid accumulation, because the conversion unit is a water-based slurry. Therefore in the algal HTL pathway, the overall economics for hydrocarbon biofuels are more strongly influenced by improvements in productivity rather than extractable lipid contents (Biddy et al., 2013).

14.4.2 BIOCHEMICAL CONVERSION

Biochemical conversion is the method of conversion of biomass into biofuel with the help of microorganisms or microbial enzymes, which utilizes the metabolite products stored in the biomass. Mainly, the reserved carbohydrates are converted to bioethanol or biogas through the fermentation process. Biogas or biomethanation, the fermentation process, and biohydrogen production are the standard biochemical conversion technologies for algal biomass (Zabed et al., 2020). Biomethanation is the conversion of biomass to biogas with the help of methanogenic bacteria under anaerobic conditions. The methane yield of biomass from *Chlorella sp.* (is 286 ml/g volatile solids) compared to other species of microalgae (Lakaniemi et al., 2011). Heating and pre-processed algal biomass is reported as capable of producing a higher concentration of methane than untreated ones (Khan et al., 2018a). The methane content of algal biogas is greatly influenced by the carbon-nitrogen ratio in the feedstock, feeding rate, temperature, pH, duration, and solid contents in the biomass (Klassen et al., 2017).

In the alcoholic fermentation process, the carbohydrate content of algal biomass is converted to ethanol with the help of yeast, bacterial, and fungal enzymes. The most common enzymes used for such a purpose include amylases, cellulases, and invertases, which can hydrolyze the sugar component of the algal biomass and convert it into bioethanol. Algae from phaeophyceaen microalgal species and many species of green algae such as *Scenedesmus sp.*, *Chlorella sp.*, and *Dunaliella sp.* are well known as a useful biomass resource for bioethanol productions (Khan et al., 2018b).

Biohydrogen production is also a biochemical process: the photobiological reaction in the absence of oxygen with the help of hydrogenase enzymes that occur in certain species of algae. This hydrogen fuel is considered eco-friendly, clean, and very energy efficient, and hence is a right fuel for the future energy consortium. The eukaryotic alga *Chlamydomonas reinhardtii* is the most commonly used microalgae in biohydrogen productions (Oncel et al., 2015). The continuous biohydrogen production under non-limiting conditions was achieved using a green alga, *Chlamydomonas reinhardtii*, and it became possible through genetic engineering technology (Ben-Zvi et al., 2019). The other species of common microalgae genetically engineered for biohydrogen production are *Chlorella vulgaris*, *Chlorella sorokiniana*, *Scenedesmussp.*, *Thalassiosira pseudonana*, *Dunaliella salina*, *Volvox carteri*, and so on (Radakovits et al., 2010; Sharma and Arya, 2017). The simultaneous saccharification and fermentation process of the biomass of *Chlorella* species provides the highest yield of biohydrogen production (170–172 mL-H_2 g-VS^{-1}) (Giang et al., 2019) currently. The common mechanisms used for biohydrogen production in algae are direct photolysis, indirect photolysis, and photo fermentation processes (Das and Veziroglu, 2008; Prince and Kheshgi, 2005).

14.4.3 TRANSESTERIFICATION

The transesterification method is the conversion of algal lipids into biodiesel. In this conversion process, the triglycerides of algal oil react with alcohol in presence of a catalyst such as alkali, acid or lipase into alkyl esters (Figure 14.2). The yield of biodiesel from algal lipids depends upon the fatty acid compositions of oil, the reaction efficiency of catalysts, and the methods used for biodiesel production (Semwal et al., 2011; Wu et al., 2012). There are two types of algal transesterification methods currently in use, such as direct transesterification and indirect transesterification. In direct transesterification methods, the wet biomass is treated directly with catalyst without the oil extraction (Sorgatto et al., 2019). The biodiesel yield is reported as very low by this method compared to indirect methods (Naghdi et al., 2014). However, it helps to reduce the time, energy, and reagents. Pretreatments of biomass

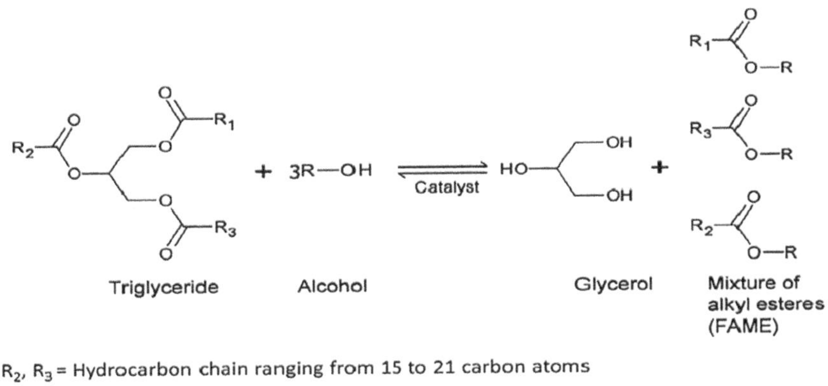
Triglyceride
+ 3R—OH
Alcohol
Catalyst
Glycerol
Mixture of
alkyl esteres
(FAME)
R1, R2, R3 = Hydrocarbon chain ranging from 15 to 21 carbon atoms

FIGURE 14.2 Transesterification reaction (Kasteren and nisworo, 2007).

with ultrasound irradiation and microwave digestion methods can increase the biodiesel yield to 90%–95% for algae biomass (Refaat et al., 2008). In indirect transesterification methods, the previously extracted oil can be transesterified to produce biodiesel, but this will only contain a relatively lesser amount of non-lipid components. In this regard, the biodiesel yield has been reported as higher than that of direct methods (Naghdi et al., 2014).

The supercritical methanol direct transesterification is a non-catalytic method for algal biodiesel production. This method has been reported as a highly efficient, eco-friendly and less energy-intensive technique for conversion of algal lipid or oil into fatty acid methyl esters, which converts 98% of fatty acids to biodiesel (Srivastava et al., 2018). In enzymyatic biodiesel conversion by using lipases, the enzymes non-specifically catalyze triglyceritdes into diverse kinds of fatty acids such as free fatty acids, tri, di, and mono glycerol fatty acids, and finally all such fatty acids formed are converted to their corresponding methyl esters up to 80%–90% levels.Yücel et al., 2013). Lipase enzymes are very costly compared to the acid and base catalase; however, the enzyme immobilization techniques are found to reduce their operational cost. The maximum biodiesel conversion yield of immobilized lipases has been reported as 95% (Surendhiran et al., 2015).

14.4.4 Photosynthetic Microbial Fuel Cell

A photosynthetic fuel cell is a system of conversion of photosynthetic energy of algae to bioelectric energy economically and sustainably. The integrated system of microbial fuel cell mechanism consists of algal cathode occupied by the photosynthetic algae for oxygen generation for the oxygen reduction reaction, causing the increment in the electron transfer from the anode occupied by the bacteria or as a substrate for bacteria (Gajda et al., 2015; Shukla and Kumar, 2018). The output of electricity is directly correlated with the dissolved oxygen level produced by the microalgae in the cathode chamber (Ling et al., 2019). *Chlorella*-based microbial fuel cell system is reported as the best system with the maximum current density of 539 mA/m^2, at normal solar illumination period (Strik et al., 2008). The most commonly reported species of microalgae for microbial fuel cell systems are *Scenedesmus sp.*, *Chlorella vulgaris*, *Chlamydomonas reinhardtii*, and so on (Angioni et al., 2018).

14.5 RESEARCH ON UP-GRADATION OF THE QUALITY OF ALGAE-BASED BIOFUEL

Intensive research on transformations of algal lipid into fine biofuels for various transportation purposes is currently going on in different laboratories. Up-gradation of algal biofuel quality is a bio-refinery process by which the extracted oil or bio-crude or biomass of algae can be converted to high-quality, fine-end usable fuels. In the up-gradation process, the chemical components present in algal bio-oils or biomass is converted into better compounds of high-quality fuel. The end products can be a liquid or solid or gas fuel, depending on the physicochemical factors employed in the conversion process such as temperature, pressure, and catalytic compounds applied in the process. Biocrude or oil obtained from algae is a complex mixture of many oxygenated hydrocarbons, aromatics, organic oxygenates, and nitrogenous compounds (Babich et al., 2011; Zhou et al., 2020). In general, high oxygen and nitrogen content in an algal biocrude adversely affect the fuel value, decreasing its heating value and affecting the combustion process (Kukushkin et al., 2015). Therefore, the up-gradation process of algal products into refined liquid fuels requires oxygen removal for improving the fuel-grades.

At different stages of the manufacturing, different kinds of biophysical methods are used in the thermochemical or biochemical conversion process of algal oil for up-gradation of its fuel quality. The most commonly used technology in the laboratory, as well as the refinery industry for refining of algal oil, includes catalytic hydro treatment, hydrodeoxygenation, and catalytic cracking (Bezergianni, 2016; Speight, 2015). Hydrotreatment and hydrodeoxygenation methods have been performed at high hydrogen pressure with the help of noble metal and transition metal catalysts (Chang et al., 2015). Catalytic cracking methods help to produce low molecular weight hydrocarbons from high molecular weight compounds with the help of acid catalysts, under high pressure (Zhou and Hu, 2020). These fundamental processes are collectively achieved through the different conversion reaction methods such as the Fischer–Tropsch conversion, hydrotreated conversion, hydrocracking conversion, and hydroisomerization conversions.

14.5.1 FISCHER–TROPSCH CONVERSION

The pyrolysis process can convert algal biomass directly into a synthetic gas known as syncrude that consists of a mixture of carbon monoxide, hydrogen, CO_2, methane, and short-chain hydrocarbons (Sikarwar et al., 2017). This syncrude mixture can be upgraded and converted to synthetic fuel using the Fischer–Tropsch conversion process. The Fischer–Tropsch reaction converts syncrude into liquid hydrocarbons in the presence of metal catalysts at high temperatures (150°C–300°C) and pressure (1–4 MPa) (Baldino et al., 2019). The reaction, such as hydrotreating, hydrocracking, hydroisomerization, and fractionation, leads to the formation of Fischer–Tropsch transportation jet fuels.

14.5.2 Hydrotreated Conversion

The hydrotreating conversion method is the most suitable for the conversion of extracted algal lipids to hydrocarbon-rich refined fuel. In this method, the crude lipids extracted from the microalgae undergo reaction with organic compounds under high temperature (250°C–374°C) and pressure (4–22 MPa) in the presence of hydrogen to remove oxygen along with other heteroatoms such as nitrogen, sulfur, and chlorine (Elliott et al., 2015). The hydrotreated conversion mainly helps to remove oxygen from the feedstock and thereby to make alkanes out of the lipid chains, propane out of the glycerol backbone, and water out of all oxygen molecules (Davis et al., 2011). Finally, this process helps to obtain hydrotreated transportation grade fuels with less amount of sulfur, nitrogen, or metal content.

The principal components of algal oil may be obtained after the hydrotreating process (Robota et al., 2013). Hydroprocessing of *Chlorella* biocrudes with sulfided CoMo and NiMo catalysts at 350°C and 405°C at hydrogen pressure (60–66 bar) shows a yield of 41%–69% of hydrocarbon-rich biofuels (Biller et al., 2015). Hydrotreatment of algal biocrude under 380°C with the commercially available catalysts Al_2O_3 supported NiMo show that the yield of 60% with deoxygenation levels similar to that of crude oil (Patel et al., 2017). *Chlamydomonas* biomass treated with ethanol, solid acid (SO_4^{2-}/ZrO_2), and catalyst Mo_2C/Biochar at 340°C temperature and 3.44 MPa pressure of hydrogen gives the yield of 93.7% hydrocarbons (Zhang et al., 2017). Hydrotreating of wet and dry algal biomass lipids under 250°C–370°C with Nickel- Molybdenum/carrier aluminum oxide catalyst (10%–30%) reported to give the yield of 73.02% (dry algae) and 76.47% (wet algae) high-quality hydrocarbon biofuels (Shi et al., 2018). Currently, the hydroteatment technology is found an easy to be operated method in conventional refineries.

14.5.3 Hydrocracking Conversion

The large hydrocarbon molecules of algal biocrude or bio-oil are broken down to shorter hydrocarbon chains by breaking the carbon-carbon bonds in the presence of hydrogen and catalyst to produce more desirable short-chain hydrocarbon products. The process of hydrocracking achieved by high pressure (12–20 MPa) and high temperature (350°C–400°C) produce bio-gasoline (Murata et al., 2014). The most widely used catalyst for hydrocracking process is zeolite/silica-alumina and metal oxides such as nickel/molybdenum (Bello-Zakari, 2015; Forghani, 2014; Shi et al., 2012). Through the hydrocracking process, the desired product can be achieved either of diesel fuel or jet fuel-range hydrocarbons. The lower carbon chain hydrocarbons are converted to gases due to inevitable over-cracking (Jones et al., 2014).

The hydrocracking process can increase the hydrogen/carbon ratio, and at the same time, it can help to remove the heteroatoms such as S, N, O, and metals from the bio-oil feedstocks (Anand and Sinha, 2012). According to Hillen et al. (1982) the hydrocracking of *Botryococcus braunii* oil using cobalt–molybdenum catalysts yield low molecular weight linear alkanes and isomers such as 67% gasoline, 15% aviation turbine fuel fraction, 15% diesel fuel fraction, and 3% residual oil fractions. Similarly, the hydrocracking of *Botryococcus braunii* oil with Pt–Re/SiO_2–Al_2O_3 catalysts, under the temperature 310°C–340°C, can produce aviation fuel-range hydrocarbons (C10–C15) with the yield of 50.2% and the diesel-range hydrocarbons (C16–C20) of 16.7% yield (Murata et al., 2014). Studies on cerium exchanged zeolite beta catalyst shows higher selectivity towards the C10–C14 hydrocarbons at high temperature (400°C) and pressure (400 psi), and it can yield 85% of aviation-grade fuel from hydrocracking of algal fuel (Bala and Chidambaram, 2016). In terms of hydrogen consumption and the quantity of oxygen content in the final product process, the hydrocracking methods remain a better option for up-gradation of algal biocrude into final transportation fuels (Ren et al., 2018).

14.5.4 Hydroisomerization Conversion

The hydroisomerization process is the conversion of normal hydrocarbons of bio-crude or bio-oil to branched ones, having the same carbon number with the help of catalysts (Mäki-Arvela et al., 2018). The selective hydroisomerization is most commonly employed for the improvement of the octane number for the gasoline pool (C5–C6) and also for the dewaxing of long-chain hydrocarbons as an improvement in its cetane number and cold flow properties (Mériaudeau et al., 1997). The most common catalysts used for the hydroisomerization reactions are zeolites, crystalline silica-aluminophosphates, Pt and zeolite beta (Mériaudeau et al., 1997). Sulfided Nimo catalysts supported on hierarchical mesoporous H-ZSM-5 are proved as an ideal catalyst for obtaining high yield of jet fuel from algal oil (Verma et al., 2011). The presence of n-alkanes in fuel increases its pour point, cloud point, and cold filter clogging point, which can be improved by controlling the extent of branched alkane formation during the hydroisomerization process (Hsieh et al., 2014).

According to Verma et al. (2011) 77% of kerosene range hydrocarbon yield can be obtained from algal oil by using sulfided Nimo catalyst supported with semi-crystalline ZSM-5 zeolite and the process can obtain high isomerization selectivity. However, algal triglycerides can be converted to diesel and hydrotreated ester-based n-alkanes jet fuel fractions using 0.5% Pt/US-Y zeolite catalyst (Robota et al., 2013). Experimental studies on the hydroisomerization conversion of algal oil to jet fuel have shown that Pt/ZSM-12 based catalyst is a significant catalyst for the conversion of n-C15 paraffin into multi-branched isomers and mono branched isomers (60% of yield) of jet fuel specifications (Ju et al., 2016). Kruger et al. (2017) have successfully hydroisomerized the algal oil and converts to diesel with a cloud point of below -10°C by using Pt/SAPO-11 catalyst. In the hydroisomerization process, the final yield and quality of fuel are mainly dependent upon the catalytic properties of catalysts, catalyst feed molar ratio, types of a reactor using, and the degree of chain length of fatty acids or oils (Mäki-Arvela et al., 2018).

Bioconversion of biomass and up-gradation of biofuel quality is highly dependent upon the post-conversion pathways. Physicochemical and biochemical treatments are effective for up-gradation of biofuels; however, the cost of process needs to be reduced or could be integrated with other treatments. Among the conversion of biomass to biofuel technology, HTL stands as a viable process; however, extensive research is needed to develop better catalysts and to increase conversion process in a low-cost manner. Optimization of the products, development of the high-value co-products and minimization of waste will help to reduce the overall production cost of the final fuel. The diversified aspects of algal biofuel production systems research are currently a significant area of fuel research worldwide. Algal biofuels are considered as an effective panacea for reducing greenhouse gas emissions as a part of the national and international mission.

14.6 CONCLUSION

Algal bioenergy for better quality biofuels is considered a critical component in our national energy mission and also an international energy mission to reduce global greenhouse gas emissions associated with increasing energy demands. Algal technological research is now more focused on carbon sequestration, combined with fuel production, improvement in fuel efficiency, phytoremediation of eutrophicated waters, and other commercially valuable products from a single species. Naturally, the progress and sustenance of algal industries depend on the identification of better species or strains, which can provide the required algal species diversity essential to the artificial recombination of desirable qualities in the suitable species. Such data on many novel species provide opportunities for intensive research in various biotechnological applications of algae. The industrial application of algae-bacterial symbiosis and their application in wastewater treatment can be suitable to make commercial and eco-friendly green technology in the future. The thermochemical and biochemical conversion technology of algal biomass improves the biofuel economy. Recent research outputs show, among the biomass conversion technology, HTL as a viable process; however, extensive research is needed to develop catalysts and to increase conversion process in a low-cost manner. Other significant hurdles currently addressed by the industry on third- and fourth-generation biofuel are the lack of efficient technology for the mass production of algal biomass at available natural environments. Therefore, the biomass, biofuel, and industrial potential of many high yielding species of microalgae could not be correctly studied and examined. The current trends in the algal biotechnology, genetic engineering, and bioconversion include specialized research programs that show the large-scale algae-based treatment for minimizing the current technical or economic barriers of microalgae-based eco-friendly biofuel production.

REFERENCES

Abba, Zubainatu, Hazel Monica Matias-Peralta, Siti Fatimah, Zahrah Mohammad, Yousif Abdalla Abakr, Isah Yakub Mohammed, and Muhammad Muhammad Nmaya. 2017. "Fatty Acids Composition of Microalga Botryococcus Sp. Cultured in Synthetic Medium". *Journal of Science and Technology* 9 (4): 117–121.

Abishek, M.P., J. Patel, A. Prem Rajan 2014. Algae Oil: A Sustainable Renewable Fuel of Future. *Biotechnol. Res. Int.* 1–8. doi:10.1155/2014/272814

Adarme-Vega, T Catalina, David KY Lim, Matthew Timmins, Felicitas Vernen, Yan Li, and Peer M Schenk. 2012. "Microalgal Biofactories: A Promising Approach towards Sustainable Omega-3 Fatty Acid Production." *Microbial Cell Factories* 11 (1): 96. https://doi.org/10.1186/1475-2859-11-96.

Anand, Mohit, and Anil K. Sinha. 2012. "Temperature-Dependent Reaction Pathways for the Anomalous Hydrocracking of Triglycerides in the Presence of Sulfided Co-Mo-Catalyst." *Bioresource Technology* 126: 148–155. https://doi.org/10.1016/j.biortech.2012.08.105.

Anders, S Carlsson, B van Beilen Jan, Ralf Möller, and Clayton David. 2007. *Micro- and Macro-Algae: Utility for Industrial Applications.* CPL Press, UK: 1–86. https://doi.org/10.1037/11474-008.

Angioni, S., L. Millia, P. Mustarelli, E. Doria, M. E. Temporiti, B. Mannucci, F. Corana, and E. Quartarone. 2018. "Photosynthetic Microbial Fuel Cell with Polybenzimidazole Membrane: Synergy between Bacteria and Algae for Wastewater Removal and Biorefinery." *Heliyon* 4 (3): e00560. https://doi.org/10.1016/j.heliyon.2018.e00560.

Aratboni, Alishah Hossein, Nahid Rafiei, Raul Garcia-Granados, Abbas Alemzadeh, and José Rubén Morones-Ramírez. 2019. "Biomass and Lipid Induction Strategies in Microalgae for Biofuel Production and Other Applications." *Microbial Cell Factories* 18 (1): 1–17. https://doi.org/10.1186/s12934-019-1228-4.

Babich, I. V., M. van der Hulst, L. Lefferts, J. A. Moulijn, P. O'Connor, and K. Seshan. 2011. "Catalytic Pyrolysis of Microalgae to High-Quality Liquid Bio-Fuels." *Biomass and Bioenergy* 35: 3199–3207. https://doi.org/10.1016/j.biombioe.2011.04.043.

Bala, Dharshini D., and Dev Chidambaram. 2016. "Production of Renewable Aviation Fuel Range Alkanes from Algae Oil." *RSC Advances* 6 (18): 14626–14634. https://doi.org/10.1039/c5ra23145k.

Baldino, Authors Chelsea, Rosalie Berg, Nikita Pavlenko, and Stephanie Searle. 2019. "Advanced Alternative Fuel Pathways: Technology Overview and Status." *The International Council on Clean Transportation* (13): 1–31. https://theicct.org/publications/advanced-alternative-fuel-pathways.

Bello-Zakari, B., 2015. "Hydro RO Processing Microalgae Derived Hydrothermal Liquefaction Bio- Crude For Middle Distillate Fuels Production – A Review." *Nigerian Journal of Technology* 34 (4): 737–749. https://doi.org/http://dx.doi.org/10.4314/njt.v34i4.11.

Ben-Zvi, Oren, Eyal Dafni, Yael Feldman, and Iftach Yacoby. 2019. "Re-Routing Photosynthetic Energy for Continuous Hydrogen Production in Vivo." *Biotechnology for Biofuels* 12 (266): 1–13. https://doi.org/10.1186/s13068-019-1608-3.

Benemann, John R., 2000. "Hydrogen Production by Microalgae." *Journal of Applied Phycology* 12: 291–300. https://doi.org/10.1023/A:1008175112704.

Bezergianni, S., 2016. *Catalytic Hydroprocessing of Liquid Biomass for Biofuels Production.* Intech, 299–326. doi:http://dx.doi.org/10.5772/57353

Biddy, Mary, Ryan Davis, Susanne Jones, and Yunhua Zhu. 2013. "Whole Algae Hydrothermal Liquefaction Technology Pathway." *Technical Report Efficiency &Renewable Energy, operated by the Alliance for Sustainable Energy, LLC. NREL/TP-5100-58051 PNNL-22314,* no. March: 1–10. www.nrel.gov.

Biller, Patrick, Brajendra K Sharma, Bidhya Kunwar, and Andrew B Ross. 2015. "Hydroprocessing of Bio-Crude from Continuous Hydrothermal Liquefaction of Microalgae." *Fuel* 159: 197–205. https://doi.org/http://dx.doi.org/10.1016/j.fuel.2015.06.077.

Bhola, V., F. Swalaha, R. Ranjith Kumar, M. Singh, F. Bux, 2014. Overview of the potential of microalgae for CO_2 sequestration. *International Journal of Environmental Science and Technology* 11, 2103–2118. doi:10.1007/s13762-013-0487-6

Bounnit, Touria, Imen Saadaoui, Rihab Rasheed, Kira Schipper, Maryam Al Muraikhi, and Hareb Al Jabri. 2020. "Sustainable Production of Nannochloris Atomus Biomass towards Biodiesel Production." *Sustainability (Switzerland)* 12 (5): 1–21. https://doi.org/10.3390/su12052008.

Cai, Ting, Stephen Y. Park, and Yebo Li. 2013. "Nutrient Recovery from Wastewater Streams by Microalgae: Status and Prospects." *Renewable and Sustainable Energy Reviews* 19: 360–369. https://doi.org/10.1016/j.rser.2012.11.030.

Cakmak, Yavuz Selim, Murat Kaya, and Meltem Asan-Ozusaglam. 2014. "Biochemical Composition and Bioactivity Screening of Various Extracts from Dunaliella Salina, a Green Microalga." *EXCLI Journal* 13: 679–690. https://doi.org/10.17877/DE290R-6669.

Chang, Z., P. Duan, Y. Xu, 2015. Catalytic hydropyrolysis of microalgae: Influence of operating variables on the formation and composition of bio-oil. *Bioresources Technology* 184, 349–354. doi:10.1016/j.biortech.2014.08.014

Chen, Zhipeng, Lingfeng Wang, Shuang Qiu, and Shijian Ge. 2018. "Determination of Microalgal Lipid Content and Fatty Acid for Biofuel Production." *BioMed Research International*, 2018: 1–18. https://doi.org/10.1155/2018/1503126.

Chen, J., J. Li, Q. Li, S. Wang, L. Wang, H. Liu, C. Fan. 2020. Engineering a chemoenzymatic cascade for sustainable photobiological hydrogen production with green algae. *Energy Environmental Science* 1–7. doi:10.1039/d0ee00993h

Chia, S.R., H.C. Ong, K.W. Chew, P.L. Show, S.M. Phang, T.C. Ling, D. Nagarajan, D.J. Lee, J.S. Chang. 2018. Sustainable approaches for algae utilisation in bioenergy production. *Renewable Energy* 129, 838–852. doi:10.1016/j.renene.2017.04.001

Chiaramonti, David, Matteo Prussi, Marco Buffi, David Casini, and Andrea Maria Rizzo. 2015. "Thermochemical Conversion of Microalgae: Challenges and Opportunities." *Energy Procedia* 75: 819–826. https://doi.org/10.1016/j.egypro.2015.07.142.

Chinnasamy, Senthil, Ashish Bhatnagar, Ryan W. Hunt, and K. C. Das. 2010. "Microalgae Cultivation in a Wastewater Dominated by Carpet Mill Effluents for Biofuel Applications." *Bioresource Technology* 101 (9): 3097–3105. https://doi.org/10.1016/j. biortech.2009.12.026.

Chisti, Y. 2007. "Biodiesel from Micro Algae." *Biotechnology Advances* 25: 294–306.

Chou, Hsiang Hui, Hsiang Yen Su, Xiang Di Song, Te Jin Chow, Chun Yen Chen, Jo Shu Chang, and Tse Min Lee. 2019. "Isolation and Characterisation of Chlorella sp. Mutants with Enhanced Thermo- And CO_2 Tolerances for CO_2 Sequestration and Utilisation of Flue Gases." *Biotechnology for Biofuels* 12 (251): 1–14. https://doi.org/10.1186/ s13068-019-1590-9.

Cobos, Marianela, Jae D. Paredes, J. Dylan Maddox, Gabriel Vargas-Arana, Leenin Flores, Carla P. Aguilar, Jorge L. Marapara, and Juan C. Castro. 2017. "Isolation and Characterisation of Native Microalgae from the Peruvian Amazon with Potential for Biodiesel Production." *Energies* 10 (224): 1–16. https://doi.org/10.3390/en10020224.

Cong, Nguyen Van, Do Thi Hoa Vien, and Dang Diem Hong. 2018. "Fatty Acid Profile And Nutrition Values Of Microalga (*Thalassiosira pseudonana*) Commonly Used In White Shrimp (*Litopenaeus vannamei*) Culturing." *Vietnam Journal of Science and Technology* 56 (4A): 138–145. https://doi.org/10.15625/2525-2518/56/4a/12810.

Craggs, R., J. Park, S. Heubeck, and D. Sutherland. 2014. "High Rate Algal Pond Systems for Low-Energy Wastewater Treatment, Nutrient Recovery and Energy Production." *New Zealand Journal of Botany* 52 (1): 60–73. https://doi.org/10.1080/0028825X.2013.861855.

Damiani, M. Cecilia, Cecilia A. Popovich, Diana Constenla, and Patricia I. Leonardi. 2010. "Lipid Analysis in Haematococcus Pluvialis to Assess Its Potential Use as a Biodiesel Feedstock." *Bioresource Technology* 101 (11): 3801–3807. https://doi.org/10.1016/j. biortech.2009.12.136.

Dammak, Mouna, Sandra Mareike Haase, Ramzi Miladi, Faten Ben Amor, Mohamed Barkallah, David Gosset, Chantal Pichon, et al. 2016. "Enhanced Lipid and Biomass Production by a Newly Isolated and Identified Marine Microalga." *Lipids in Health and Disease* 15 (1): 1–13. https://doi.org/10.1186/s12944-016-0375-4.

Darzins, Al, Philip Pienkos, and Les Edye. 2010. "Current Status and Potential for Algal Biofuels Production." *A REPORT TO IEA BIOENERGY TASK 39*, no. August: 146.

Das, Debabrata, and T. Nejat Veziroglu. 2008. "Advances in Biological Hydrogen Production Processes." *International Journal of Hydrogen Energy* 33 (21): 6046–6057. https://doi. org/10.1016/j.ijhydene.2008.07.098.

Davis, Ryan, Andy Aden, and Philip T. Pienkos. 2011. "Techno-Economic Analysis of Autotrophic Microalgae for Fuel Production." *Applied Energy* 88 (10): 3524–3531. https://doi.org/10.1016/j.apenergy.2011.04.018.

Deviram, Gvns, K V Pradeep, and R Gyana Prasuna. 2011. "Purification of Waste Water Using Algal Species." *European Journal of Experimental Biology* 1 (3): 216–222.

Dong, Liang, Dong Li, and Chun Li. 2020. "Characteristics of Lipid Biosynthesis of Chlorella Pyrenoidosa under Stress Conditions." *Bioprocess and Biosystems Engineering* 43 (5): 877–884. https://doi.org/10.1007/s00449-020-02284-x.

Elliott, Douglas C., Patrick Biller, Andrew B. Ross, Andrew J. Schmidt, and Susanne B. Jones. 2015. "Hydrothermal Liquefaction of Biomass: Developments from Batch to Continuous Process." *Bioresource Technology* 178: 147–156. https://doi.org/10.1016/j.biortech.2014.09.132.

Fabris, M., R.M. Abbriano, M. Pernice, D.L. Sutherland, A.S. Commault, C.C. Hall, L. Labeeuw, J.I. McCauley, U. Kuzhiuparambil, P. Ray, T. Kahlke, P.J. Ralph. 2020. Emerging Technologies in Algal Biotechnology: Toward the Establishment of a Sustainable, Algae-Based Bioeconomy. *Frontiers in Plant Science* 11, 1–22. doi:10.3389/fpls.2020.00279

Feinberg, D.A., and A.M. Hill. 1984. "Fuel from Microalgae Lipid Products." *Solar Energy*, no. SERI/TP-231-2348 UC Category: 61a DE84004506: 1–17.

Forghani, Amir Ahmad. 2014. "Catalytic Hydro-Cracking of Bio-Oil to Bio-Fuel." Thesis, University of Adelaide, Austalia, no. 1200751: 40–104.

Fu, Weiqi, David R. Nelson, Alexandra Mystikou, Sarah Daakour, and Kourosh Salehi-Ashtiani. 2019. "Advances in Microalgal Research and Engineering Development." *Current Opinion in Biotechnology* 59: 157–164. https://doi.org/10.1016/j.copbio.2019.05.013.

Gajda, Iwona, John Greenman, Chris Melhuish, and Ioannis Ieropoulos. 2015. "Self-Sustainable Electricity Production from Algae Grown in a Microbial Fuel Cell System." *Biomass and Bioenergy* 82: 87–93. https://doi.org/10.1016/j.biombioe.2015.05.017.

Gendy, Tahani S., and Seham a. El-Temtamy. 2013. "Commercialization Potential Aspects of Microalgae for Biofuel Production: An Overview." *Egyptian Journal of Petroleum* 22 (1): 43–51. https://doi.org/10.1016/j.ejpe.2012.07.001.

Giang, Tran Thi, Siriporn Lunprom, Qiang Liao, Alissara Reungsang, and Apilak Salakkam. 2019. "Enhancing Hydrogen Production from Chlorella sp. Biomass by Pre-Hydrolysis with Simultaneous Saccharification and Fermentation (PSSF)." *Energies* 12 (908): 1–14. https://doi.org/10.3390/en12050908.

Gnouma, Asma, Emna Sehli, Walid Medhioub, Rym Ben Dhieb, Mahmoud Masri, Norbert Mehlmer, Wissem Slimani, et al. 2018. "Strain Selection of Microalgae Isolated from Tunisian Coast: Characterisation of the Lipid Profile for Potential Biodiesel Production." *Bioprocess and Biosystems Engineering* 41 (10): 1449–1459. https://doi.org/10.1007/s00449-018-1973-5.

Goecke, Franz, Víctor Hernández, Magalis Bittner, Mariela González, José Becerra, and Mario Silva. 2010. "Fatty Acid Composition of Three Species of Codium (Bryopsidales, Chlorophyta) in Chile." *Revista de Biología Marina y Oceanografía* 45 (2): 325–330. https://doi.org/10.4067/S0718-19572010000200014.

Gollakota, A. R.K., Nanda Kishore, and Sai Gu. 2018. "A Review on Hydrothermal Liquefaction of Biomass." *Renewable and Sustainable Energy Reviews* 81: 1378–1392. https://doi.org/10.1016/j.rser.2017.05.178.

Gour, Rakesh Singh, Aseem Chawla, Harvinder Singh, Rajinder Singh Chauhan, and Anil Kant. 2016. "Characterisation and Screening of Native Scenedesmus sp. Isolates Suitable for Biofuel Feedstock." *PLoS ONE* 11 (5): 1–16. https://doi.org/10.1371/journal.pone.0155321.

Guihéneuf, Freddy, Manuela Fouqueray, Virginie Mimouni, Lionel Ulmann, Boris Jacquette, and Gérard Tremblin. 2010. "Effect of UV Stress on the Fatty Acid and Lipid Class Composition in Two Marine Microalgae Pavlova Lutheri (Pavlovophyceae) and Odontella Aurita (Bacillariophyceae)." *Journal of Applied Phycology* 22 (5): 629–638. https://doi.org/10.1007/s10811-010-9503-0.

Guzmán, Héctor Mendoza, Adelina de la Jara Valido, Laura Carmona Duarte, and Karen Freijanes Presmanes. 2010. "Estimate by Means of Flow Cytometry of Variation in Composition of Fatty Acids from Tetraselmis Suecica in Response to Culture Conditions." *Aquaculture International* 18 (2): 189–199. https://doi.org/10.1007/s10499-008-9235-1.

Hassan, Masjuki Hj, and Md Abul Kalam. 2013. "An Overview of Biofuel as a Renewable Energy Source: Development and Challenges." *Procedia Engineering* 56: 39–53. https://doi.org/10.1016/j.proeng.2013.03.087.

Hawrot-Paw, Małgorzata, Adam Koniuszy, Małgorzata Gałczynska, Grzegorz Zajac, and Joanna Szyszlak-Bargłowicz. 2020. "Production of Microalgal Biomass Using Aquaculture Wastewater as Growth Medium." *Water (Switzerland)* 12 (106): 1–11. https://doi.org/10.3390/w12010106.

Hena, S, S Fatimah, and S Tabassum. 2015. "Cultivation of Algae Consortium in a Dairy Farm Wastewater for Biodiesel Production." *Water Resources and Industry* 10: 1–14. https://doi.org/10.1016/j.wri.2015.02.002.

Higgins, Brendan T., Ingrid Gennity, Patrick S. Fitzgerald, Shannon J. Ceballos, Oliver Fiehn, and Jean S. VanderGheynst. 2018. "Algal–Bacterial Synergy in Treatment of Winery Wastewater." *Npj Clean Water* 1 (6): 1–10. https://doi.org/10.1038/s41545-018-0005-y.

Hillen, LW, G Pollard, LV Wake, and N White. 1982. "Hydrocracking of the Oils of Botryococcus Braunii to Transport Fuels." *Biotechnology and Bioengineering* 24 (1982): 193–205.

Hoek, CVD, D. Mann, and HM Jahns. 1995. "*Algae: An Introduction to Phycology.*" Cambridge University Press, 627.

Hsieh, Peter Y., Jason A. Widegren, Tara J. Fortin, and Thomas J. Bruno. 2014. "Chemical and Thermophysical Characterisation of an Algae-Based Hydrotreated Renewable Diesel Fuel." *Energy and Fuels* 28: 3192–3205. https://doi.org/10.1021/ef500237t.

Hu, Yulin, Shanghuan Feng, Zhongshun Yuan, Chunbao Xu, and Amarjeet Bassi. 2017. "Investigation of Aqueous Phase Recycling for Improving Bio-Crude Oil Yield in Hydrothermal Liquefaction of Algae." *Bioresource Technology* 239: 151–169. https://doi.org/10.1016/j.biortech.2017.05.033.

Ikawa, Miyoshi. 2004. "Algal Polyunsaturated Fatty Acids and Effects on Plankton Ecology and Other Organisms." *UNH Center for Freshwater Biology Research* 6 (2): 17–44.

Islam, Muhammad Aminul, Marie Magnusson, Richard J. Brown, Godwin a. Ayoko, Md Nurun Nabi, and Kirsten Heimann. 2013. "Microalgal Species Selection for Biodiesel Production Based on Fuel Properties Derived from Fatty Acid Profiles." *Energies* 6 (11): 5676–5702. https://doi.org/10.3390/en6115676.

James, Gabriel O., Charles H. Hocart, Warwick Hillier, Hancai Chen, Farzaneh Kordbacheh, G. Dean Price, and Michael A. Djordjevic. 2011. "Fatty Acid Profiling of Chlamydomonas Reinhardtii under Nitrogen Deprivation." *Bioresource Technology* 102: 3343–3351. https://doi.org/10.1016/j.biortech.2010.11.051.

Jesús-Campos, Damaristelma de, José Antonio López-Elías, Luis Ángel Medina-Juarez, Gisela Carvallo-Ruiz, Diana Fimbres-Olivarria, and Corina Hayano-Kanashiro. 2020. "Chemical Composition, Fatty Acid Profile and Molecular Changes Derived from Nitrogen Stress in the Diatom Chaetoceros Muelleri." *Aquaculture Reports* 16: 100281. https://doi.org/10.1016/j.aqrep.2020.100281.

Ji, Xiyan, Mengqi Jiang, Jibiao Zhang, Xuyao Jiang, and Zheng Zheng. 2018. "The Interactions of Algae-Bacteria Symbiotic System and Its Effects on Nutrients Removal from Synthetic Wastewater." *Bioresource Technology* 247: 44–50. https://doi.org/10.1016/j.biortech.2017.09.074.

Jo, S.-W., J.-M. Do, H. Na, J.W. Hong, I.-S. Kim, H.-S. Yoon. 2020. Assessment of biomass potentials of microalgal communities in open pond raceways using mass cultivation. *PeerJ* 8, e9418. doi:10.7717/peerj.9418

Jones, Susanne, Yunhua Zhu, D Anderson, Richard T. Hallen, and Douglas C. Elliott. 2014. "Process Design and Economics for the Conversion of Algal Biomass to Hydrocarbons : Whole Algae Hydrothermal Liquefaction and Upgrading." *US Department of Energy Report*, no. PNNL-23227: 1–69. https://doi.org/10.2172/1126336.

Josephine, A, C Niveditha, A Radhika, A Brindha Shali, T S Kumar, G Dharani, and R Kirubagaran. 2015. "Analytical Evaluation of Different Carbon Sources and Growth Stimulators on the Biomass and Lipid Production of Chlorella Vulgaris e Implications for Biofuels." *Biomass and Bioenergy* 75: 170–179. http://dx.doi.org/10.1016/j.biombioe.2015.02.016.

Ju, Chao, Yuping Zhou, Mingli He, Qiuying Wu, and Yunming Fang. 2016. "Improvement of Selectivity from Lipid to Jet Fuel by Rational Integration of Feedstock Properties and Catalytic Strategy." *Renewable Energy* 97: 1–7. https://doi.org/10.1016/j.renene.2016.05.075.

Kamani, M.H., I. Eş, J.M. Lorenzo, F. Remize, E. Roselló-Soto, F.J. Barba, J. Clark, A. Mousavi Khaneghah. 2019. Advances in plant materials, food by-products, and algae conversion into biofuels: Use of environmentally friendly technologies. *Green Chemistry* 21, 3213–3231. doi:10.1039/c8gc03860k

van Kasteren, J. M. N., and A. P. Nisworo. 2007. "A Process Model to Estimate the Cost of Industrial Scale Biodiesel Production from Waste Cooking Oil by Supercritical Transesterification." *Resources, Conservation and Recycling* 50: 442–458. https://doi.org/10.1016/j.resconrec.2006.07.005.

Khalekuzzaman, M., Muhammed Alamgir, M. Bashirul Islam, and Mehedi Hasan. 2019. "A Simplistic Approach of Algal Biofuels Production from Wastewater Using a Hybrid Anaerobic Baffled Reactor and Photobioreactor (HABR-PBR) System." *PLoS ONE* 14 (12): 1–24. https://doi.org/10.1371/journal.pone.0225458.

Khan, Muhammad Imran, Jin Hyuk Shin, and Jong Deog Kim. 2018a. "The Promising Future of Microalgae: Current Status, Challenges, and Optimisation of a Sustainable and Renewable Industry for Biofuels, Feed, and Other Products." *Microbial Cell Factories* 17 (36): 1–21. https://doi.org/10.1186/s12934-018-0879-x.

Khan, Shakeel A., Mir Z. Hussain Rashmi, S. Prasad, and U. C. Banerjee. 2009. "Prospects of Biodiesel Production from Microalgae in India." *Renewable and Sustainable Energy Reviews* 13 (9): 2361–2372. https://doi.org/10.1016/j.rser.2009.04.005.

Khan, M.I., J.H. Shin, and J.D. Kim. 2018b. The promising future of microalgae: Current status, challenges, and optimisation of a sustainable and renewable industry for biofuels, feed, and other products. *Microbe Cell Factors* 17, 1–21. doi:10.1186/s12934-018-0879-x

Kim, Jungmin, Gursong Yoo, Hansol Lee, Juntaek Lim, Kyochan Kim, Chul Woong, Min S Park, and Ji-won Yang. 2013. "Methods of Downstream Processing for the Production of Biodiesel from Microalgae." *Biotechnology Advances* 31 (6): 862–876. https://doi.org/10.1016/j.biotechadv.2013.04.006.

Klassen, Viktor, Olga Blifernez-Klassen, Daniel Wibberg, Anika Winkler, Jörn Kalinowski, Clemens Posten, and Olaf Kruse. 2017. "Highly Efficient Methane Generation from Untreated Microalgae Biomass." *Biotechnology for Biofuels* 10 (186): 1–12. https://doi.org/10.1186/s13068-017-0871-4.

Knothe, G. 2006. "Analysing Biodiesel: Standards and Other Methods." *Jaocs* 83 (10): 823–833. https://doi.org/10.1007/s11746-006-5033-y.

Kruger, Jacob S., Earl D. Christensen, Tao Dong, Stefanie Van Wychen, Gina M. Fioroni, Philip T. Pienkos, and Robert L. McCormick. 2017. "Bleaching and Hydroprocessing of Algal Biomass-Derived Lipids to Produce Renewable Diesel Fuel." *Energy and Fuels* 31 (10): 10946–10953. https://doi.org/10.1021/acs.energyfuels.7b01867.

Kukushkin, R. G., O. A. Bulavchenko, V. V. Kaichev, and V. A. Yakovlev. 2015. "Influence of Mo on Catalytic Activity of Ni-Based Catalysts in Hydrodeoxygenation of Esters." *Applied Catalysis B: Environmental* 163: 531–538. https://doi.org/10.1016/j.apcatb.2014.08.001.

Lakaniemi, Aino Maija, Christopher J. Hulatt, David N. Thomas, Olli H. Tuovinen, and Jaakko A. Puhakka. 2011. "Biogenic Hydrogen and Methane Production from Chlorella Vulgaris and Dunaliella Tertiolecta Biomass." *Biotechnology for Biofuels* 4: 1–12. https://doi.org/10.1186/1754-6834-4-34.

Lapuerta, Magín, José Rodríguez-Fernández, and Emilio Font de Mora. 2009. "Correlation for the Estimation of the Cetane Number of Biodiesel Fuels and Implications on the Iodine Number." *Energy Policy* 37 (11): 4337–4344. https://doi.org/10.1016/j.enpol.2009.05.049.

Laurens, L. M. L., N. Nagle, R. Davis, N. Sweeney, S. Van Wychen, A. Lowell, and P. T. Pienkos. 2015. "Acid-Catalyzed Algal Biomass Pretreatment for Integrated Lipid and Carbohydrate-Based Biofuels Production." *Green Chemistry* 17 (2): 1145–1158. https://doi.org/10.1039/c4gc01612b.

Ling, Jiayin, Yanbin Xu, Chuansheng Lü, Weikang Lai, Guangyan Xie, Li Zheng, Manjunatha P. Talawar, Qingping Du, and Gangyi Li. 2019. "Enhancing Stability of Microalgae Biocathode by a Partially Submerged Carbon Cloth Electrode for Bioenergy Production from Wastewater." *Energies* 12 (17): 3229. https://doi.org/10.3390/en12173229.

Lackner, M., 2016. *Handbook of Climate Change Mitigation and Adaptation*. doi:10.1007/978-1-4614-6431-0

Lopes, J. C. A., L. Boros, M. A. Kráhenbúhl, A. J. A. Meirelles, J. L. Daridon, J. Pauly, I. M. Marrucho, and J. A. P. Coutinho. 2008. "Prediction of Cloud Points of Biodiesel." *Energy and Fuels* 22 (2): 747–752. https://doi.org/10.1021/ef700436d.

Lü, J., C. Sheahan, P. Fu. 2011. Metabolic engineering of algae for fourth-generation biofuels production. *Energy Environ. Sci.* 4, 2451–2466. doi:10.1039/c0ee00593b

Makareviciene, V, V Skorupskaite, D Levisauskas, V Andruleviciute, and K Kazancev. 2014. "The Optimisation of Biodiesel Fuel Production from Microalgae Oil Using Response Surface Methodology." *International Journal of Green Energy* 11 (5): 527–541. https://doi.org/Doi 10.1080/15435075.2013.777911.

Mäki-Arvela, Päivi, Taimoor A.Kaka Khel, Muhammad Azkaar, Simon Engblom, and Dmitry Yu Murzin. 2018. "Catalytic Hydroisomerization of Long-Chain Hydrocarbons for the Production of Fuels." *Catalysts* 8 (534): 1–27. https://doi.org/10.3390/catal8110534.

Mériaudeau, P., V. A. Tuan, V. T. Nghiem, S. Y. Lai, L. N. Hung, and C. Naccache. 1997. "SAPO-11, SAPO-31, and SAPO-41 Molecular Sieves: Synthesis, Characterization, and Catalytic Properties in n-Octane Hydroisomerization." *Journal of Catalysis* 169 (1): 55–66. https://doi.org/10.1006/jcat.1997.1647.

Molazadeh, Marziyeh, Hossein Ahmadzadeh, Hamid R. Pourianfar, Stephen Lyon, and Pabulo Henrique Rampelotto. 2019. "The Use of Microalgae for Coupling Wastewater Treatment with CO_2 Biofixation." *Frontiers in Bioengineering and Biotechnology* 7 (42): 1–12. https://doi.org/10.3389/fbioe.2019.00042.

Mona, S., S.S. Kumar, V. Kumar, K. Parveen, N. Saini, B. Deepak, and A. Pugazhendhi. 2020. Green technology for sustainable biohydrogen production (waste to energy): A review. *Science Total Environment* 728, 138481. doi:10.1016/j.scitotenv.2020.138481

Murata, K., Y. Liu, M.M. Watanabe, M. Inaba, I. Takahara. 2014. Hydrocracking of algae oil into aviation fuel-range hydrocarbons using a Pt-Re catalyst. *Energy and Fuels* 28, 6999–7006. doi:10.1021/ef5018994

Naghdi, Forough Ghasemi, Skye R. Thomas-Hall, Reuben Durairatnam, Steven Pratt, and Peer M. Schenk. 2014. "Comparative Effects of Biomass Pre-Treatments for Direct and Indirect Transesterification to Enhance Microalgal Lipid Recovery." *Frontiers in Energy Research* 2 (57): 1–10. https://doi.org/10.3389/fenrg.2014.00057.

Nascimento, Iracema Andrade, Sheyla Santa Izabel Marques, Iago Teles Dominguez Cabanelas, Gilson Correia Carvalho, Maurício A. Nascimento, Carolina Oliveira Souza, Janice Isabel Druzian, Javid Hussain, and Wei Liao. 2014. "Microalgae Versus Land Crops as Feedstock for Biodiesel: Productivity, Quality, and Standard Compliance." *BioEnergy Research* 7: 1002–1013. https://doi.org/10.1007/s12155-014-9440-x.

Nascimento, Iracema Andrade, Sheyla Santa Izabel Marques, Iago Teles Dominguez Cabanelas, Solange Andrade Pereira, Janice Isabel Druzian, Carolina Oliveira de Souza, Daniele Vital Vich, Gilson Correia de Carvalho, and Maurício Andrade Nascimento. 2013. "Screening Microalgae Strains for Biodiesel Production: Lipid Productivity and Estimation of Fuel Quality Based on Fatty Acids Profiles as Selective Criteria." *BioEnergy Research* 6 (1): 1–13. https://doi.org/10.1007/s12155-012-9222-2.

Ngangkham, Momocha, Sachitra Kumar Ratha, Radha Prasanna, Anil Kumar Saxena, Dolly Wattal Dhar, Chandragiri Sarika, and Rachapudi Badari Narayana Prasad. 2012. "Biochemical Modulation of Growth, Lipid Quality and Productivity in Mixotrophic Cultures of Chlorella Sorokiniana." *SpringerPlus* 1 (33): 1–13. https://doi.org/10.1186/2193-1801-1-33.

Odjadjare, E.C., T. Mutanda, and A.O. Olaniran. 2017. Potential biotechnological application of microalgae: a critical review. *Critical Review Biotechnology* 37, 37–52. doi:10.3109/07388551.2015.1108956

Ohse, Silvana, Roberto Bianchini Derner, Renata Ávila Ozório, Rafaela Gordo Corrêa, Eliana Badiale Furlong, and Paulo Cesar Roberto Cunha. 2015. "Lipid Content and Fatty Acid Profiles in Ten Species of Microalgae." *Idesia (Arica)* 33 (1): 93–101. https://doi.org/10.4067/s0718-34292015000100010.

Olaizola, Miguel. 2003. "Commercial Development of Microalgal Biotechnology: From the Test Tube to the Marketplace." *Biomolecular Engineering* 20: 459–466. https://doi.org/10.1016/S1389-0344(03)00076-5.

Oncel, S.S., A. Kose, C. Faraloni, E. Imamoglu, M. Elibol, G. Torzillo, and F. Vardar Sukan. 2015. Biohydrogen production from model microalgae Chlamydomonas reinhardtii: A simulation of environmental conditions for outdoor experiments. *International Journal of Hydrogen Energy* 40, 7502–7510. doi:10.1016/j.ijhydene.2014.12.121

Park, J. B. K., R. J. Craggs, and A. N. Shilton. 2011. "Wastewater Treatment High Rate Algal Ponds for Biofuel Production." *Bioresource Technology* 102 (1): 35–42. https://doi.org/10.1016/j.biortech.2010.06.158.

Patel, Bhavish, Pedro Arcelus-Arrillaga, Arash Izadpanah, and Klaus Hellgardt. 2017. "Catalytic Hydrotreatment of Algal Biocrude from Fast Hydrothermal Liquefaction." *Renewable Energy* 101: 1094–1101. https://doi.org/10.1016/j.renene.2016.09.056.

Patel, D.S., A. Pandey, S. Srivastava, A.N. Sawarkar, and S. Kumar. 2020. Ultrasound-intensified biodiesel production from algal biomass: a review. *Environmental Chemical Letters* 1–21. doi:10.1007/s10311-020-01080-z

Piasecka, Agata, Izabela Krzemińska, and Jerzy Tys. 2017. "Enrichment of Parachlorella Kessleri Biomass with Bioproducts: Oil and Protein by Utilisation of Beet Molasses." *Journal of Applied Phycology* 29 (4): 1735–1743. https://doi.org/10.1007/s10811-017-1081-y.

Pittman, Jon K., Andrew P. Dean, and Olumayowa Osundeko. 2011. "The Potential of Sustainable Algal Biofuel Production Using Wastewater Resources." *Bioresource Technology* 102 (1): 17–25. https://doi.org/10.1016/j.biortech.2010.06.035.

Praveen, Prashant, and Kai Chee Loh. 2016. "Nitrogen and Phosphorus Removal from Tertiary Wastewater in an Osmotic Membrane Photobioreactor." *Bioresource Technology* 206: 180–187. https://doi.org/10.1016/j.biortech.2016.01.102.

Prince, Roger C., and Haroon S. Kheshgi. 2005. "The Photobiological Production of Hydrogen: Potential Efficiency and Effectiveness as a Renewable Fuel." *Critical Reviews in Microbiology* 31: 19–31. https://doi.org/10.1080/10408410590912961.

Pugliese, A., L. Biondi, P. Bartocci, F. Fantozzi. 2020. Selenastrum Capricornutum a New Strain of Algae for Biodiesel Production. *Fermentation* 6, 46. doi:10.3390/fermentation6020046

Pulz, Otto, and Wolfgang Gross. 2004. "Valuable Products from Biotechnology of Microalgae." *Applied Microbiology and Biotechnology* 65 (6): 635–648. https://doi.org/10.1007/s00253-004-1647-x.

Radakovits, Randor, Robert E. Jinkerson, Al Darzins, and Matthew C. Posewitz. 2010. "Genetic Engineering of Algae for Enhanced Biofuel Production." *Eukaryotic Cell* 9 (4): 486–501. https://doi.org/10.1128/EC.00364-09.

Ramanan, Rishiram, Byung-hyuk Kim, Dae-hyun Cho, Hee-mock Oh, and Hee-sik Kim. 2016. "Algae – Bacteria Interactions : Evolution, Ecology and Emerging Applications." *Biotechnology Advances* 34 (1): 14–29. https://doi.org/10.1016/j.biotechadv.2015.12.003.

Rakesh, S., J. Tharunkumar, and B. Sri. 2020. Sustainable Microalgae Harvesting Strategies for the Production of Biofuel and Oleochemicals. *Highlights Bioscience* 3, 1–8. doi:10.36462/H.BioSci.20209

Ray, J.G., P. Santhakumaran, and S. Kookal. 2020. Phytoplankton communities of eutrophic freshwater bodies (Kerala, India) in relation to the physicochemical water quality parameters. *Environmental Develpment Sustainable* doi:10.1007/s10668-019-00579-y

Refaat, A. A., S. T. El Sheltawy, and K. U. Sadek. 2008. "Optimum Reaction Time, Performance and Exhaust Emissions of Biodiesel Produced by Microwave Irradiation." *International Journal of Environmental Science and Technology* 5 (3): 315–322. https://doi.org/10.1007/BF03326026.

Ren, Rui, Xue Han, Haiping Zhang, Hongfei Lin, Jianshe Zhao, Ying Zheng, and Hui Wang. 2018. "High Yield Bio-Oil Production by Hydrothermal Liquefaction of a Hydrocarbon-Rich Microalgae and Biocrude Upgrading." *Carbon Resources Conversion* 1: 153–159. https://doi.org/10.1016/j.crcon.2018.07.008.

Robota, Heinz J., Jhoanna C. Alger, and Linda Shafer. 2013. "Converting Algal Triglycerides to Diesel and HEFA Jet Fuel Fractions." *Energy and Fuels* 27: 985–996. https://doi.org/10.1021/ef301977b.

Saad, Marwa G, Noura S Dosoky, Mohamed S Zoromba, and Hesham M Shafik. 2019. "Algal Biofuels : Current Status and Key Challenges." *Energiesgies* 12 (1920): 1–22.

Sahoo, Dinabandhu, Geetanjali Elangbam, and Salam Sonia Devi. 2012. "Using Algae for Carbon Dioxide Capture and Bio-Fuel Production to Combat Climate Change." *Phykos* 42 (1): 32–38.

Santhakumaran, P., S.K. Kookal and J.G. Ray. 2018. "Biomass Yield and Biochemical Profile of Fourteen Species of Fast-Growing Green Algae from Eutrophic Bloomed Freshwaters of Kerala, South India." *Biomass and Bioenergy* 119: 155–165. https://doi.org/10.1016/j.biombioe.2018.09.021.

Santhakumaran, Prasanthkumar, Santhosh Kumar Kookal, Linu Mathew, and Joseph George Ray. 2020. "Experimental Evaluation of the Culture Parameters for Optimum Yield of Lipids and Other Nutraceutically Valuable Compounds in Chloroidium Saccharophillum (Kruger) Comb. Nov." *Renewable Energy* 147: 1082–1097. https://doi.org/10.1016/j.renene.2019.09.071.

Satpati, Gour Gopal, Prakash Chandra Gorain, and Ruma Pal. 2016. "Efficacy of EDTA and Phosphorous on Biomass Yield and Total Lipid Accumulation in Two Green Microalgae with Special Emphasis on Neutral Lipid Detection by Flow Cytometry" 2016: 14–17.

Semwal, Surbhi, Ajay K. Arora, Rajendra P. Badoni, and Deepak K. Tuli. 2011. "Biodiesel Production Using Heterogeneous Catalysts." *Bioresource Technology* 102 (3): 2151–2161. https://doi.org/10.1016/j.biortech.2010.10.080.

Shaikh, Kashif Mohd, Asha Arumugam Nesamma, Malik Zainul Abdin, and Pannaga Pavan Jutur. 2019. "Molecular Profiling of an Oleaginous Trebouxiophycean Alga Parachlorella Kessleri Subjected to Nutrient Deprivation for Enhanced Biofuel Production." *Biotechnology for Biofuels* 12 (1): 1–15. https://doi.org/10.1186/s13068-019-1521-9.

Shalaby, Emad A. 2011. "Algal Biomass and Biodiesel Production." *Biodiesel - Feedstocks and Processing Technologies*, 111–131. www.intechopen.com.

Sharma, Archita, and Shailendra Kumar Arya. 2017. "Hydrogen from Algal Biomass: A Review of Production Process." *Biotechnology Reports* 15 (June): 63–69. https://doi.org/10.1016/j.btre.2017.06.001.

Sharma, Kalpesh K., Holger Schuhmann, and Peer M. Schenk. 2012. "High Lipid Induction in Microalgae for Biodiesel Production." *Energies* 5: 1532–1553. https://doi.org/10.3390/en5051532.

Sharma, Prabin Kumar, Manalisha Saharia, Richa Srivstava, Sanjeev Kumar, and Lingaraj Sahoo. 2018. "Tailoring Microalgae for Efficient Biofuel Production." *Frontiers in Marine Science* 5 (NOV): 1–19. https://doi.org/10.3389/fmars.2018.00382.

Sheehan, J., T. Dunahay, J. Benemann, and P. Roessler. 1998. "A Look Back at the US Department of Energy's Aquatic Species." *National Renewable Energy Laboratory, Report No. TP-580-24190* NREL/TP-58: 1–328. https://doi.org/10.2172/15003040.

Shen, Y., Z.J. Pei, Q. Wu, and E. Mao. 2009. Microalgae Mass Production Methods. *American Society of Agricultural and Biological Engineering* 52, 1275–1287.

Shi, Fan, Ping Wang, Yuhua Duan, Dirk Link, and Bryan Morreale. 2012. "Recent Developments in the Production of Liquid Fuels via Catalytic Conversion of Microalgae: Experiments and Simulations." *RSC Advances* 2 (26): 9727–9747. https://doi.org/10.1039/c2ra21594b.

Shi, Ze, Bingwei Zhao, Shun Tang, and Xiaoyi Yang. 2018. "Hydrotreating Lipids for Aviation Biofuels Derived from Extraction of Wet and Dry Algae." *Journal of Cleaner Production* 204: 906–915. https://doi.org/10.1016/j.jclepro.2018.08.351.

Shukla, Madhulika, and Sachin Kumar. 2018. "Algal Growth in Photosynthetic Algal Microbial Fuel Cell and Its Subsequent Utilisation for Biofuels." *Renewable and Sustainable Energy Reviews* 82: 402–414. https://doi.org/10.1016/j.rser.2017.09.067.

Shuping, Zou, Wu Yulong, Yang Mingde, Imdad Kaleem, Li Chun, and Junmao Tong. 2010. "Production and Characterisation of Bio-Oil from Hydrothermal Liquefaction of Microalgae Dunaliella Tertiolecta Cake." *Energy* 35 (12): 5406–5411. https://doi.org/10.1016/j.energy.2010.07.013.

Sikarwar, Vineet Singh, Ming Zhao, Paul S. Fennell, Nilay Shah, and Edward J. Anthony. 2017. "Progress in Biofuel Production from Gasification." *Progress in Energy and Combustion Science* 61: 189–248. https://doi.org/10.1016/j.pecs.2017.04.001.

Singh, Amrik, Amit Pal, and Sagar Maji. 2017. "Biodiesel Production from Microalgae Oil through Conventional and Ultrasonic Methods." *Energy Sources, Part A: Recovery, Utilisation, and Environmental Effects* 39 (8): 806–810. https://doi.org/10.1080/15567036.2016.1263260.

Sivasubramanian, V., V. Subramanian, B.G. Raghavan, and R. Ranjithkumar. 2009. "Large Scale Phycoremediation of Acidic Effluent from an Alginate Industry." *ScienceAsia* 35 (3): 220. https://doi.org/10.2306/scienceasia1513-1874.2009.35.220.

Sorgatto, Vanessa Ghiggi, Julio Cesar de Carvalho, Eduardo Bittencourt Sydney, Adriane Bianchi Pedroni Medeiros, Luciana Porto de Souza Vandenberghe, and Carlos Ricardo Soccol. 2019. "Microscale Direct Transesterification of Microbial Biomass with Ethanol for Screening of Microorganisms by Its Fatty Acid Content." *Brazilian Archives of Biology and Technology* 62: e19180178. https://doi.org/10.1590/1678-4324-2019180178.

Speight, J.G. 2015. "Fouling During Hydrotreating." *Fouling in Refineries* 303–328. doi:10.1016/b978-0-12-800777-8.00012-7

Srivastava, Garima, Atanu Kumar Paul, and Vaibhav V. Goud. 2018. "Optimisation of Non-Catalytic Transesterification of Microalgae Oil to Biodiesel under Supercritical Methanol Condition." *Energy Conversion and Management* 156: 269–278. https://doi.org/10.1016/j.enconman.2017.10.093.

Stansell, Graham Robert, Vincent Myles Gray, and Stuart David Sym. 2012. "Microalgal Fatty Acid Composition: Implications for Biodiesel Quality." *Journal of Applied Phycology* 24: 791–801.

Strik, David PBTB, Hilde Terlouw, Hubertus V.M. Hamelers, and Cees J.N. Buisman. 2008. "Renewable Sustainable Biocatalyzed Electricity Production in a Photosynthetic Algal Microbial Fuel Cell (PAMFC)." *Applied Microbiology and Biotechnology* 81 (4): 659–668. https://doi.org/10.1007/s00253-008-1679-8.

Su, Chia-Hung, Liang Jung Chien, James Gomes, Yu-Sheng Lin, Yuan Kun Yu, Jhang Song Liou, and Rong Jhih Syu. 2011. "Factors Affecting Lipid Accumulation by Nannochloropsis oculata in a Two-Stage Cultivation Process." *Journal of Applied Phycology* 23: 903–908. https://doi.org/10.1007/s10811-010-9609-4.

Sun, Xian, Yu Cao, Hui Xu, Yan Liu, Jianrui Sun, Dairong Qiao, and Yi Cao. 2014. "Effect of Nitrogen-Starvation, Light Intensity and Iron on Triacylglyceride/Carbohydrate Production and Fatty Acid Profile of Neochloris Oleoabundans HK-129 by a Two-Stage Process." *Bioresource Technology* 155: 204–212. https://doi.org/10.1016/j.biortech.2013.12.109.

Surendhiran, Duraiarasan, Abdul Razack Sirajunnisa, and Mani Vijay. 2015. "An Alternative Method for Production of Microalgal Biodiesel Using Novel Bacillus Lipase." *3 Biotech* 5: 715–725. https://doi.org/10.1007/s13205-014-0271-4.

Tayari, Sara, Reza Abedi, and Ali Abedi. 2020. "Investigation on Physicochemical Properties of Wastewater Grown Microalgae Methyl Ester and Its Effects on CI Engine." *Environmental and Climate Technologies* 24 (1): 72–87. https://doi.org/10.2478/rtuect-2020-0005.

Thao, Tran Yen, Dinh Thi Nhat Linh, Vo Chi Si, Taylor W. Carter, and Russell T. Hill. 2017. "Isolation and Selection of Microalgal Strains from Natural Water Sources in Viet Nam with Potential for Edible Oil Production." *Marine Drugs* 15 (7): 1–14. https://doi.org/10.3390/md15070194.

Tong, Dongmei, Changwei Hu, Kanghua Jiang, and Yuesong Li. 2011. "Cetane Number Prediction of Biodiesel from the Composition of the Fatty Acid Methyl Esters." *JAOCS: Journal of the American Oil Chemists' Society* 88 (3): 415–423. https://doi.org/10.1007/s11746-010-1672-0.

Valdez, Peter J., Michael C. Nelson, Henry Y. Wang, Xiaoxia Nina Lin, and Phillip E. Savage. 2012. "Hydrothermal Liquefaction of Nannochloropsis Sp.: Systematic Study of Process Variables and Analysis of the Product Fractions." *Biomass and Bioenergy* 46: 317–331. https://doi.org/10.1016/j.biombioe.2012.08.009.

Venkataraman, L.V., M.R. Somasekarappa, I. Somasekeran, and T. Lalitha. 1982. "Simplified Method of Raising Inoculum of Blue-Green Alga Spirulina Platensis for Rural Application." *Phykos* 21: 56–62.

Verma, Deepak, Rohit Kumar, Bharat S. Rana, and Anil K. Sinha. 2011. "Aviation Fuel Production from Lipids by a Single-Step Route Using Hierarchical Mesoporous Zeolites." *Energy and Environmental Science* 4 (5): 1667–1671. https://doi.org/10.1039/c0ee00744g.

Wang, Xin Wei, Jun Rong Liang, Chun Shan Luo, Chang Ping Chen, and Ya Hui Gao. 2014. "Biomass, Total Lipid Production, and Fatty Acid Composition of the Marine Diatom Chaetoceros Muelleri in Response to Different CO_2 Levels." *Bioresource Technology* 161: 124–130. https://doi.org/10.1016/j.biortech.2014.03.012.

Wu, Xiaodan, Rongsheng Ruan, Zhenyi Du, and Yuhuan Liu. 2012. "Current Status and Prospects of Biodiesel Production from Microalgae." *Energies* 5: 2667–2682. https://doi.org/10.3390/en5082667.

Wu, T., L. Li, X. Jiang, Y. Yang, Y. Song, L. Chen, X. Xu, Y. Shen, and Y. Gu, 2019. Sequencing and comparative analysis of three Chlorella genomes provide insights into strain-specific adaptation to wastewater. *Science Report* 9, 1–12. doi:10.1038/s41598-019-45511-6

Yoshida, M., Y. Tanabe, N. Yonezawa, and M.M. Watanabe. 2012. Energy innovation potential of oleaginous microalgae. *Biofuels* 3, 761–781. doi:10.4155/bfs.12.63

Yücel, S., Terzioğlu Pınar, and Özçimen Didem. 2013. "Biodiesel - Feedstocks, Production and Applications." *Intech Open Science*, 209–250. https://doi.org/http://dx.doi.org/10.5772/57353.

Zabed, H.M., S. Akter, J. Yun, G. Zhang, Y. Zhang, and X. Qi. 2020. Biogas from microalgae: Technologies, challenges and opportunities. *Renewable Sustainable Energy Review* 117, 109503. doi:10.1016/j.rser.2019.109503

Zhang, Bo, Lijun Wang, Rui Li, Quazi Mahzabin Rahman, and Abolghasem Shahbazi. 2017. "Catalytic Conversion of Chlamydomonas to Hydrocarbons via the Ethanol-Assisted Liquefaction and Hydrotreating Processes." *Energy and Fuels* 31 (11): 12223–12231. https://doi.org/10.1021/acs.energyfuels.7b02080.

Zhou, Yingdong, Yaguang Chen, Mingyu Li, and Changwei Hu. 2020. "Production of High-Quality Biofuel via Ethanol Liquefaction of Pretreated Natural Microalgae." *Renewable Energy* 147: 293–301. https://doi.org/10.1016/j.renene.2019.08.136.

Zhou, Yingdong, and Changwei Hu. 2020. "Catalytic Thermochemical Conversion of Algae and Upgrading of Algal Oil for the Production of High-Grade Liquid Fuel: A Review." *Catalysts* 10 (145): 1–24. https://doi.org/10.3390/catal10020145.

15 Compilation of Characterization and Electrochemical Behavior from Novel Biomass as Porous Electrode Materials for Energy Storage Devices

Divya Palaniswamy and Rajalakshmi Reguramnan

Avinashilingam Institute for Home Science and Higher Education for Women, India

CONTENTS

15.1 INTRODUCTION

The depletion of global warming and limited resources of fossil fuels have increased to force the world toward the development of clean and green energy resources. For the production of hybrid electric vehicles and portable electronic devices, there has been an increasing demand for ecofriendly, non-toxic and sustainable high-power and energy sources. To reduce the environmental issues, many research efforts have been started to utilize the biomass as porous electrode materials in the application of supercapacitors. Biomass generation will increase four-fold and solar energy generation will rapidly grow seven-fold by 2040, it has been predicated [1]. Therefore, the development of energy storage devices, namely supercapacitors and batteries, by using carbon-based electrode materials from biomass were essential.

Biomass as an energy source consists of plant or animal or marine organism-derived materials due to their renewable source of earth, unique structure, ecofriendly and low cost. The richness of the carbon element and well-oriented channels for the metabolic process make these biomass-based electrode materials suitable for rapid ion transportation during electrochemical charge-discharge applications. Biomass-derived materials have been exploited for various applications, such as CO_2 capture [2, 3], hydrogen storage [4–6], water treatment [7, 8], catalysis, and energy storage [9–13].

Supercapacitors (SCs) are also known as electrochemical capacitors or ultra-capacitors and have increased in attention because of their pulse power supply, long life cycle, basic principle and high dynamics of charge propagation [14–16]. SCs have higher power density than batteries and higher energy density than regular capacitors. SCs are widely used in the application of consumer electronics, memory backup systems, braking accelerations, industrial power and energy management [17]. The main market for the next few years is hybrid electric vehicles as well as subways and tramways and also in fuel cells.

The most common electrode materials for SCs such as carbon-based materials in the form of activated carbon, carbon nanotubes, and carbon aerogels for EDLCs are more attractive and are rich in dimensionality [18] metal oxides and conducting polymers for pseudocapacitors. By altering the electrochemical performance of the material, the microstructure and surface chemistry of carbon can be tuned or adjusted easily [19]. The electrode materials should have the required properties such as good conductivity, high temperature stability, inertness, high corrosion resistance and high surface area. High surface area is a very important parameter because the energy in electrochemical storage devices is stored on the surface and the pore size promotes the transport and storage of charges. These properties of specific carbon-based materials make them effective for use in supercapacitors and batteries in terms of power density and energy storage capability [20].

The present work focuses on the compilation of using functional carbon materials from two seaweeds of SW-700/TC-700 and two plant leaves of PAL-1000/SLL-1000 as the porous green electrode materials for the performance of supercapacitors. The prepared functional carbon of SW-700/TC-700/PAL-1000/SLL-1000 was synthesized by single step without any physical or chemical activation. The seaweeds of *Sargassum Wightii* (SW), *Turbinaria Conoides* (TC) are the tropical brown algae that belong to the "Sargassaceae" family. It is found on rocky substrates and in marine waters. The SW and TC have more bioactive properties with antioxidant activities. The dried leaves of *Phyllanthus acidus* (PAL) and *Solanum Lycopersicum* (SLL) are also known as "Amla plant" and "Tomato plant" respectively. The leaves of these two plants act as a potential source for antiviral and antioxidant with more flavonoids content. The prepared SW-700/TC-700/PAL-1000/SLL-1000 electrodes were characterized and fabricated for the electrochemical performance of cyclic voltammetry (CV), galvanostatic charge/discharge (GCD), and electrochemical impedance spectroscopy (EIS) analysis. The functional carbon of SW-700/TC-700/PAL-1000/SLL-1000 was compared and evaluated. The results reveals that the SW-700/TC-700 shows highest capacitance at 354 and 416 F/g at 0.5 A/g and 1 A/g respectively even at a lower temperature of 700°C and PAL-1000/SLL-1000 shows best capacitance of 347 F/g and 345 F/g at 0.5 A/g respectively at a higher temperature of 1000°C.

15.2 EXPERIMENTAL

15.2.1 SYNTHESIS OF FUNCTIONAL CARBON FROM SELECTED DRIED SEAWEEDS AND PLANT LEAVES

The selected two seaweeds of SW/TC were collected in large quantity from the Rameswaram coastal seashore area and the two plant leaves of PAL/SLL were collected in large quantity from the Garden, Coimbatore, TamilNadu. It is thoroughly cleaned and dry at 60°C in an oven. This biomass of SW/TC/PAL/SLL were grounded to become fine powder and stored in dry atmosphere. For the preparation of functional carbon, 10 g of dried SW and TC powder at the lower temperature of 700°C for 3 h [21] and 10 g of dried PAL and SLL powder at the higher temperature of 1000°C for 5 h was heated in an alumina crucible in argon atmosphere at a heating rate of 10°C per min in a tubular furnace [22]. The resultant black solid was then washed with 5M HCl and dried for 12 h at 60°C. The just-carbonized sample for SW and TC at 700°C were denoted as SW/TC-700, and PAL and SLL at 1000°C were denoted as PAL/SLL-1000.The schematic diagram of the prepared functional carbon of SW-700/TC-700/PAL-1000/SLL-1000 were depicted in Figure 15.1.

15.2.2 CHARACTERIZATION OF SYNTHESIZED FUNCTIONAL CARBON – SW-700/ TC-700/PAL-1000/SLL-1000

The synthesized functional carbon from SW-700/TC-700/PAL-1000/SLL-1000 was characterized by X-ray powder diffraction using the ULTIMA IV Model X-ray Diffractometer with CuKα radiation (α = 1.5406 Å) in the range of 10°–90° at 0.02 steps with a count time of 0.2 s. Fourier-transform infrared (FT-IR) spectra were recorded in the model of NICOLET iS10 spectrometer from 500 to 4000 cm^{-1}. Raman spectroscopy was assessed using Lab RAM HR 520 from JY Horiba. The morphology studies were clearly observed by field emission scanning electron microscopy (FE-SEM) using Yes Carl Zeiss, UK with an acceleration voltage of 15 kV equipped with an EDS and qualitatively measured by Bruker Nano GmbH Berlin, Germany (Esprit 1.9). The carbon which are dispersed, and tiny nano-metric pores were studied

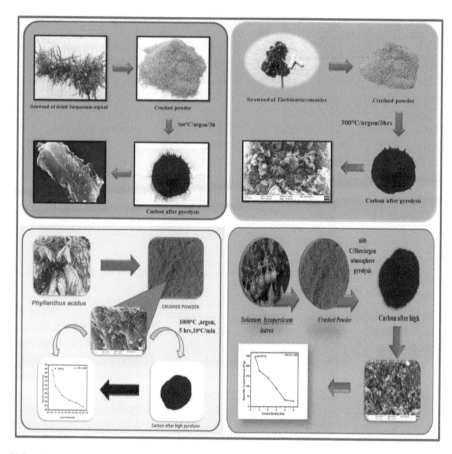

FIGURE 15.1   Pictorial representation for the preparation of functional carbon material from SW/TC/PAL/SLL.

using TEM (120 kV, FEI Tecnai G2). The surface area and pore size analysis were evaluated by BET using Quantachrome Instruments of ASiQwin (Version.3).

15.2.3   Electrode Fabrication of SW-700/TC-700/PAL-1000/SLL-1000

To examine the electrochemical properties of the synthesized functional carbon electrode materials by using active materials, Super-P as carbon conductive additive and polytetra-fluoroethylene (PTFE) as a binder in a weight ratio of 80:15:5 dissolved in NMP solvent. After ensuring good dispersion, the slurry was coated on a graphite sheet with the surface area of 1 cm^2 and dried in a hot vacuum oven at 100°C for 24 h for solvent evaporation. 1M H_2SO_4 was used as the electrolyte in a three-electrode system. For three-electrode configurations, a synthesized functional carbon material is used as working electrode, saturated calomel electrode (SCE) as reference electrode and platinum as counter electrode. The fabricated functional carbon electrode materials of SW-700, TC-700, PAL-1000, and SLL-1000 were tested and assessed by cyclic voltammetry (CV) and galvanostatic charge/discharge (GCD) analysis, EIS, and cyclic stability analysis by using CH Instrument, Model CHI608E, USA.

15.3   RESULTS AND DISCUSSION

15.3.1   Compilation of Synthesized Functional Carbon Materials of SW-700/TC-700/PAL-1000/SLL-1000 by Characterization Techniques

The XRD patterns of prepared functional carbon from SW-700/TC-700/PAL-1000/ SLL-1000 with crude samples of SW/TC/PAL/SLL are compared and represented in Figure 15.2. It clearly indicates a two sharp peaks around $2\theta = 28.5°,40.7°$ for the crude sample of seaweeds SW/TC and broad peaks at $2\theta = 14.8°, 21.4°, 24.4°/2\theta = 14.7°, 20.9°, 24.3°$ for the crude sample of plant leaves PAL/SLL was noticed. The crude samples of SW/TC and prepared functional carbon material of SW-700/ TC-700 are crystalline in nature [27, 28]. The crude samples of PAL/SLL and synthesized functional carbon material of PAL-1000/SLL-1000 30 are amphorous in nature with the disordered structure in Table 15.1

The FT-IR patterns of synthesized functional carbon from SW-700/TC-700/PAL-1000/SLL-1000 are compiled and shown in Figure 15.3. The results indicate that it contains --OH stretching group around 3448 cm^{-1}, carbonyl C=O or may be C=C stretching groups assigned to the characteristic olefinic group which represents that the graphitization is more around 1629 cm^{-1}, and the 1086 cm^{-1} attributes to the C-O stretching group, but the resultant peaks are shifted to lower wavenumbers and the crude sample of SW/TC/PAL/SLL illustrates the presence of C-H stretching and CO_2

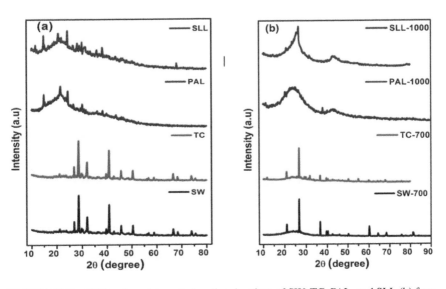

FIGURE 15.2   XRD pattern (a) crude functional carbon of SW, TC, PAL, and SLL (b) functional carbon of SW-700/TC-700/PAL-1000/SLL-1000.

TABLE 15.1

XRD Profile of SW-700/TC-700/PAL-1000/SLL-1000 with Crude Samples of SW/TC/PAL/SLL

S. No	Crude Samples	2θ Values for Crude Samples	Inference	Prepared Carbon Precursors	2θ Values for Prepared Carbon Precursors	Inference	Nature
1	SW	28.5°, 40.7°	Two sharp peaks	SW-700	26.9°, 36.7°	Two sharp peaks	Crystalline
2	TC	28.5°, 40.7°	Two sharp peaks	TC-700	26.9°	Sharp peak	Crystalline
3	PAL	24.4°	Broad peak	PAL-1000	25°	Broad peak	Amphorous
4	SLL	24.3°	Broad peak	SLL-1000	24.8°	Broad peak	Amphorous

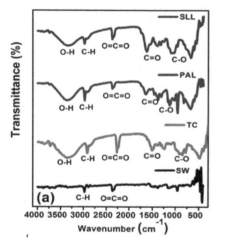 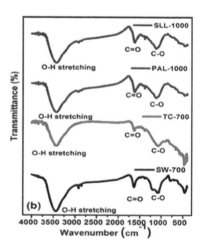

FIGURE 15.3 FT-IR pattern (a) crude functional carbon of SW, TC, PAL, and SLL. (b) functional carbon of SW-700, TC-700, PAL-1000 and SLL-1000.

groups listed in Table 15.2. It is important that the intensities of peaks resultant to few oxygen-containing functional groups were present in the samples of SW-700/TC-700/PAL-1000/SLL-1000. These are significantly reduced and more graphitic when increased to higher temperature [23–25, 27, 28, 30].

The elements and their relative parameters in synthesized SW-700/TC-700/PAL-1000/SLL-1000 samples were determined by EDS as depicted in Table 15.3. It represents that the carbon is the most prominent ingredient and implies that all the samples are well carbonized. The presence of oxygen (O) infers that there are lots of functional groups containing oxygen for the samples of SW/TC-700 and PAL-SLL-1000. Furthermore, the mineral substances such as potassium (K), silicates,

TABLE 15.2
FT-IR Profile of SW-700/TC-700/PAL-1000/SLL- 1000 with Crude Samples of SW/TC/PAL/SLL

Precursors

Crude SW	SW-700	Crude TC	TC-700	Crude PAL	PAL-1000	Crude SLL	SLL-1000	Frequency Assignment
-	3448	3389	3430	3384	3448	3390	3448	O-H stretching
-	1634	1635	1632	1637	1632	1619	1632	C=O stretching
-	1086	1033	1084	1166	1076	1065	1076	C-O stretching
2968	-	2977	-	2979	-	2979	-	C-H stretching
2360	-	2360	-	2347	-	2360	-	O=C=O

TABLE 15.3
EDS Analysis of SW-700/TC-700/PAL-1000/SLL-1000

Samples	C (%)	O (%)	Cl (%)	Mg (%)	Ca (%)
SW-700	89.11	9.67	0.43	0.36	0.00
TC-700	83.40	9.79	3.05	0.75	0.38
PAL-1000	87.53	9.82	1.53	0.41	0.19
SLL-1000	86.98	11.19	0.53	0.00	0.23

calcium, and magnesium in minor traces can be detected as well in the samples of SW-700/TC-700/PAL-1000/SLL-1000. During the carbonization process, it can be inferred that the samples contain a large amount of carbon content of >83%. It shows a significant role in the physical properties and chemical nature [26, 27, 28, 30].

Raman spectrum is used to determine the nature and amount of graphitization of prepared functional carbon materials of SW-700/TC-700/PAL-1000/SLL-1000 as represented in Figure 15.4. From the compilation table, it shows that the typical D and G bands exist with the I_D/I_G ratio of >1. This infers that SW-700/TC-700/PAL-1000/SLL-1000 depicts the amount of graphitization is more with higher defects or disorders in the prepared samples. The Lorentzian fit was done to determine the positions and widths of the D and G bands in the Raman shift spectra. From the statistical analysis of Lorentz fit, the data of R^2 were found to be equal to 1 for the sample of TC-700/PAL-1000/SLL-1000 which indicates that the regression predictions of Lorentz type perfectly fit the data. But in the sample for TC-700 [27, 28, 30], the Lorentz fit doesn't fit since the R^2 is very low value due to the position and width of the two bands which are not identical as displayed in Table 15.4.

The surface morphology and the porous nature of the prepared functional carbon of SW-700/TC-700/PAL-1000/SLL-1000 carbon materials are shown in Figure 15.5.

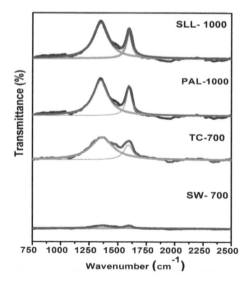

FIGURE 15.4   Raman Spectrum of SW-700/TC-700/PAL-1000/SLL-1000.

TABLE 15.4
RAMAN Profile of SW-700/TC-700/PAL-1000/SLL-1000

Samples	G Band (cm⁻¹)	D Band (cm⁻¹)	I_D/I_G ratio	$_R2$	Lorentz Type Fit
SW-700	1348	1595	1.01	0.04	Doesn't fit
TC-700	1353	1591	1.49	0.93	Fit
PAL-1000	1348	1595	1.35	0.94	Fit
SLL-1000	1348	1595	1.35	0.94	Fit

The formation of large and thick carbon particles is observed. This clearly confirms that the carbonization process in the argon atmosphere could convert SW-700/TC-700/PAL-1000/SLL-1000 sample into valuable carbon material of 89.1%, 83.4%, 87.53% & 86.9% respectively. The pore formation on the surface of the carbon at higher or lower temperature displays thicker and larger carbon particles which leads to the formation of more rough surfaces with micro size particles in Figure 15.5c–d [24]. Finally, the prepared functional carbon material reveals the formation of small or tiny pores and the high rough surface of carbon materials is found by comparing the FE-SEM images of SW-700/TC-700/PAL-1000/SLL-1000 [27, 28, 30].

The TEM images of prepared functional carbon of SW-700/TC-700/PAL-1000/SLL-1000 are indicated in Figure 15.6. The TEM image of the carbonized sample of SW-700 indicates the honeycomb-like structure which represents the ray of hollow cells formed between thin vertical walls. The cells are often columnar and hexagonal in shape and also the formation of nanosheets like layers by the carbon matrix can be seen in Figure 15.6a. The TC-700/PAL-1000 functional carbon material depicts the structure which is only made of interconnected structures or porous spheres that

FIGURE 15.5   FE-SEM Pictures (a) SW-700, (b) TC-700, (c) PAL-1000, and (d) SLL-1000.

overlap each other with irregular shapes of tiny pores. The highly porous network and nanosheets like layers of the carbon matrix can be seen in Figure 15.6b, c. The PAL-1000 looks like a carbon smoke-like structure with an interconnected network [27]. For the material of SLL-1000, the TEM images of Figure 15.6d show the evidence of amphorous 2D rod or tubular-like nano structure with a hollow center and thin tube walls. The inside and outside diameters were found to be approximately 33.2 nm and 10 nm, respectively. The surface of these rods has rich mesopores which indicates the interconnected mesopores were dispersed well within the amorphous carbon network which is significant for the fast ion transfers for energy storage systems. This type of 2D materials (micro or meso porous) not only offers ionic transfer for high-power delivery, it also shows rich interfacial active sites and high charge carrier density for high energy storage and also contributes rich active sites electrochemically in the electrochemical reactions which affords high electrode-electrolyte interfaces and low ionic resistances [28].

By comparing SW-700/TC-700/PAL-1000/SLL-1000, there is a presence of micro and meso pores in all the above-mentioned samples. This type of micro/meso-pores is distributed well within the carbon matrix by interconnected structures or

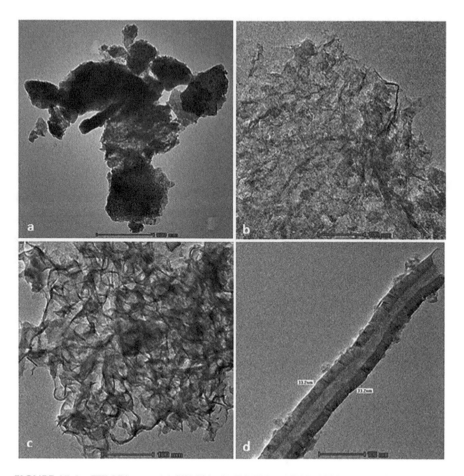

FIGURE 15.6   TEM Pictures (a) SW-700, (b) TC-700, (c) PAL-1000, and (d) SLL-1000.

porous spheres that overlap each other with irregular shapes of tiny pores. It not only contributes a large surface area but also provides more adsorbate routes by large transport networks to micropores and it could be favorable for energy storage systems to achieve high-power and energy density.

BET analysis is used to determine the surface area and the porosity of the prepared functional carbon of SW-700/TC-700/PAL-1000/SLL-1000, the BET of nitrogen adsorption-desorption isotherms and pore size distribution (PSD) are recorded, compared, and shown in Figure 15.7. From the results, it shows the presence of hysteresis loop infers Type - IV curve due to the capillary condensation. The pore size is an important parameter that determines charge storage capacity. Even at low surface, high capacitance can be achieved [22]. The results of SW-700/TC-700/PAL-1000/SLL-1000 reveal the range of size at >3 nm with the pore volume of 0.1–0.3 cm^3/g in Table 15.5. which confirms the presence of mesopores with graphitic carbon contributes excellent electrical conductivity capable of overcoming the basic kinetic limits of the electrochemical process in porous electrodes.

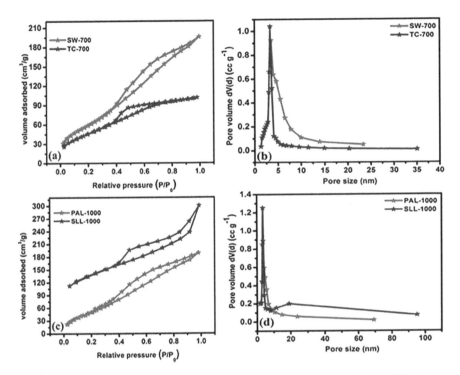

FIGURE 15.7 BET Pattern (a) Adsorption–Desorption Isotherm of SW/TC-700 (b) Pore Size Distribution of SW/TC-700 (c) Adsorption–Desorption Isotherm of PAL/SLL-1000 (b) Pore Size Distribution of PAL/SLL-1000.

TABLE 15.5
BET Analysis of SW-700/TC-700/PAL-1000/SLL-1000

S.No	Carbon Precursors	Surface Area (m²/g)	Pore Volume (cm³/g)	Range (nm)	Type of Curve
1	SW-700	231.6	0.269	3.56	Type - IV
2	TC-700	173.8	0.171	3.37	Type - IV
3	PAL-1000	216.2	0.272	3.15	Type - IV
4	SLL-1000	325.1	0.381	3.52	Type - IV

15.3.2 COMPARISON OF FABRICATED FUNCTIONAL CARBON ELECTRODES OF SW-700/TC-700/PAL-1000/SLL-1000 BY ELECTROCHEMICAL MEASUREMENTS

The CV profile of prepared functional carbon electrodes of SW-700/TC-700/PAL-1000/SLL-1000 in the potential range of –0.2–0.8 V for a sweep rate of 10–50 mV/s are shown in Figure 15.8. The prepared functional carbon of SW-700/TC-700 represents a double-layer formation perfectly due to the reversible adsorption and

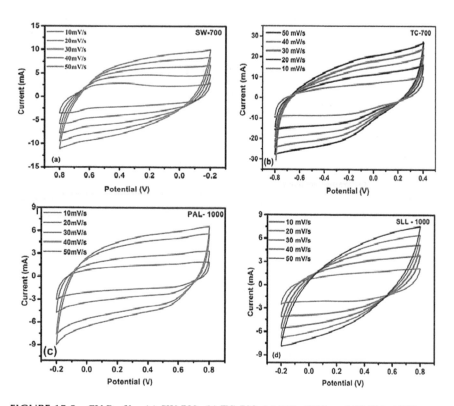

FIGURE 15.8   CV Profiles (a) SW-700, (b) TC-700, (c) PAL-1000, and (d) SLL-1000.

desorption of the ions through the carbon surface, which leads to the indication of microporous character of the material. Even at a high sweep rate, the CV curvatures maintain their character or structure without any alteration. So it clearly shows that the SW-700/TC-700 sample was found to be major electrical double-layer capacitance (EDLC)-like behavior with a pseudo capacitive nature due to the lower content of oxygen. At a lower sweep rate, the CV curves furnish the rectangular shape, clearly revealing the good reversible electrochemical behavior of EDLC [26, 27, 30].

The CV profile of PAL-1000/SLL-1000 reveals a super stable capacitive behavior. The CV curves contain rectangular shapes between a positive and negative scanning rate which specifies the very good reversible characteristic of EDLC. It also shows the absence of redox peaks (no hump) which indicates the EDLC perfectly during the charge/discharge process. From the above CV results, it is understood that when the pyrolysis temperature increases (700°C–1000°C), the capacitive behavior also increases gradually which supports the result of the galvanostatic charge/discharge (GCD) experiment. Even at an extremely high scan rates the capacitive behavior is maintained. This performance at high scan rates shows the high-power capability of the microporous PAL/SLL material [22].

The CD profile of prepared functional carbon electrode samples of SW-700/TC-700/PAL-1000/SLL-1000 at 0.5–5 A/g are depicted in Figure 15.9. The SW-700

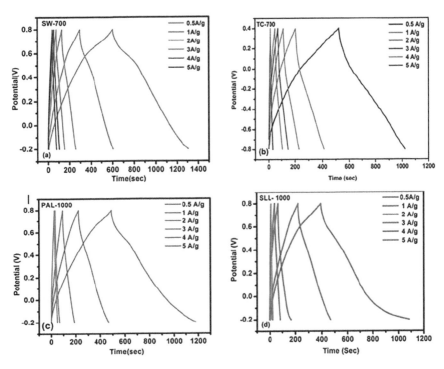

FIGURE 15.9   CD Profiles (a) SW-700, (b) TC-700, (c) PAL-1000, and (d) SLL-1000.

reveals highest specific capacitance of 354 F/g at 0.5 A/g (Figure 15.9a). As the current was increased to 1, 2, 3, 4, and 5 A/g, the specific capacitance reduced gradually to 308, 256, 228, 204, and 185 F/g, respectively. The TC-700 shows the highest specific capacitance of 416 F/g at the current density of 1 A/g (Figure 15.9b). When the current density was increased to 2, 3, 4, and 5 A/g, the specific capacitance decreased gradually to 343.3,290,250 and 216.6 F/g respectively. The PAL-1000 reveals a highest capacitance of 347 F/g at 0.5 A/g (Figure 15.9c). The specific capacitance decreases gradually to 247,188,159,144 and 120 F/g, respectively as the current density was increased to 1, 2, 3, 4, and 5 A/g. The SLL-1000 reveals a higher specific capacitance of 345 F/g at 0.5 A/g (Figure 15.9d). Therefore, the specific capacitance decreases gradually to 247,200, 126,47 and 39 F/g, respectively, when the current density was increased to 1, 2, 3, 4, and 5 A/g. Generally, the electrolyte will diffuse through the entire micro- and mesoporous surfaces at a low charging rate, so the specific capacitance will be improved significantly or increases. However, at high charging rates, the electrolyte can access only the mesoporous surface, so that the specific capacitance will be reduced gradually. The specific capacitance of SW-700, TC-700, PAL-1000, SLL-1000 at 0.5 to 5 A/g are represented in Figure 15.10 and Table 15.6.

To analyze the ion transport kinetics between the electrode material and the electrolyte ion, EIS measurements were carried out for the electrode of SW-700, TC-700, PAL- 1000, SLL-1000 in 1M H_2SO_4 (Figure 15.11a, b). The Nyquist plot of SW-700/

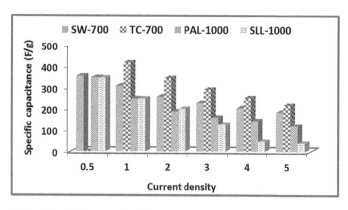

FIGURE 15.10 Bar Diagram of Specific Capacitance of SW-700/TC-700/PAL-1000/ SLL-1000.

TABLE 15.6

Compilation of GCD Analysis of Functional Carbon of SW-700/TC-700/ PAL-1000/SLL-1000

	Specific Capacitance (F/g)					
Current Density (A/g)	0.5	1	2	3	4	5
SW-700	354	308	256	228	204	185
TC-700	-	416	343.3	290	250	216.6
PAL-1000	347	247	188	159	144	120
SLL-1000	345	247	200	126	47	39

TC-700/PAL-1000/SLL-1000 represents the imaginary component plot (Z′) with the real component of impedance (Z″) which indicates the response of the electrode and electrolyte system. In order to realize the numerous resistance aspects in the interface of the electrode-electrolyte and charge transfer process, EIS results were fitted with an equivalent circuit model and signified with the physical components such as double-layer capacitance (C_{dl}), solution resistance (R_s), and charge transfer resistance (R_{ct}).

The Nyquist plot reveals a vertical straight line equivalent to the imaginary axis (Z″) at the region of lower frequency and reveals the domination of the ideal capacitance nature at the electrode-electrolyte interface which shows pure EDLC behavior and the width of the semicircle impedance loop represents charge transfer resistance in the electrode materials [22]. The statistical values from the fitted profile reveals the low R_s values of 1.4 Ω, 1.60 Ω, 1.15 Ω, 1 Ω for SW-700/TC-700/ PAL-1000/SLL-1000 respectively and R_{ct} values of 5.8 Ω, 8.34 Ω, 11 Ω, 2 Ω for SW-700/TC-700/PAL 1000/SLL-1000 respectively due to fast diffusion of electrolyte ions into the porous carbon electrode matrix and also may be loss of pseudocapacitance [22].

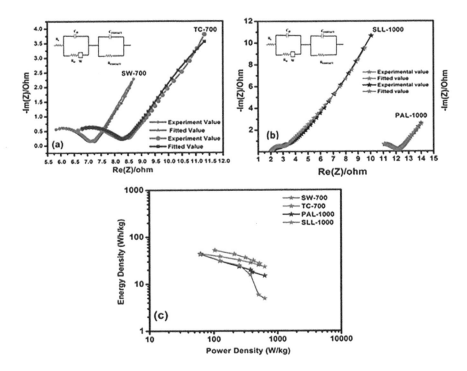

FIGURE 15.11 (a) EIS Spectrum of SW-700/TC-700 Electrodes and its Equivalent Circuit Model (inset) (b) EIS Spectrum of PAL-1000/SLL-1000 Electrodes and its Corresponding Circuit Model (inset) (c) Ragone plot of SW-700/TC-700/PAL-1000/SLL-1000 using 1M H_2SO_4.

The energy and power density of storage systems are the main factors to differentiate them for the fuel cells, supercapacitor and batteries. The Ragone plot (power Vs. energy densities) for SW-700/TC-700/PAL-1000/SLL-1000 in Figure 15.11c. When the current density increases, the energy density decreases and power density increases. The reason is only some part of the surface pores were accessible to the electrolyte which leads to fast discharge at high current density, whereas all the pores including those near to the surfaces and also in bulk were accessible to the electrolyte which leads to slow discharge at low current density [22]. In other words, the highly porous nature in nm level resulted in an increase in the number of sites which are accessible to electrolyte ions and enhances the electron transfer process between the electrode and electrolyte [38, 41] which gives rise to a higher energy density of 43.38 Wh/kg [27–30].

The SW-700 electrode material exhibits a high energy density 44.25 Wh/kg at a power density of 62.5 W/kg, and still maintains 23.13 Wh/kg at a high-power density of 625 W/kg [27]. The TC-700 shows high energy density 52 Wh/kg at a power density of 104 W/kg and still maintains 27.08 Wh/kg at a high-power density of 520.8 W/kg in 1M H_2SO_4 which is closer to batteries [28].

The prepared functional carbon electrode material of PAL-1000 displays a higher energy density 43.38 Wh/kg at a 625 W/kg of power density and still maintains

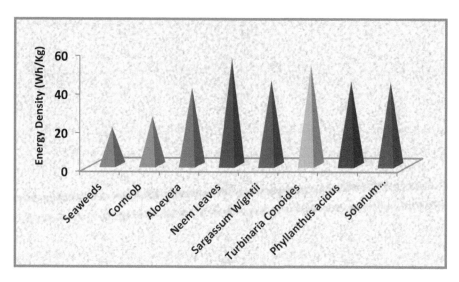

FIGURE 15.12 The Energy Density Graph of Functional Carbon Material of SW-700/TC-700/PAL 1000/SLL-1000 with Various Biomass-derived Carbon Precursors.

15 Wh/kg even at a high-power density of 625 W/kg and SLL-1000-based electrode materials reveal a high energy density of 43.13 Wh/kg at a power density of 61.34 W/kg and it still maintains 4.9 Wh/kg (energy density) at 624.7 W/kg (high-power density) in 1M H_2SO_4 which is greater in the application of supercapacitors [30]. The energy density of various biomass precursors (seaweeds, corncob, aloevera and neem leaves) is compared with the biomass-based functional carbon material of SW-700/TC-700/PAL-1000/SLL-1000 as depicted in Figure 15.12. The specific capacitance, energy and power densities of functional carbon material of SW-700/TC-700/PAL-1000/SLL-1000 are shown in Table 15.7.

The long-term cyclic stability is another important parameter that determines the supercapacitors application [23, 24]. The prepared functional carbon electrode material of SW-700/TC-700/PAL-1000/SLL-1000 affords an initial capacitance of 89.5%, 85.3%, 88.7% and 87.3% respectively retained over 5000 cycles at 5 A/g in Figure 15.13. The stability of these electrodes shows a promising and suitable candidate for supercapacitor application. Therefore, the carbonization temperature for the functional carbon electrodes derived from seaweeds of SW/TC and plant leaves of PAL/SLL specifies high capacitance, superior energy and power, and best stability even at a lower temperature of 700°C and 1000°C respectively for further investigations.

Table 15.8 summarizes the different carbon materials prepared from natural waste biomass with/without activation and their behavior in supercapacitors in aqueous medium. From this Table 15.8, it clearly affords that in our case, the functional carbon material derived from selected biomass of SW/TC/PAL/SLL without any activation offers high specific capacitance of 354 F/g, 416 F/g, 347 F/g and 345 F/g respectively in aqueous medium compared to other biomass material derived carbon

TABLE 15.7

Specific Capacitance, Energy and Power Densities of Functional Carbon Material of SW- 700/TC-700/PAL-1000/SLL-1000

Samples	Current Density (A/g)	0.5	1	2	3	4	5
SW-700	Specific capacitance (F/g)	354	308	256	228	204	185
	Energy density (Wh/Kg)	44.25	38.5	32	28.5	25.5	23.13
	Power density (W/Kg)	62.5	125	250	375	500	625
TC-700	Specific capacitance (F/g)	-	416	343.3	290	250	216.6
	Energy density (Wh/Kg)	-	52	43	36.3	31.3	27.1
	Power density (W/Kg)	-	104	208.3	312.5	416.7	520.8
PAL- 1000	Specific capacitance (F/g)	347	247	188	159	144	120
	Energy density (Wh/Kg)	43.38	30.88	23.5	19.88	18	15
	Power density (W/Kg)	62.51	125	250	375	400	625
SLL- 1000	Specific capacitance (F/g)	345	247	200	126	47	39
	Energy density (Wh/Kg)	43.13	30.9	25	15.8	5.9	4.9
	Power density (W/Kg)	61.34	125.0	250	375	502	624.7

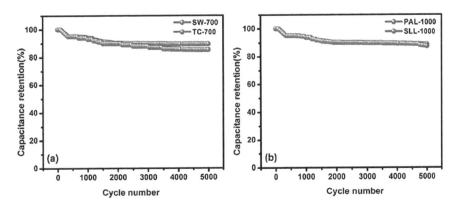

FIGURE 15.13   Capacitance Retention as a Function of Cycle Number (a) SW-700/TC-700 (b) PAL-1000/SLL-1000.

electrode precursors without activation (banana peel waste, seaweeds, watermelon, wood carbon monolith and neem dead leaves etc.) and also some of the biomass-based carbon materials followed by activation are listed in Table 15.8. From these results, it is established that the high specific capacitance, high energy and power density can be achieved for the SW/TC/PAL/SLL even without any physical or chemical activation and the observed results are on par with the results of various biomass-based carbon materials with and without activation as listed in Table 15.8. So, the synthesized biomass-based carbon materials from selected biomass of SW/TC/PAL/SLL without activation delivers an excellent performance for the application of supercapacitors.

TABLE 15.8

Compilation of the Various Carbon Material Properties from Biomass for Supercapacitors (from the literature survey)

S. No	Carbon Materials	Activation Agent	Specific Capacitance	Measurements Were Done At	Electrolyte	Reference
1	Waste Coffee beans	$ZnCl_2$	368	50 mA/g	1M H_2SO_4	[31]
2	Pomegranate rind	KOH	~268	0.1 A/g	1M H_2SO_4	[32]
3	Sugarcane bagasse	$ZnCl_2$	300	250 mA/g	1M H_2SO_4	[33]
4	Cassava peel waste	KOH	264	-	0.5 H_2SO_4	[34]
5	Argan Seed shell (Arganiaspinosa)	KOH/melamine	355	125 mA/g	1M H_2SO_4	[35]
6	Soya bean residue	KOH	250–260	-	1M H_2SO_4	[36]
7	Paulownia flower	KOH	297	1 A/g	1M H_2SO_4	[37]
8	Cherry stones	KOH	230	1 mA/cm_2	2M H_2SO_4	[38]
9	Fir wood	KOH and CO_2	197	-0.1 to 0.9 V	H_2SO_4	[39]
10	Fir wood	H_2O	140	25mV/s	0.5 H_2SO_4	[39]
11	Fir wood	KOH	180	10 mV/s	0.5 H_2SO_4	[39]
12	Pistachio Shell	KOH	120	10 mV/s	0.5 H_2SO_4	[40]
13	Tamarind Fruit shell	KOH	412	1.56 A/g	1M H_2SO_4	[41]
14	Camellia oleifera	$ZnCl_2$	374	0.2 A/g	1M H_2SO_4	[42]
15	Grapefruit peel	KOH	311	0.1 A/g	1M H_2SO_4	[43]
16	Rice straw	KOH	332	0.5 A/g	1M H_2SO_4	[24]
17	Bean dregs	KOH	482	1 A/g	1M H_2SO_4	[44]

S. No	Carbon Materials	Activation Agent	Specific Capacitance	Measurements Were Done At	Electrolyte	Reference
18	Beer leaves	KOH	188	1 mA/cm^2	0.1M H$_2$SO$_4$	[45]
19	Hallow fibers from wood waste	HCHO and HCl	295	0.5 A/g	1M H$_2$SO$_4$	[46]
20	Banana Peel waste	No activation	68	1 mv/s	1M H$_2$SO$_4$	[47]
21	Seaweeds	No activation	264	200 mA/g	1M H$_2$SO$_4$	[21]
22	Watermelon	No activation	333.1	1 A/g	6M KOH	[48]
23	Wood carbon monolith	No activation	234	5–50 mA/cm	2M KOH	[49]
24	Neem dead leaves	No activation	400	500 mA/g	1M H$_2$SO$_4$	[22]
25	Sargassum Wightii seaweeds	No activation	354	0.5 A/g	1M H$_2$SO$_4$	[26]
26	Turbinaria conoides seaweeds	No activation	416	1 A/g	1M H$_2$SO$_4$	[27]
27	Solanum lycopersicum leaves	No activation	345	0.5 A/g	1M H$_2$SO$_4$	[30]

15.4 CONCLUSION

The functional carbon materials of SW-700/TC-700/PAL-1000/SLL-1000 were prepared by a single carbonization process without using any parameters of physical or chemical activation. The PAL/SLL-1000 samples reveal more amorphous in nature and seaweeds of SW/TC-700 are more crystalline using XRD analysis. The presence of a few oxygen-containing functional groups was significantly reduced on the carbon surface as confirmed using FT-IR spectroscopy and the elements present in the prepared samples of SW-700/TC-700/PAL-1000/SLL-1000 contains more carbon content of above 83%. The intensity of an I_D/I_G ratio for SW-700/TC-700/PAL-1000/SLL-1000 directs the high amount of graphitization from Raman spectra. The PSD of SW-700/TC-700/PAL-1000/SLL-1000 reveals a low surface area at the range of 3 nm which confirms the presence of mesoporous with Type IV isotherm which is a good observation with TEM results and affords a best charge carrier density for high energy storage. The prepared functional carbon electrode materials of SW-700/TC-700 and PAL-1000/SLL-1000 represents superior capacitance, high energy and power density, more cyclic stability even at low temperature of 700°C for seaweeds and at high temperature of 1000°C for plant leaves respectively. The SW-700/TC-700/PAL-1000/SLL-1000 reveals low ionic resistance due to the fast diffusion of electrolyte ions and the limited content of hetero atom of oxygen. The prepared functional carbon of SW/TC-700/PAL/SLL-1000 from raw biomass of SW/TC/PAL/SLL acts as excellent porous electrode materials, which are non-toxic, renewable and low cost for the high performance of supercapacitors.

CONFLICT OF INTEREST

The author declares that there is no conflict of interests regarding the publication of this paper.

REFERENCES

1. U.S. Energy Information Administration, Annual Energy Outlook with Projections to 2035 (2012).
2. S. Choi, J. H. Drese and C. W. Jones, *ChemSusChem*, 2, 796–854 (2009).
3. N. Balahmar, A. C. Mitchell and R. Mokaya, *Adv. Energy Mater*, 5, 1500867 (2015).
4. M. Sevilla and R. Mokaya, *Energy Environ. Sci.* 7, 1250–1280 (2014).
5. A. F. Dalebrook, W. Gan, M. Grasemann, S. Moret and G. Laurenczy, *Chem. Commun*, 49, 8735–8751 (2013).
6. C. Zhang, Z. Geng, M. Cai, J. Zhang, X. Liu, H. Xin and J. Ma, *Int. J. Hydrogen Energy*, 38, 9243–9250 (2013).
7. Y. Matsumura, M. Sasaki, K. Okuda, S. Takami, S. Ohara, M. Umetsu and T. Adschiri, *Combust. Sci. Technol.*, 178, 509–536 (2006).
8. X. Liao, C. Chen, Z. Wang, R. Wan, C.-H. Chang, X. Zhang and S. Xie, *Process Biochem*, 48, 312–316 (2013).
9. M. M. Titirici, R. J. White, N. Brun, V. L. Budarin, D. S. Su, F. del Monte, J. H. Clark and M. J. MacLachlan, *Chem. Soc. Rev*, 44, 250–290 (2015).
10. E. M. Lotfabad, J. Ding, K. Cui, A. Kohandehghan, W. P. Kalisvaart, M. Hazelton and D. Mitlin, *ACS Nano*, 8, 7115–7129 (2014).

11. N. Liu, K. Huo, M. T. McDowell, J. Zhao and Y. Cui, *Sci. Rep*, 3, 1919 (2013).

12. A. Jain, V. Aravindan, S. Jayaraman, P. S. Kumar, R. Balasubramanian, S. Ramakrishna, S. Madhavi and M. Srinivasan, *Sci. Rep.*, 3, 3002 (2013).

13. D. Larcher and J. Tarascon, *Nat. Chem*, 7, 19–29 (2015).

14. A. Burke, *J. Power Sources*, 91, 37–50 (2000).

15. M. Winter and R. J. Brodd, *Chem. Rev*, 104, 4245–4269 (2004).

16. L. L. Zhang and X. S. Zhao, *Chem. Soc. Rev*, 38, 2520–2531 (2009).

17. J. R. Miller and P. Simon, Materials Science: Electrochemical Capacitors for Energy Management. *Science*, 321, 651–652 (2008).

18. L. L. Zhang and X. S. Zhao, *Chem. Soc. Rev.*, 38 (9), 2520–2531 (2009).

19. F. Beguin, V. Presser, A. Balducci and E. Frackowiak, *Adv. Mater.*, 26 14, 2219–2251, (2014).

20. P. Simon and Y. Gogotsi, *Nat. Mater.*, 7 11, 845–885, (2008).

21. E. Raymundo-Pinero, M. Cadek and F. Beguin, *Adv. Funct. Mater*, 19, 1032 (2009).

22. M. Biswal, A. Banerjee, M. Deo, and S. Ogale 6(4), 1249–1259 (2013).

23. M. Karnan, K. Subramani, N. Sudhan, N. Ilayaraja, and M. Sathish, *ACS Appl. Mater. Interfaces.* 8, 35191–35202 (2016).

24. N. Sudhan, K. Subramani, M. Karnan, N. Ilayaraja, and M. Sathish, *Energy Fuels*, 31, 977–985 (2017).

25. T. E. Rufford, D. Hulicova-Jurcakova, Z. Zhu, and G. Q. Lu, *Electrochem. Commun.*, 10,1594 (2008).

26. P. Divya, A. Prithiba, and R. Rajalakshmi, *IOP Conference Series, Mater. Sci. Eng.*, 2, 640–650 (2019).

27. P. Divya, and R. Rajalakshmi, *Mater. Today: Proceedings*, 27, 44–53 (2020).

28. Z. Bi, Q. Kong, Y. Cao, G. Sun, F. Su, X. Wei, C.M. Chen, *J. Mater. Chem. A*, 7, 16028, (2019).

29. W. Xiaozhong, W. Xing, J. Florek, J. Zhou, G. Wang, S. Zhuo, Q. Xue, Z. Yan and F. Kleitz, *J. Mater. Chem*, 44 (2014).

30. P. Divya, and R. Rajalakshmi , *J. Energy Storage*, 27, 101149 (2020).

31. T. E. Rufford, D. Hulicova-Jurcakova, Z. Zhu and G. Q. Lu, *Electrochem. Commun.*, 10(10), 1594–1597 (2008).

32. F. Qin, K. Zhang, J. Li, Y. Lai, H. Lu, W. Liu and J. Fang, *J. Solid State Electrochem.*, 20(2), 469–477 (2015).

33. T. E. Rufford, D. Hulicova-Jurcakova, K. Khosla, Z. Zhu and G. Q. Lu, *J. Power Sources*, 195, 912 (2010).

34. A. E. Ismanto, S. Wang, F. E. Soetaredjo, and S. Ismadji, *Bioresource Technol.*, 101(10), 3534–3540 (2010).

35. A. Elmouwahidi, Z. Zapata-Benabithe, F. Carrasco-Marín, and C. Moreno-Castilla, *Bioresource Technol.*, 111, 185–190 (2012).

36. G. A. Ferrero, A. B. Fuertes, and M. Sevilla, *Sci. Rep.*, 5(1), (2015).

37. J. Chang, Z. Gao, X. Wang, D. Wu, F. Xu, X. Wang, and K. Jiang, *Electrochimica Acta*, 157, 290–298 (2015).

38. M. Olivares-Marín, J. A. Fernández, M. J. Lázaro, C. Fernández-González, A. Macías-García, V. Gómez-Serrano, and T. A. Centeno, *Mater. Chem. Phys.*, 114(1), 323–327 (2009).

39. F.-C. Wu, R.-L. Tseng, C.-C. Hu, and C.-C. Wang, *J. Power Sources*, 159(2), 1532–1542 (2006).

40. F. Wu, R. Tseng, C. Hu and C. Wang, *J. Power Sources*, 144, 302 (2005).

41. S. T. Senthilkumar, R. K. Selvan, J. S. Melo, and C. Sanjeeviraja, *ACS Appl. Mater. Interfaces*, 5(21), 10541–10550 (2013).

42. J. Zhang, L. Gong, K. Sun, J. Jiang, and X. Zhang, *J. Solid State Electrochem.*, 16(6), 2179–2186 (2012).

43. Y.-Y. Wang, B.-H. Hou, H.-Y. Lü, C.-L. Lü, and X.-L. Wu, *Chemistry Select*, 1(7), 1441–1447 (2016).

44. C. Ruan, K. Ai, and L. Lu, *RSC Adv.*, 4(58), 30887 (2014).

45. S. G. Lee, K. H. Park, W. G. Shim, M. S. Balathanigaimani, and H. Moon. *J. Industrial Eng. Chem.*, 17(3) (2011).

46. X. Ma, C. Ding, D. Li, M. Wu, and Y. Yu, *Cellulose*, 25(8), 4743–4755 (2018).

47. E. Taer, R. Taslim, Z. Aini, S. D. Hartati, and W.S. Mustika, *AIP Conference Proc.*, 1801, 040004 (2017).

48. X.-L. Wu, T. Wen, H.-L. Guo, S. Yang, X. Wang, and A.-W. Xu, *ACS Nano*, 7(4), 3589–3597 (2013).

49. M.-C. Liu, L.-B. Kong, P. Zhang, Y.-C. Luo, and L. Kang, *Electrochimica Acta*, 60, 443–448 (2012).

Index

Page numbers in **bold** indicate tables, page numbers in *italic* indicate figures.

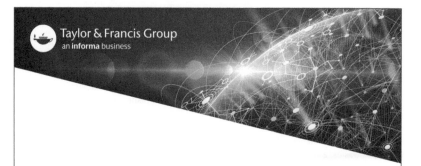